Mathematics Educ...

Volume 13

Series Editors

Dragana Martinovic, University of Windsor, Windsor, ON, Canada
Viktor Freiman, Faculte des sciences de l'education, Université de Moncton, Moncton, NB, Canada

Editorial Board

Marcelo Borba, State University of São Paulo, São Paulo, Brazil
Rosa Maria Bottino, CNR – Istituto Tecnologie Didattiche, Genova, Italy
Paul Drijvers, Utrecht University, Utrecht, The Netherlands
Celia Hoyles, University of London, London, UK
Zekeriya Karadag, Giresun Üniversitesi, Giresun, Turkey
Stephen Lerman, London South Bank University, London, UK
Richard Lesh, Indiana University, Bloomington, USA
Allen Leung, Hong Kong Baptist University, Kowloon Tong, Hong Kong
Tom Lowrie, University of Canberra, Bruce, Australia
John Mason, Open University, Buckinghamshire, UK
Sergey Pozdnyakov, Saint-Petersburg State Electro Technical University, Saint-Petersburg, Russia
Ornella Robutti, Università di Torino, Torino, Italy
Anna Sfard, USA & University of Haifa, Michigan State University, Haifa, Israel
Bharath Sriraman, University of Montana, Missoula, USA
Anne Watson, University of Oxford, Oxford, UK

The Mathematics Education in the Digital Era (MEDE) series explores ways in which digital technologies support mathematics teaching and the learning of Net Gen'ers, paying attention also to educational debates. Each volume will address one specific issue in mathematics education (e.g., visual mathematics and cyber-learning; inclusive and community based e-learning; teaching in the digital era), in an attempt to explore fundamental assumptions about teaching and learning mathematics in the presence of digital technologies. This series aims to attract diverse readers including: researchers in mathematics education, mathematicians, cognitive scientists and computer scientists, graduate students in education, policy-makers, educational software developers, administrators and teachers-practitioners. Among other things, the high quality scientific work published in this series will address questions related to the suitability of pedagogies and digital technologies for new generations of mathematics students. The series will also provide readers with deeper insight into how innovative teaching and assessment practices emerge, make their way into the classroom, and shape the learning of young students who have grown up with technology. The series will also look at how to bridge theory and practice to enhance the different learning styles of today's students and turn their motivation and natural interest in technology into an additional support for meaningful mathematics learning. The series provides the opportunity for the dissemination of findings that address the effects of digital technologies on learning outcomes and their integration into effective teaching practices; the potential of mathematics educational software for the transformation of instruction and curricula; and the power of the e-learning of mathematics, as inclusive and community-based, yet personalized and hands-on.

Title forthcoming:

Proof Technology in Mathematics Research and Teaching: Gila Hanna, David Reid and Michael de Villiers (eds.), due mid-2019

Submit your proposal:

Book proposals for this series may be submitted per email to Springer or the Series Editors. - Springer: Natalie Rieborn at Natalie.Rieborn@springer.com - Series Editors: Dragana Martinovic at dragana@uwindsor.ca and Viktor Freiman at viktor.freiman@umoncton.ca

More information about this series at http://www.springer.com/series/10170

Gilles Aldon · Jana Trgalová
Editors

Technology in Mathematics Teaching

Selected Papers of the 13th ICTMT Conference

Springer

Editors
Gilles Aldon
S2HEP
École Normale Supérieure de Lyon
Lyon, France

Jana Trgalová
S2HEP
Université Claude Bernard Lyon 1
Villeurbanne, France

ISSN 2211-8136 ISSN 2211-8144 (electronic)
Mathematics Education in the Digital Era
ISBN 978-3-030-19743-8 ISBN 978-3-030-19741-4 (eBook)
https://doi.org/10.1007/978-3-030-19741-4

© Springer Nature Switzerland AG 2019
This work is subject to copyright. All rights are reserved by the Publisher, whether the whole or part of the material is concerned, specifically the rights of translation, reprinting, reuse of illustrations, recitation, broadcasting, reproduction on microfilms or in any other physical way, and transmission or information storage and retrieval, electronic adaptation, computer software, or by similar or dissimilar methodology now known or hereafter developed.
The use of general descriptive names, registered names, trademarks, service marks, etc. in this publication does not imply, even in the absence of a specific statement, that such names are exempt from the relevant protective laws and regulations and therefore free for general use.
The publisher, the authors and the editors are safe to assume that the advice and information in this book are believed to be true and accurate at the date of publication. Neither the publisher nor the authors or the editors give a warranty, expressed or implied, with respect to the material contained herein or for any errors or omissions that may have been made. The publisher remains neutral with regard to jurisdictional claims in published maps and institutional affiliations.

This Springer imprint is published by the registered company Springer Nature Switzerland AG
The registered company address is: Gewerbestrasse 11, 6330 Cham, Switzerland

An Introductory Walk in an Inspiring Book

Introduction

It is both a pleasure and an honour to introduce this volume of work originating from the *13th International Conference for Technology in Mathematics Teaching* (ICTMT) since the ICTMT conferences mirror our two journeys in the field of mathematics education/didactics of mathematics. ICTMT started life, in Birmingham UK in 1993, as a European version of courses and conferences organized by Frank Demana and Bert Waits, pioneers in introducing mathematical hardware and software into collegiate mathematics in the USA, having the intuition that 'the use of calculator—or computer-based graphing approach—dramatically changes results in the classroom […] transformed into a mathematics laboratory' (Demana, Waits, & Clemens, 1992 p. vii). In ICTMT, as in our professional lives, the meeting of mathematics, digital technology and education is a meeting point for mathematicians, researchers, teachers and technologists to listen to each other and learn from each other. The subfield of mathematics education focused on digital technologies includes practical and theoretical evaluations of new technologies for the learning and teaching of mathematics. It is imperative, in our opinion, that practice and theory, though they may be considered separately at times, are never fully separated; they are dialectically related and the synergy of their interaction drives our subfield forward. ICTMT has, from the outset, respected this synergy.

Back in 1993, we had what might be called a *naïve optimism* for positive changes that digital technology could bring to learning (and teaching) mathematics; and this *optimism* is reflected in the volume reporting on ICTMT 1 (Burton & Jaworski, 1995). Since 1993, many (what could be called) 'implementation difficulties' in the integration of digital technologies have appeared. We consider two of these in two chapters in our book (Monaghan, Trouche, & Borwein, 2016): *Integrating tools as an ordinary component of the curriculum in mathematics education*: and *The calculator debate*. A factor in these implementation difficulties is, we believe, the 'culture of mathematics'. This is a culture that we have the greatest respect for but this culture was formed, over thousands of years, before digital technologies arrived and adjusting to a rapidly developing class of new tools is not easy; but this culture is gradually changing. We had the good fortune to write our book with the research mathematician Jonathan M. Borwein, an experimental mathematician who sadly died shortly after the book was published. In our tribute to Jon (Kortenkamp, Monaghan, & Trouche, 2016, p. 135) we say:

> Jon valued the rich cultural heritage of our discipline but this did not make him a slave to this culture ... The mathematical culture that Jon worked in included: computer languages, mathematical software, web applications (e.g., Sloane's Encyclopedia of Integer Sequences), and web databases. He viewed the use of these tools as data mining and that in his work the "boundaries between mathematics and the natural sciences and between inductive and deductive reasoning are blurred".

We see ICTMT as one of the many vehicles for the further advancing the culture of mathematics.

Sections 1–4 below follow the themes of this volume.

Digital Technology and Assessment

At the time of the first ICTMT conference, assessment in mathematics classrooms meant, to all intents and purposes, summative assessment of students' mathematics skills. The term 'formative assessment' (FA) was coined in 1967 but it was not until Black & Wiliam's (1998) classic review of literature on assessment and classroom learning, that FA, as an explicit activity in the classroom, really 'took off'. Black & Wiliam (1998, p. 8) note:

> the term formative assessment does not have a tightly defined and widely accepted meaning. In this review, it is to be interpreted as encompassing all those activities undertaken by teachers, and/or by their students, which provide information to be used as feedback to modify the teaching and learning activities in which they are engaged.

The chapters under Theme 1, *Digital technologies and assessment*, relate to this definition and, of course, the potential of digital technologies to aid FA.

The chapter by Cusi, Morselli & Sabena builds on a recent article (Cusi, Morselli, & Sabena, 2017) on the design and implementation of digital resources to promote FA in connected classrooms. An important element of their work is digital polls. Polls, canvassing students' opinions on their understanding of the mathematics they are engaged with, have been a feature of FA of many classrooms and lecture theatres since the turn of the millennium. But digital tools have the potential to provide the teacher and students with more information than *I do/do not understand*. In the work described in this chapter, digital poll worksheets are designed for students to make metacognitive and affective judgments. Such polls allow for emergent re-design of lessons but, of course, increase the complexity of managing the classroom.

The chapter by Olsher is more speculative than the chapter by Cusi et al., but no less interesting and with the constant development of digital tools, it is very important to speculate. Olsher focuses on a possible adjunct to a dynamic geometry environment (DGE) that could complement and/or inform a teacher's categorization of students' geometric conjectures; as Olsher notes, a digital filter could be linked to a DGE to provide both student and teacher with feedback on the generality of students' solutions as well as automatically determining their correctness.

FA is a multidimensional construct and an important dimension is students' self-assessment. Ruchniewicz' chapter reports work in progress of a digital tool designed for student self-assessment in the context of students constructing velocity–time graphs from descriptions of situations, e.g. 'Niklas rides home …'. Typical student misconceptions are built into the digital tool which can provide feedback to students, not just on the correctness of the solution but information that facilitates self-assessment. Critical evaluation of student use of the tool suggests areas for development.

Barzel and Ball's chapter appropriately concludes this theme on digital technologies and assessment with an overview and analysis of digital technologies for formative self-assessment. An important issue raised is agency, who/what directs learning? Traditionally, it has been the teacher and textbook. These agents remain but digital tools are rising agents in mathematics classrooms and these tools, arguably, promote student agency. What exciting times we live and work in.

The field of technology in mathematics education lies at the meeting of mathematics, technology and education and the subfield concerned with formative assessment can draw from this meeting of fields. Formative assessment tools in general use include Plickers (https://www.plickers.com/) and Socrative (https://www.socrative.com/) and mathematics teachers are appropriating these general tools for mathematics classroom (see Umameh & Monaghan, 2017). The chapters in this theme, however, largely focus on formative assessment tailored for the special features of our subject, e.g. students' geometric conjectures, mathematical concept development and switching between representations of mathematical objects.

The chapters in this theme provide a window on an emerging field of practice-focused scholarship. Important work outside the window include Bokhove's constructs of *timing and fading, crises and feedback variation* (see Bokhove & Drijvers, 2012) and Sangwin's work on the automation of marking students' computer-based mathematics (see Sangwin & Köcher, 2016). The issue of digital tools in assessment is important, as Ridgeway et al. (2004, p. 4) wrote 14 years ago:

> The issue for e-assessment is not if it will happen, but rather, what, when and how it will happen. E-assessment is a stimulus for rethinking the whole curriculum, as well as all current assessment systems.

Innovative Technologies and Approaches to Mathematics Education

The second theme of this book concerns innovative technologies and approaches in our field of inquiry, mathematics education. But what is 'innovation'? Innovation, whether it concerns technology (artefacts) or approaches (using artefacts) or both, is popularly regarded as *something new*. We accept this with the proviso that *the new* comes from reflections on our past experiences. We outline our basis for this statement before considering the chapters under this theme. Our basis comes from an essay by Wartofsky (1979), which argues for an historical epistemology of perception.

How do we see the possibility to create something new or to do something in a new way? At any point in sociocultural development, human action (*praxis*) involves communicating and producing, using extant artefacts and representations, 'the very use of tools for a certain purpose is what determines how such tools will be seen' (ibid., p. 205). But if we venture to reflect on actual use, then we may enter an 'imaginative praxis, the perceptual modes are derived from and related to a given historical mode of perception, but are no longer bound to it' (ibid., p. 209). This, *derived from—related to—but not bound to*, can be seen in the three chapters under this second theme of this book.

The chapter by Dimmel & Bock considers *Handwaver*, a gesture-based virtual environment that allows users to use their hands to create and modify mathematical objects. Gestures have always been a part of doing mathematics, though widespread recognition of this did not come, until after the publication of Lakoff & Nunes (2000). Virtual digital environments originated outside of mathematics. The designers of *Handwaver* certainly engaged in 'imaginative praxis' to bring gestures and virtual environments together. The development of *Handwaver* is linked to research-based ideas about productive mathematical activity and is thus poised to be more than a fun gimmick in future mathematics lessons.

The chapter by Kobylanski reports on WIMS, an interactive multipurpose server for mathematics as well as other school subjects. WIMS is multipurpose in the sense that it provides exercises, games and subject-specific tools, amongst other things. With regard to 'imaginative praxis' it takes non-digital cultural practices such as exercises and repackages them digitally. The mathematics side of WIMS is designed to interface with existing mathematics software for graph plotting, numerical analysis and symbolic manipulation. WIMS has been around for two decades, so there is an argument that it is no longer an innovation—but this does depend on what we regard as the unit of sociocultural time. It also raises questions about cross-cultural use of an innovation, as WIMS is popular in France but not elsewhere.

The chapter by El-Demerdash, Trgalová, Labs and Mercat looks at software to support creativity in mathematics. This focus on creativity is especially pertinent to the ideas of Wartofsky as creativity and innovation are two (of many) interrelated dimensions of 'imaginative praxis'. The chapter is based on work done in a

European Union funded project on creative mathematical thinking. Within this project a set of resources have been written under the guidance of scholars from computer science, mathematics and mathematics education. The chapter focuses on interrelated resources linked to a dynamic geometry package (*Cinderella*). We see again Wartofsky's 'imaginative praxis' as mathematical problems in the topic of loci are re-presented in a digital media.

Digital technology is arguably the *sine qua non* of innovation in recent decades but *actioning* innovation (enabling it to move from imaginative to actual praxis) is arguably the most important goal of mathematics education research. Actioning the digital environments of the three chapters described above depends on mathematics teachers. This leads into the next theme of the book on *digital technology and teacher professional development*, but before we leave the theme of innovation, a word of caution from twentieth-century history.

In Chapter 15 of our book on tools and mathematics (Monaghan, Trouche, & Borwein, 2016), we ask 'Is there anything special about teachers' use of digital tools?' and answer 'no and yes'. Our 'no' response includes a consideration of Cuban (1986), an historical analysis of educational innovation in the USA following technology innovation; educational use of film and radio, television and computers in classrooms from 1920 to the 1980s. In each case, Cuban traces initial optimism for the potential of the artefact for education turning, over time, to infrequent use due to lack of skills, costs, accessibility and difficulties in adapting established perceptions of what a lesson should be, to the new media.

There is a sense in which we, and the authors in this book, are part of a subcommunity of the mathematics education research community charged with providing research-informed valuations of the potential of digital technology for improving the learning and teaching of mathematics. It is important that we take 'actioning innovation' seriously to prevent positively valued digital innovations remaining solely the preserve of imaginative praxis.

Digital Technologies and Teacher Professional Development

The mathematics teacher professional development (PD) needed for a fruitful integration of digital technologies was largely underestimated in the last century, as evidenced by Lagrange *et al.* (2003, p. 259), on the basis of a systematic literature review:

> In the years 1994–1998, questions about the teacher necessarily brought about more general problems with few solutions. There was a tendency to focus on teachers' development and an implicit assumption that the transfer of innovative situations of use, possibly supported by outcomes of research, would provide the teacher with sufficient material for an easy integration. Aware of the complexity of teaching and learning situations with ICT, researchers are now more cautious

The 17th ICMI study, dedicated in 2006 to 'mathematics education and technology' confirmed this awareness, showing up

> to what point knowledge had progressed in the last two decades, allowing to understand better how digital technologies modify teacher professional work, requiring new competencies, up to what point the usual discourse accompanying the promotion of technology has been misleading and counterproductive, the educational resources and training strategies poorly appropriate. (Artigue, 2010, p. 471).

The third section of the book, focusing on digital technologies and teacher professional development, offers a new opportunity for taking stock of the progression of the knowledge on this critical point.

The chapter by Tabach & Trgalová proposes an analysis of knowledge and skills mathematics teachers need for ICT integration, focusing on the issue of standards for teaching mathematics with technology. For this purpose, they develop a theoretical frame mainly based on three pillars: *instrumental genesis* (Rabardel, 2002); *Pedagogical Technology Knowledge* (PTK, Thomas, & Hong, 2005); and teacher *orientations* and *goals* (Schoenfeld, 2011). They apply this frame for analysing documents produced by three international structures (UNESCO, NCTM and M-TPACK framework) and two national educational institutions (Australia and France). They draw, from this analysis, two main results: the existing standards are too general, 'as they are neither school level, nor subject matter specific'; and teachers' personal orientation towards integrating technology are not sufficiently considered. These results lead the authors to theoretical and prospective considerations: from a theoretical point of view, they propose an adaptation of the PTK framework, 'emphasizing the determining role of the components of the mathematical knowledge for teaching with technology that are related to teacher orientations, personal and professional instrumental genesis'. From a prospective point of view, the authors call for the mathematics education community to take the standards issue seriously.

Considering standards, in this domain, raises new questions: is it possible to design stable standards in a time of very rapid technological changes? Is it possible to design the same standards for countries with very different cultural and social backgrounds? And, which aspects of mathematics teacher education should be developed for reaching given standards?

The following chapter, by Drijvers, van den Bogaart and Tolboom, addresses this issue of teacher education, in the domain of STEM, under the form of open online modules for blended learning. Blended learning, integrating face-to-face and online learning, is more and more used in teacher education (see Trouche *et al.* 2013). Teacher education organization requires specific skills for teacher educators (Gueudet *et al.* 2012), combining technical, mathematical, didactical and design competencies. The chapter appropriately proposes a report on a design research project in which teacher educators engaged in a co-design process of developing and field testing two open online learning units for mathematics and science didactics. This report evidences some major results: the time needed by the educators for designing the units, and more particularly to incorporate new training modes in current educational practices; a desire to design flexible building blocks rather than ready-to-use courses; and the need for associating in the design each actor of the learning process (teachers, educators, as well as students).

The last chapter of this section, by Aldon, Arzarello, Panero, Robutti, Taranto and Trgalová, reports on an international research about two MOOCs for in-service mathematics teacher education, developed, respectively, by an Italian and a French team. Compared to the previous chapter, this teacher education organization works essentially online, and caters for a large number of teachers (several hundred). These conditions raise new needs, on both practical and theoretical levels. The practical need is related to teachers' collaboration: the use of sophisticated platforms for supporting the online courses does not reduce but rather, increases the need for teachers to collaborate. Teachers' engagement in each MOOC seems to be directly linked to the development of *communities of practice* (Wenger 1998). On a theoretical level, monitoring MOOCs with conceptual intentions—introducing didactical concepts—calls for rethinking the transposition of theories from communities of research to practitioners, what the authors call the *meta-didactical transposition* (Arzarello *et al.* 2014).

Finally, this section reminds us the reflection we proposed in introducing the ICTMT conference: 'New forms of activity with digital tools enabled by connecting these tools also provide teachers with opportunities to engage in new forms of action in their classrooms, and with their colleagues, giving new opportunities for teacher professional development' (Monaghan & Trouche, 2016, p. 381).

Teaching and Learning Experiences with Digital Technologies

The fourth theme is dedicated to experiences with digital technologies. Entering this theme, we had in mind Dewey's statement, evidencing 'the organic connection between education and personal experience' (Dewey 1938, p. 25). Contrasting (already in 1938…) 'traditional' and 'progressive' schools, he stated (p. 25) that 'the belief that all genuine education comes about through experience does not mean that all experiences are genuinely or equally educative'. His criterion, for developing genuine and educative experience, could be retained, today: 'The central problem of an education based upon experience is to select the kind of present experiences that live fruitfully and creatively in subsequent experiences' (ibid. p. 27–28): a criterion to keep in mind for considering current experiences with digital technologies!

This theme contains three chapters, each of them analysing experiences with different digital technologies: 3D modelling software, dynamic algebra and geometry software and a computer-based learning environment.

The chapter by Uygan and Turgut analyses how through the use of a 3D modelling software, a subject experiences and interprets concrete, kinaesthetic and dynamic images. Anchored in a multimodal and embodied cognition paradigm, the authors state that 'reasoning on mathematical objects not only includes the acts of thinking, constructing and expressing meaning per se, but also those acts interlaced with our gestures, mimics and sometimes with specific sketches'. The research question is here: 'what kind of spatial-semiotic resources emerge when an eighth-grade student solves spatial tasks using 3D modelling software?' the notion of 'resources' referring here to techniques, or strategies. For answering this question, a student, who is performing moderately well in mathematics but mastering the use of basic computer tools, is chosen. He has to solve tasks (constructing and representing building made of cubes) in the 3D modelling software environment. A task-based interview was conducted with him. One of the main results is that the student's gestures remain limited to the use of the software: 'There did not appear to be any gestures independent of the artefact (mouse and keyboard), such as hand movements, tracing with a finger and so on'. Further to this, the authors identified two components structuring the student's activity: 'a spatial-analytic strategy that is related to constructing a certain part of a 3D building independently from other parts and a spatial-holistic strategy which is about constructing a building as a whole by considering the relationships between all parts within it'. It would be interesting to know how this kind of activity could live 'fruitfully and creatively in subsequent experiences', allowing the student to develop his own thinking through (and outside of) the use of the artefact.

The chapter by Lisarelli focuses on occasions for students to experience the dependence relation and to explore functions as covariation through the use of a dynamic algebra and geometry software. We have, in our own research (Guin & Trouche 1999), experienced to which extent the act of representing functions graphically has as much potential to produce confusion as enlightenment. Anchored in both the theory of semiotic mediation and instrumental theory, the author explores the semiotic potential of the representation of functions with two parallel axes, supported by a dynamic environment allowing the combined movement of two ticks bounded to the two axes. She presents excerpts from a pilot study involving students working in Pairs. These excerpts evidence how students' descriptions are rich in references to movement, time and space. This experience appears to foster, for the students, building mathematical meanings related to functional dependence, as a relation between two co-varying quantities: one depending on the other one. Here also, the need for thinking 'subsequence experiences' is evoked: 'choosing functions that can support the coming to light of other relevant properties of functions, in order to gain a deeper insight into possible exploits of the semiotic potential of functions' representation with parallel axes'.

The chapter by Jedtke and Greefrath analyses the role of feedback, in a computer-based learning environment, for steering students' positive experience in a context of quadratic functions problems. Regarding problems of this kind, one of the major learning objectives is to develop students' ability to flexibly switch between distinct representations of quadratic functions (verbal, graphical, tabular, symbolic). Following institutional requirement in Germany, the authors aimed to design a digital educational medium having the property of "multimediality, interactivity, networkability, changeability, and divisibility". They focus on the nature of the feedback, exploring the following question: 'Does a learning path for the topic of quadratic functions that incorporates feedback featuring additional explanations and hints have a more positive influence on the mathematics performance of students than the same learning path with feedback that merely states the correct solution?' To answer this question, they analyse the work of students groups into Pairs. This analysis evidences students' interest for working in Pairs, and having feedback directed towards understanding more than correctness.

Finally, for developing 'genuine and educative experience' in mathematics education with digital technologies, we retain from a cross-reading of these three chapters, some critical dimensions to be taken into account: the collective dimension (working in pairs is appreciated); the design dimension (simultaneously thinking of task design and of the design of the environment itself); the time dimension (for exploring, sharing inspiration, analysing feedback…). The word 'experience' is actually used in two senses: 'having experience in the use of something' (as Atakan in Uygan & Turgut's chapter), and 'experience something' (as 'experience difficulties', in Jedtke and Greefrath's chapter). Extending Dewey's idea (see above), we could perhaps state that the central problem for teaching and learning mathematics related to digital experiences is to be able to link current experiences with previous ones, and to develop them fruitfully and creatively, in subsequent experiences.

Conclusion

In closing this foreword to selection of papers from ICTMT 13, we look to our community and beyond. We view our community as research active (or, at least, research aware) members of the subfield of mathematics educators working with digital technologies. This is, in our opinion, the primary audience for this book. Members of this community also belong, as we mentioned in our Introduction, to other communities, for instance, mathematicians and computer scientists. This book is also relevant for members of these communities. These sentiments hark back to a UNESCO publication (Cornu & Ralston, 1992) which revisited themes of the 1985 ICMI study *The Influence of Computers and Informatics on Mathematics and Its Teaching*: the effect of computers on mathematics; the impact of computers and computer science on the Mathematics curriculum; and computers as an aid to teaching and learning mathematics.

It is important that our community interacts with these other communities but the scope for interaction is wider still. In closing our book on tools and mathematics, we asked 'Who is empowered by knowledge on tool use in mathematics?' In addressing this question, we considered groups of people who deserve to know about advances in mathematics and mathematics education related to the use of digital technologies. Groups mentioned, beyond those considered above, include: policymakers; teacher educators; international agencies; technical developers;

employers; school staff; online teacher associations; and students and their families. We do not expect these groups to read this book but it is important, to scale-up that which is believed to improve students' education, for these groups to be empowered, to be aware of advances in our subfield.

Our final comments return to our digital–mathematics–education community, with special thanks to Jana and to Gilles, for their invitation to open the ICTMT 13 conference and the invitation to open this book.

March 27   John Monaghan

Luc Trouche

References

Artigue, M. (2010). The future of teaching and learning mathematics with digital technologies. In C. Hoyles & J.-B. Lagrange, *Mathematics education and technology—Rethinking the terrain* (pp. 463–475). New York (N.Y.): Springer.

Arzarello, F., Robutti, O., Sabena, C., Cusi, A., Garuti, R., Malara, N., et al. (2014). Meta-didactical transposition: A theoretical model for teacher education programs. In A. Clark-Wilson, O. Robutti, & N. Sinclair (Eds.), *The mathematics teacher in the digital era: An international perspective on technology focused professional development* (Vol. 2, pp. 347–372). Dordrecht, The Netherlands: Springer.

Black, P., & Wiliam, D. (1998). Assessment and classroom learning. *Assessment in Education: Principles, Policy & Practice, 5*(1), 7–74.

Bokhove, C., & Drijvers, P. (2012). Effects of feedback in an online algebra intervention. *Technology, Knowledge and Learning, 17*(1–2), 43–59.

Burton, L., & Jaworski, B. (Eds.), (1995). *Technology in mathematics teaching: A bridge between teaching and learning.* Lund, Sweden: Chartwell-Bratt.

Cornu, B., & Ralston, A. (1992). The influence of computers and informatics on mathematics and its teaching. Paris: UNESCO, *Science and Technology Education* 44 (second edition of the 1985 ICMI study). http://unesdoc.unesco.org/images/0009/000937/093772eo.pdf.

Cuban, L. (1986). *Teachers and machines: The classroom use of technology since 1920.* New York: Teachers College Press.

Cusi, A., Morselli, F., & Sabena, C. (2017). Promoting formative assessment in a connected classroom environment: Design and implementation of digital resources. *ZDM—Mathematics Education, 49*(5), 755–767.

Demana, F., Waits, K. W., & Clemens, S. R. (1992). *College algebra & trigonometry. A graphing approach.* USA: Addison-Wesley Publishing Company.

Dewey, J. (1938). *Experience & education.* Kappa Delta Pi Lecture Series. NY: Collier MacMillan Publishers.

Guin, D., & Trouche, L. (1999). The complex process of converting tools into mathematical instruments. The case of calculators. *The International Journal of Computers for Mathematical Learning, 3*(3), 195–227.

Gueudet, G., Sacristan, A. I., Soury-Lavergne, S., & Trouche, L. (2012). Online path in mathematics teacher training: New *resources* and new *skills* for teacher educators. *ZDM—Mathematics Education, 44*(6), 717–731.

Kortenkamp, U., Monaghan, J., & Trouche, L. (2016). Jonathan M Borwein (1951–2016): Exploring, experiencing and experimenting in mathematics—An inspiring journey in mathematics. *Educational Studies in Mathematics, 93*(2), 131–136.

Lagrange, J.-B., Artigue, M., Laborde, C., & Trouche, L. (2003). Technology and mathematics education: A multidimensional study of the evolution of research and innovation. In A. J. Bishop, M. A. Clements, C. Keitel, J. Kilpatrick, & F. K. S. Leung (Eds.), *Second international handbook of mathematics education* (pp. 239–271). Dordrecht: Kluwer Academic Publishers.

Lakoff, G., & Núñez, R. E. (2000). *Where mathematics comes from: How the embodied mind brings mathematics into being.* New York: Basic Books.

Monaghan, J., & Trouche, L. (2016). Mathematics teachers and digital tools. In J. Monaghan, L. Trouche, & J. M. Borwein, *Tools and mathematics: Instruments for learning* (pp. 357–384). Berlin: Springer International Publishing.

Monaghan, J., Trouche, L., & Borwein, J. M. (2016). *Tools and mathematics: Instruments for learning.* Berlin: Springer International Publishing.

Rabardel, P. (2002). *People and technology—A cognitive approach to contemporary instruments.* Université Paris 8. Retrieved October 15, 2013. https://hal-univ-paris8.archives-ouvertes.fr/file/index/docid/1020705/filename/people_and_technology.pdf.

Ridgeway, J., McCusker, S., & Pead, D. (2004). *Literature review of E–assessment.* Futurelab Series 10 Futurelab. ISBN: 0-9544695-8-5.

Sangwin, C. J., & Köcher, N. (2016). Automation of mathematics examinations. *Computers & Education, 94*, 215–227.

Schoenfeld, A. H. (2011). Toward professional development for teachers grounded in a theory of decision making. *ZDM—Mathematics Education 43*(4), 457–469.

Thomas, M. O. J., & Hong, Y. Y. (2005). Teacher factors in integration of graphic calculators into mathematics learning. In H. L. Chick, & J. L. Vincent (Eds.), *Proceedings of the 29th Conference of the International Group for the Psychology of Mathematics Education* (Vol. 4, pp. 257–264). Melbourne: University of Melbourne.

Trouche, L., Drijvers, P., Gueudet, G., & Sacristan, A. I. (2013). Technology-driven developments and policy implications for mathematics education. In A. J. Bishop, M. A. Clements, C. Keitel, J. Kilpatrick, & F. K. S. Leung (Eds.), *Third international handbook of mathematics education* (pp. 753–790). Springer.

Umameh, M., & Monaghan, J. (2017). A classification of resources used by mathematics teachers in an English high school. In G. Aldon, & J. Trgalová (Eds.), *Proceedings of the 13th International Conference on Technology in Mathematics Teaching*(pp. 276–283). Available on-line.

Wartofsky, M. (1979). Perception, representations, and the forms of action: Towards an historical epistemology (written 1973). In M. Wartofsky, *Models: Representation and the scientific understanding* (pp. 188–210). Dordrecht: D. Reidel.

Wenger, E. (1998). *Communities of practice: Learning, meaning, identities.* Cambridge University Press.

Contents

Part I Digital Technologies and Assessment

The Use of Polls to Enhance Formative Assessment Processes in Mathematics Classroom Discussions................................ 7
Annalisa Cusi, Francesca Morselli and Cristina Sabena

Making Good Practice Common Using Computer-Aided Formative Assessment 31
Shai Olsher

Technology Supporting Student Self-Assessment in the Field of Functions—A Design-Based Research Study 49
Hana Ruchniewicz and Bärbel Barzel

Students' Self-Awareness of Their Mathematical Thinking: Can Self-Assessment Be Supported Through CAS-Integrated Learning Apps on Smartphones?............................... 75
Bärbel Barzel, Lynda Ball and Marcel Klinger

Part II Innovative Technologies and Approaches to Mathematics Education: Old and New Challenges

Dynamic Mathematical Figures with Immersive Spatial Displays: The Case of Handwaver....................................... 99
Justin Dimmel and Camden Bock

WIMS: Innovative Pedagogy with 21 Year Old Interactive Exercise Software... 123
Magdalena Kobylanski

Design and Evaluation of Digital Resources for the Development of Creative Mathematical Thinking: A Case of Teaching the Concept of Locus 145
Mohamed El-Demerdash, Jana Trgalová, Oliver Labs and Christian Mercat

Part III Mathematics Teachers' Education for Technological Integration: Necessary Knowledge and Possible Online Means for its Development. Introduction to the Section

The Knowledge and Skills that Mathematics Teachers Need for ICT Integration: The Issue of Standards 183
Michal Tabach and Jana Trgalová

Co-Design and Use of Open Online Materials for Mathematics and Science Didactics Courses in Teacher Education: Product and Process 205
Theo van den Bogaart, Paul Drijvers and Jos Tolboom

MOOCs for Mathematics Teacher Education to Foster Professional Development: Design Principles and Assessment 223
Gilles Aldon, Ferdinando Arzarello, Monica Panero, Ornella Robutti, Eugenia Taranto and Jana Trgalová

Part IV Teaching and Learning Experiences with Digital Technologies

Spatial-Semiotic Analysis of an Eighth Grade Student's Use of 3D Modelling Software 253
Candas Uygan and Melih Turgut

Activities Involving Dynamic Representations of Functions with Parallel Axes: A Study of Different Utilization Schemes 275
Giulia Lisarelli

A Computer-Based Learning Environment About Quadratic Functions with Different Kinds of Feedback: Pilot Study and Research Design 297
Elena Jedtke and Gilbert Greefrath

Concluding Remarks 323

Part I
Digital Technologies and Assessment

Gilles Aldon
S2HEP, EducTice, Institut Français de l'Éducation,
École Normale Supérieure de Lyon

Introduction

It is obvious that the process of assessment is fundamental in teaching and in mathematics teaching. The ICTMT conference in Lyon focused largely on assessment in a context of technology, in particular about formative assessment. Starting from the work done during the European project FASMEd,[1] the round table of this conference showed the potentialities of technology in a Formative Assessment process. The objective of this European design research project was to foster high-quality interactions in classrooms that are instrumental in raising achievement but also expanding our knowledge of technologically enhanced teaching and assessment methods addressing achievement in mathematics and science. The definition that was adopted comes from Thompson and Wiliam (2007) and Black and Wiliam (1998):

"Students and teachers using evidence of learning to adapt teaching and learning to meet immediate needs minute-to-minute and day-by-day" (Thompson & Wiliam, 2007); "All those activities undertaken by teachers, and by their students in assessing themselves, which provide information to be used as feedback to modify the teaching and learning activities in which they are engaged. Such assessment becomes 'formative assessment' when the evidence is actually used to adapt the teaching work to meet the needs" (Black & Wiliam, 1998).

The dimension of technology was theoretically conceptualized, taking into account the fundamental questions of formative assessment but also, who is actually assessing and the properties that technology can bring to the process of assessment. These potentialities have been gathered in three main groups: (a) sending and

---

[1]Formative Assessment for Science and Maths Education. The research leading to these results has received funding from the European Community's Seventh Framework Programme fp7/2007–2013 under grant agreement No [612337]. Toolkit of FASMEd: https://www.fasmed.eu.

displaying, (b) processing and analyzing, and (c) providing an interactive environment. This framework, as well as, its potentialities in term of analysis are developed in Chapter "The Use of Polls to Enhance Formative Assessment Processes in Mathematics Classroom Discussions" where Annalisa Cusi, Francesca Morselli, and Cristina Sabena highlight its use in the design and analysis of assessment lessons, and in Chapter "Technology Supporting Student Self-Assessment in the Field of Functions—A Design-Based Research Study" where Hana Ruchniewicz and Bärbel Barzel use the model in the design of self-assessment tools that improve functional thinking. FASMEd has developed a toolkit to support teachers in using technology for Formative Assessment in mathematics and science. The toolkit is a set of materials for teachers and teacher trainers which are both theoretical (what is formative assessment, how do we model formative assessment with technology?) and practical (assessment lessons, professional development sessions).

The four first chapters of this book propose a reflection about assessment and particularly about formative assessment. I will present in this introduction the main ideas leading to the relationship between Formative Assessment and teaching and learning, and add a reflection about the use of technology in such a process.

Assessment, Formative Assessment, and Assessment for Learning

Assessment is at the core of a long tradition of research all around the world, because it is the force engine of all the decisions that students and teachers can take regarding the processes of learning and teaching. In the cycle of assessment, measurement based on criteria leads to a judgment that causes decision taking in relation to the intention of assessment. Teachers take decisions from assessment through the perspective of modifying or adapting their own teaching; however, these decisions coming from assessment are important with regards to the students' course choices, that is to say they prepare and ascertain the choices that will have lifelong importance. Assessment makes sense only because it comes to one or several decisions that concern the knowledge learning at stake, the competencies in play, the teaching method that is used, the coherence of the relationship between teaching and learning, the organization of a school, the politics of education, and so on: "the process of 'evaluation' or 'assessment' as requiring the gathering of data, establishing weightings, selecting goals and criteria in order to compare performances and justify each of these. In other words, to make a judgment we must decide on what elements are important, why these are important, how each element is important in relation to the others and finally, provide a justification of all the choices made" (Taras, 2012).

These decisions are based on the function that is given to assessment by actors of education. These functions of assessment have for a long time declined as being diagnostic, formative, summative. However, it appears that summative or diagnostic assessment can be part of formative assessment (Wiliam, 2000; Wiliam & Black, 2006; Taras, 2012) through data collection, feedback, and decisions "that are likely to be better, or better founded, than the decisions they would have taken in the absence of the evidence that was elicited" (Black & Wiliam, 2009). Following

Perrenoud and Cardinet, the functions of assessment can be clarified as functions of regulation, of orientation and of certification. The function of regulation is to allow students to understand and use the feedback to enhance their learning and give the teacher the possibility to adjust his/her course to complement the students' knowledge; the function of orientation gives the learners a direction in which they can develop and enhance their competencies, while the function of certification illustrates, for society, the proof of a knowledge, or competencies level. These functions are in tension between a cognitive or didactical point of view and a social point of view as represented in Fig. 1. In that view, Formative Assessment becomes, through feedback, a model of teaching taking into account the different actors that take place in the game: the teacher, the student as a whole, and the peers through the social relationships built in the classroom. Assessment of learning gradually becomes assessment for learning. The work of Linda Allal, Lucie Mottier-Lopez (Allal & Mottier-Lopez, 2007), and other authors gives the regulation a crucial importance in the assessment process by distinguishing interactive regulation, given during a learning activity and supported by the teacher, and retroactive regulation, given after the activity alongside feedback offered to students, sometimes in a summative way. Allal and Mottier-Lopez (Op. cit.) add a proactive regulation that corresponds to an organization of teaching centered on the students' possibilities and difficulties. Briefly outlined, the landscape of assessment in its formative part asks to be completed by the "how to" issue. Particularly, in the context of the ICTMT conference, the role of technology has to be questioned. Does technology change the process of assessment or does it bring facilities in the different phases of the process? The European project FASMEd gave the opportunity during the round table to share some of the results obtained in the project, both from the point of view of the material collaboratively built with teachers and the research questions.

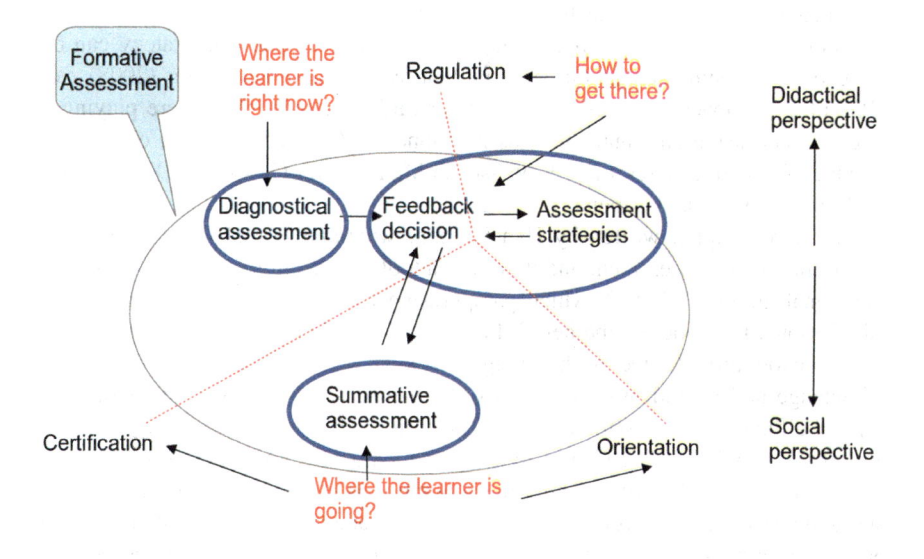

Fig. 1 The three functions of assessment in a formative assessment process

From that starting point, new issues regarding technology arise: does the technology provide help for formative assessment? How is it possible to use such technology? And more precisely, going back to the research questions of FASMEd, how do teachers process formative assessment data from students using a range of technologies? How do teachers inform their future teaching using such data? How is formative assessment data used by students to inform their learning trajectories? When technology is positioned as a learning tool rather than a data logger for the teacher, what issues does this pose for the teacher in terms of them being more informed about student understanding? Three of the four chapters of this section are written by partners of this project and I will not reveal their results; instead, I will let the authors develop the theoretical framework built on both the formative assessment framework and the technological potentialities.

Assessment and Technology
In the first chapter, Annalisa Cusi, Francesca Morselli, and Cristina Sabena discuss the way in which technology may support formative assessment strategies in whole classroom activities. In the first section, they detail the theoretical framework that allows consideration of the dynamics of the classroom through the filter of formative assessment with technology. In this perspective, the authors highlight the strategies of formative assessment that polls with technology can reinforce. It is interesting to notice, through the categorization of polls and the discussion, the technology, formative assessment from isolating each actor, favors discussions and debates within the classroom. Through the orchestration of technology, formative assessment strategies allow the learning dynamic that the theoretical model describes.

The viewpoint of Shai Olsher in the second chapter is completely different. Starting from teacher practices in the process of feedback, he analyzes and discusses the categorization that the teacher uses to implement them in an automatic system. The Formative Assessment process that comes from the feedback strategy can be more or less assimilated to the strategy that Wiliam and Thompson (2008) call "Providing feedback that moves learners forward", the technology here playing a role of "processing and analyzing" as mentioned in Chapter "The Use of Polls to Enhance Formative Assessment Processes in Mathematics Classroom Discussions" by Cusi, Morselli, and Sabena.

The third chapter, written by Hana Ruchniewicz and Bärbel Barzel, takes the viewpoint of auto-assessment and describes a tool, the electronic self-assessment of functional thinking (SAFE), which paraphrasing the gamebooks, would be called "the lesson where you are the hero"! Leaving students' facing functional problems aside, the tool provides feedback, giving the student the opportunity to assess his/her knowledge and to find information in order to overcome misconceptions or conceptual difficulties. The frame of FASMEd is once again presented and technology appears as a tool allowing formative assessment strategies to be put into action.

The step from self-assessment to metacognition is not so large and Barbel, Ball, and Klinger address this issue from the study of mathematical apps that are able to provide information, methods, and results of quadratic equations. The issue that the authors tackle concerns the "kind of formative assessment" that students perform when using these apps, most of the time out of school, and

Formative Assessment from the sight of the teacher. The aim here could be to study the potentialities that these apps can provide with the help of teachers in a Formative Assessment perspective.

Conclusion

A running theme in these chapters is that technology is a tool which helps teachers to take decisions and to assess, but is not a tool that replaces the teacher. The necessary orchestration of technology that has been studied for a long time remains an important factor of the success of technologically enhanced assessment. It's not a question of doing better or faster with technology, but rather that using technology transforms the way formative assessment is proposed in the classroom, both in cognitive and mathematical aspects. The chapter of Cusi, Morselli, and Sabena shows clearly the importance of the teacher when technology gives students and their peers the opportunity to benefit, both from technology and each other. The message here is that technology leads to a deeper understanding of the subject due to facilitating the analysis of difficulties and successes, and the feedback that teachers may use to improve the students' learning; this is even if, as Olsher points out, "Orchestrating the work of students in a technological environment, referred to by Trouche as instrumental orchestration, while collecting information about students that could be used for formative assessment, presents challenges for teachers" (Olsher, in this book, Chapter "Making Good Practice Common Using Computer-Aided Formative Assessment", p. xxx). Self-assessment, as well as metacognition competencies, addressed by Ruchniewicz and Barbel in Chapter "Technology Supporting Student Self-Assessment in the Field of Functions—A Design-Based Research Study" and Ball, Barbel, and Klinger in Chapter "Students' Self-Awareness of Their Mathematical Thinking: Can Self-Assessment be Supported Through CAS-Integrated Learning Apps on Smartphones?", are surely largely improved by the use of technology: a tool built by researchers with a high degree of didactical reflection or tools available on smartphones and largely shared by students. In the two cases, it is noticed that we certainly need further research to understand better how students work and what they gain using these tools alone or with peers.

We can see through all these texts the undeniable contribution of technology to the formative assessment process. But it would certainly be premature to think that technology is transforming teachers' work habits in itself. Precautions should be taken to ensure that teachers see technology as tools to facilitate the implementation of formative assessment. Works, such as that of FaSMEd which promotes tools available to teachers to facilitate the implementation of formative assessment, and training, both pre-service and in-service, are still needed to ensure that the difficulties of introducing technology into the classroom do not exceed the benefits that have been clearly demonstrated in the texts of this part. I finish this introduction by quoting David Wright,[2] who said during the round table that technology brings a lot for Formative Assessment, but that "you have a few issues to take into account if you want to help teachers to use technology: first of all, it has to be easy to use… you can't expect teachers to radically change their practices; there have to be small steps, which fit in with what they are already doing… You need some 'quick wins', which are immediately going to bring about change, while also assuring teachers that they will have technical support to fix any problem. You also need a professional learning community, and a 'champion' who will provide an initial boost and sustain the promotion of innovative activities."

Finally, common to the four chapters is that technology may help students and teachers in the assessment process when feedback will make the former work more than the latter. There is a need to subordinate teaching to learning, and technology may help!

References

Allal, L., & Mottier Lopez, L. (2007). *Régulation des apprentissages en situation scolaire et en formation*. Louvain-la-Neuve, Belgique: De Boeck Supérieur. doi:10.3917/dbu.motti.2007.01.

Black, P., & Wiliam, D. (1998). Assessment and classroom learning. *Assessment in education: Principles, Policy & Practice, 5*(1), 7–74.

Black, P., & Wiliam, D. (2009). Developing the theory of formative assessment. *Educational Assessment, Evaluation and Accountability, 21*(1), 5–31.

Taras, M. (2012). Assessing assessment theories. *Online Educational Research Journal*.

Wiliam, D., & Black, P. (2006). Meanings and consequences: A basis for distinguishing formative and summative functions of assessment? *British Educational Research Journal, 22*(5), 537–548. doi:10.1080/0141192960220502.

Wiliam, D. (2000). Integrating summative and formative functions of assessment. Keynote address to the European Association for Educational Assessment; Prague: Czech Republic, November 2000. http://discovery.ucl.ac.uk/1507176/1/Wiliam2000IntergratingAEA-E_2000_keynoteaddress.pdf.

Wiliam, D., & Thompson, M. (2008). Integrating assessment with instruction: What will it take to make it work? In C. A. Dwyer (Ed.), *The future of assessment: Shaping teaching and learning* (pp. 53–82). Mahwah, NJ: Erlbaum.

---

[2] David Wright was the leader of the Formative AssessmentSMEd project and one of the panelists of the ICTMT 13 round table. https://ictmt13.sciencesconf.org/resource/page/id/16.

The Use of Polls to Enhance Formative Assessment Processes in Mathematics Classroom Discussions

Annalisa Cusi, Francesca Morselli and Cristina Sabena

1 Introduction and Background

Formative assessment (FA) or assessment for learning is generally conceived as a teaching method, where "evidence about student achievement is elicited, interpreted, and used by teachers, learners, or their peers, to make decisions about the next steps in instruction that are likely to be better, or better founded, than the decisions they would have taken in the absence of the evidence that was elicited" (Black & Wiliam, 2009, p. 7). Large-scale reviews have revealed the effectiveness of educational interventions that focus on the development of teaching using FA in classrooms compared to most other intervention approaches (Hattie, 2009). However, FA practices constitute a great challenge for teachers in the classroom and research has started investigating how technology can support them: this was one of the goals of the European project FaSMEd—Improving Progress for Lower Achievers through Formative Assessment in Science and Mathematics Education). Within FaSMEd, we focused in particular on the role that the so-called connected classroom technologies (CCT) may play in FA mathematics classroom processes at primary and lower secondary school levels (Cusi, Morselli, & Sabena, 2017).

CCT are networked systems of computers or handheld devices specifically designed to be used in a classroom for interactive teaching and learning. Previous research has underlined those affordances of CCT that make them effective tools for FA: monitoring students' progress, collecting the content of students' interaction

---

A. Cusi (✉)
Sapienza University of Roma, Rome, Italy
e-mail: annalisa.cusi@uniroma1.it

F. Morselli
University of Genova, Genoa, Italy

C. Sabena
University of Torino, Turin, Italy

© Springer Nature Switzerland AG 2019
G. Aldon and J. Trgalová (eds.), *Technology in Mathematics Teaching*,
Mathematics Education in the Digital Era 13,
https://doi.org/10.1007/978-3-030-19741-4_1

over long timespans and for multiple sets of classroom participants (Roschelle & Pea, 2002); providing students with immediate private feedback, supporting them with appropriate remediation and keeping them oriented on the path to deep conceptual understanding (Irving, 2006); enabling students to take a more active role in classroom discussions and encouraging them to reflect and monitor their own progress (Roschelle & Pea, 2002; Ares, 2008).

Notwithstanding the potential of these tools, many researchers have stressed that their effectiveness depends on the skill of the instructor and on his/her ability to incorporate procedures such as tracking students' progress, keeping students motivated and enhancing reflection with technologies (Irving, 2006; Kay & Le Sage, 2009). Some studies, in particular, have highlighted that CCT increase the complexity of the teacher's role with respect to 'orchestrating' the lesson (Clark-Wilson, 2010; Roschelle & Pea, 2002). Therefore, in order to bring about progress in student participation and achievement, technology must be used in conjunction with particular kinds of teaching strategies. To this respect, Beatty and Gerace (2009) developed technology-enhanced formative assessment (TEFA), a pedagogical approach for teaching science and mathematics with the aid of classroom response system (CRS). CRS[1] consist of a set of input devices for students, communicating with the software running on the instructor's computer, and enabling the instructor to pose questions to students and take a follow-up poll (Beatty & Gerace, 2009).

To help teachers implement FA, the TEFA approach (Beatty & Gerace, 2009) introduces an iterative cycle of question posing, answering, and discussing, which forms a scaffold for structuring whole-class interaction with the aid of CRS. The essential phases of the cycle are:

- pose a challenging question to the students;
- have students wrestle with the question and decide upon a response;
- use a CRS to collect responses and display a chart of the aggregated responses;
- elicit different reasons and justifications from students for the chosen responses;
- develop a student-dominated discussion of the assumptions, perceptions, ideas, and arguments involved;
- provide a summary, micro-lecture, and meta-level comments.

From the 1970s to the 2010s, many research studies on the use of CRS in educational settings have been developed, as documented in meta-reviews that highlighted benefits and challenges of CRS-integrated instruction (Fies & Marshall, 2006; Kay & Le Sage, 2009; Chien, Chang, & Chang, 2016; Hunsu, Adesope, & Bayly, 2016).

Among the benefits of CRS-integrated instruction, research indicates that CRS simultaneously provide anonymity and accountability, support collecting answers from all students in a class rather than just the few who speak up or are called upon, and enable recording data of students' individual and collective responses for subsequent analysis (Beatty & Gerace, 2009).

---

[1] In research literature different terms are used to refer to these devices, such as clickers, Audience Response Systems, Student Response Systems, Interactive Response Systems.

Research has also identified different factors that contribute in increasing the quantity and quality of class discussions, when CRS are used: asking subjects to generate explanations and justifications for their own answers to the questions posed by means of CRS (Chien et al., 2016); engaging students in peer discussion (Chien et al., 2016); and collecting students' responses and presenting them to the class without providing the correct answer (Kay & Le Sage, 2009).

Beatty and Gerace (2009) highlight that CRS may be exploited by the teacher with great flexibility, and list specific instructional purposes connected to their use. Among them are the uses of polls for:

- status check, that is to ask students their self-reported degree of confidence in their understanding of a topic;
- exit poll, that is to poll students to find out which concepts they want to spend more time on;
- assess prior knowledge, that is to elicit what students know or believe about a topic;
- provoke thinking, that is to ask a question to get students engaged within a new topic;
- elicit a misconception, that is to lead students to manifest a specific common misconception or belief that may hinder their learning;
- exercise a cognitive skill, that is to engage students in a specific cognitive activity;
- stimulate discussion with questions having multiple reasonable answers;
- review, that is to pose questions aimed at reminding students of a body of material already covered.

Meta-reviews also highlight specific aspects on which future research should focus. First of all, most of the meta-reviews stress the need for further research on the use of CRS in K-12 classrooms, since most of the examined studies used undergraduates or graduates as research samples (Kay & Le Sage, 2009; Chien et al., 2016; Hunsu et al., 2016). Others focus on the methodologies through which CRS are used in the classroom and, in particular, on the ways in which students are engaged in CRS-integrated instruction. Fies and Marshall (2006), for example, suggest deeply exploring the effects of group mode use of CRS. In tune with this idea, Kay and Le Sage (2009) suggest studying the influence of students' comparison with peers on developing a classroom community. Finally, we remark that another important feature highlighted is the need to explore how CRS could be used in order to promote meta-cognitive and self-regulatory learning strategies in classes (Hunsu et al., 2016).

Against this background, in our study we focused on the use of polls to enhance effective classroom discussions with FA purposes at primary and lower secondary school levels.

2 Formative Assessment with Technology: A Theoretical Framework

Wiliam and Thompson (2007) identified five key strategies for FA:

(A) Clarifying and sharing learning intentions and criteria for success;
(B) Engineering effective classroom discussions and other learning tasks that elicit evidence of student understanding;
(C) Providing feedback that moves learners forward;
(D) Activating students as instructional resources for one another;
(E) Activating students as the owners of their own learning.

These strategies may be activated by three agents: the teacher, the peers and the student themselves. Technology, indeed, may support the three agents in activating the FA strategies in different ways.

Within the FaSMEd project, we developed a three-dimensional framework for the design and implementation of technologically-enhanced formative assessment activities. The framework is represented in the chart[2] in Fig. 1 and extends Wiliam and Thompson's model (whose dimensions are the five key strategies for FA and the agents), adding to it the dimension related to the functionalities through which technology could support FA (Aldon, Cusi, Morselli, Panero, & Sabena, 2017; Cusi et al., 2017). These functionalities are:

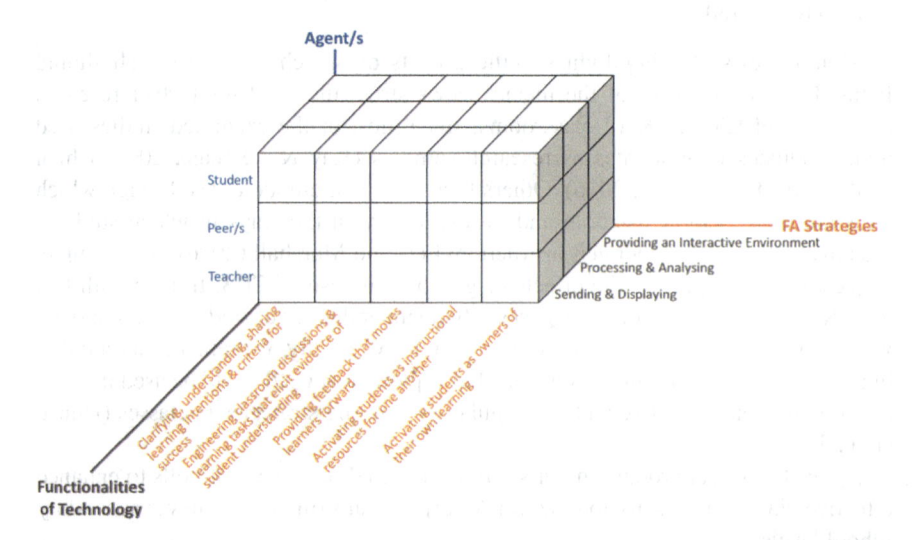

Fig. 1 Chart of the FaSMEd three-dimensional framework (the three dimensions—FA strategies, agents, functionalities of technology—are represented in the three axes of the diagram)

---

[2]We thank D. Wright (Newcastle University) for the digital version of the chart and Hana Ruchniewicz (University of Duisburg-Essen) for its adaptation.

(1) *Sending and displaying*, that is the ways in which technology supports the communication among the agents of FA processes (e.g., sending and receiving messages and files, displaying and sharing screens or documents to the whole class).
(2) *Processing and analysing*, that is the ways in which technology supports the processing and the analysis of the data collected during the lessons (e.g., through the sharing of the statistics of students' answers to polls or questionnaires, feedback given directly by the technology to the students when they are performing a test).
(3) *Providing an interactive environment*, that is when technology enables the creation of environments in which students can interact to work individually or in groups on a task or to explore mathematical/scientific contents (e.g., through the creation of interactive boards to be shared by teacher and students or the use of specific software that provides an environment where it is possible to dynamically explore specific mathematical problems).

A fundamental aspect on which FA is focused is feedback, defined by Hattie and Timperley (2007) as "information provided by an agent (e.g. teacher, peer, book, parent, self, experience) regarding aspects of one's performance or understanding" (p. 81).

In our design and in the data analysis we refer, in particular, to the four major levels of feedback introduced by Hattie and Timperley (ibid.):

- feedback about the task, which concerns how well a task is being accomplished or performed;
- feedback about the processing of the task, which concerns the processes underlying tasks or relating and extending tasks;
- feedback about self-regulation, which refers to the way students monitor, direct, and regulate actions toward the learning goal;
- feedback about the self as a person, which consists in positive (and sometimes negative) evaluations of and effects on the student.

3 Designing FA Activities Within a CCT Environment

The study documented in this paper is part of a wider design-based research, characterized by cycles of design, enactment, analysis and redesign, where the goal of designing learning environments is intertwined with that of developing new theories (DBRC, 2003). The research is carried out in authentic settings (classroom environments) focusing on "interactions that refine our understanding of the learning issues involved" (DBRC, 2003, p. 5).

In our design study, in tune with the FA framework, we focus on the crucial role of the interaction with peers and with an expert in students' learning. In line with Vygotskian perspectives (Vygotsky, 1978), we consider effective mathematical discussions (Bartolini Bussi, 1998) as fundamental activities, where the teacher plays a key role in planning and promoting fruitful occasions for FA and learning.

Moreover, we also believe that FA has to focus on affective (Hannula, 2011) and metacognitive (Schoenfeld, 1992) factors. Accordingly, we designed activities aimed at supporting students in (a) making their thinking visible (Collins, Brown, & Newmann, 1989) through the sharing of their thinking processes with the teacher and classmates by means of argumentative processes, (b) developing their ongoing reflections on the learning processes and (c) discussing their emotions during the FA activity.

An important feature of the task design is a strong argumentative component: students are always required to explain their answers in a written text. Students' argumentations are then collectively analysed according to three criteria: the correctness (Do these justifications contain any mistake?), the clearness (Is every reader able to easily understand these justifications?) and the completeness (Do these justifications contain all the information necessary to draw these conclusions?) of the justifications provided by the students.

Concerning technology, we explored the use of a CCT (provided by a software called IDM-TClass), which connects the students' tablets with the teacher's laptop, allows the students to share their productions and the teacher to easily collect the students' opinions and reflections, during or at the end of an activity, by means of the creation of instant polls.

The use of IDM-TClass was integrated within a set of activities on relations and functions as well as their representations (symbolic representations, tables, graphs) adapted from different sources (the ArAl Project: www.progettoaral.it; and the Mathematics Assessment Program: http://map.mathshell.org). For each activity, we designed a sequence of worksheets (doc files), to be sent to the students' tablets or to be displayed on the interactive whiteboard (or through the data projector).

The worksheets were designed according to three main categories: (1) worksheets introducing a problem and asking one or more questions (problem worksheets); (2) worksheets aimed at providing support to students who met difficulties in facing the given tasks (helping worksheets); and (3) worksheets prompting a poll between proposed options (poll worksheets).

Concerning the poll modality, IDM-TClass software collects all the students' choices and processes them, displaying an analytical record (collection of each answer) as well as a synthetic overview (bar chart). In reference to the analytical framework, instant polls are used through the support of the "Processing and Analysing" functionality of technology. The possibility of showing the results in real time also brings to the fore the "Sending and Displaying" functionality of technology.

In principle, the software also enables the teacher to set the time given to students before completing the poll, and offers the opportunity to provide an immediate automatic correction to the student. However, our choice is not to provide an immediate automatic correction to students, so that they can instead be engaged in a subsequent classroom discussion. In tune with Beatty and Gerace's (2009) framework, in fact, we conceive the use of polls as a way of scaffolding whole-class interaction with the aim of fostering the sharing of results and the comparison between students (FA strategy B). This is also coherent with our belief on the key role of the teacher and the importance of peer interaction.

During the design experiments, we implemented planned polls and instant polls. Planned polls were created a priori and were part of each teaching sequence. They were realized through poll worksheets, which can be used alternatively to problem worksheets. Instant polls were on the contrary created and implemented on the spot during the lesson. In design-based research, instant polls that prove fruitful in terms of FA strategies may be inserted in the repertoire of planned polls for subsequent cycles of experimentation.

4 Categories of Polls

In our design, polls are always intended as a starting point for class discussion and not for individual "revising" or "status check". After three cycles of design, implementation and analysis of the classroom activities, we classified polls according to the different foci and aims of the classroom discussions developed from them.

We identified four categories of polls, which are presented in Table 1, together with the corresponding aim and an example from our design experiments. Two examples will be presented in an extended way in the data analysis section.

Referring to the instructional purposes of polls described by Beatty and Gerace's framework (2009), polls belonging to category 1 may be related to the aims to "provoke thinking" and "exercise a cognitive skill", whereas the polls belonging to category 2 may be linked to the aims to "elicit a misconception" and "stimulate discussion with questions having multiple reasonable answers". Polls belonging to categories 3 and 4 are of different nature: even if they could be somehow related to "status check", they bring to the fore metacognitive and affective issues that are not so evident in Beatty and Gerace's list.

5 Research Questions and Methodology

Concerning polls, our investigation is guided by the following research questions:

- *Which FA strategies can be activated by means of technologically enhanced polls?*
- *What are the main characteristics of the FA discussions developed by means of technologically enhanced polls? i.e.,: How could they be initiated? How could they evolve?*

All the lessons were video-recorded, fields notes were taken and students' productions (doc files) were collected, building a large amount of data (about 450 h of class sessions, carried out in collaboration with 20 teachers).

Table 1 Categories of polls resulting from our design experiments

Category of poll	Corresponding aims	Example from our design experiments
(1) Polls on specific mathematical content: these polls ask to choose the correct answer to a problem or to a specific question	– To highlight students' understanding of specific topics developed during the lesson – To make students discuss the reasons linked to the choice of the correct answers (providing feedback about the task) – To promote a discussion on the solving strategies, in order to enable students to share and compare them (providing feedback about the processing of the task)	Every morning Tommaso walks a straight road from his home to a bus stop, a distance of 160 m. The graph shows his journey on one particular day *Every morning Tommaso walks along a straight road from his home to a bus stop, a distance of 160 meters. The graph shows his journey on one particular day.* (3) After how many seconds does Tommaso reach the bus stop? (a) After 120 s; (b) After 50+70+100+120 seconds, that is after 340 seconds; (c) After 100 seconds; (d) After 50 seconds. After how many seconds does Tommaso reach the bus stop? (a) After 120 s (b) After 50 + 70 + 100 + 120 s, that is after 340 s (c) After 100 s (d) After 50 s

(continued)

Table 1 (continued)

Category of poll	Corresponding aims	Example from our design experiments
(2) Polls on argumentation: these polls ask students to compare different justifications about the answers to a given problem	– To promote a discussion at a meta-mathematical level, focused on the way the answer is justified. Thanks to this kind of discussion, students are led to identify the criteria to assess the answers. Hence, students receive and provide feedback on the processing of the task and feedback about self-regulation	See Sect. 7
(3) Polls on metacognitive aspects: these polls are focused on the difficulties students meet when facing specific kind of tasks or on the best strategies to be used to face specific tasks	– To promote metacognitive reflections that could help students in identifying the available tools to face similar tasks in the future and in becoming aware of how to monitor themselves while facing this kind of task (providing feedback about self-regulation)	See Sect. 8
(4) Polls on affective aspects: these polls are focused on students' feelings when facing a specific kind of task or when a particular methodology is adopted during the lessons	– To bring to the fore the affective dimension, supporting students in becoming aware of their way of posing themselves during this kind of activity (providing feedback about self-regulation)	How did you feel when your answer was displayed on the interactive whiteboard? A) Uneasy B) Happy C) Calm D) Worried

In line with design-based research, the study is carried out through close collaboration between researchers and teachers, who share the aim of improving practice, taking into account both contextual constraints and research aims.

At least one researcher was always present in the classroom as participant observer during the design experiments.

The analysis of the video-recordings of class discussion was developed according to the following methodology:

- a preliminary selection of class sessions was made on the basis of researchers' direct observations;
- the selected classroom discussion episodes were transcribed and analysed separately by the researchers, who coded the transcripts in terms of FA strategies;
- problematic codes were discussed together so that researchers could come to an agreement.

6 Data Analysis

Hereunder we present two examples taken from our experiments. Both examples refer to a task sequence on time-distance graphs adapted from the task sequence "Interpreting time-distance graphs", from the Mathematics Assessment Program (http://map.mathshell.org/materials/lessons.php). From the original source based on paper and pencil materials for grade 8, we adapted the activities and created a set of 19 worksheets to be used with students from grades 5–7 (ages 10–12).

The sequence starts with a short text about the walk of a student, Tommaso, from home to the bus stop. This text is accompanied by a time-distance graph, as illustrated in Fig. 2.

Students' interpretation of this graph is guided through questions, posed to them within problem, helping and poll worksheets. After interpretation of the time-distance graph according to the given story, the activity develops through the matching of different graphs and the corresponding stories and the construction of graphs associated to specific stories. Since this was students' first encounter with time-distance graphs, we designed an introductory activity based on the use of a motion sensor, in which students could explore in a laboratory way the construction of the graph after a motion experience along a straight line. The students worked in two ways: first, they walked and observed the graph of their motion provided by the motion sensor; after, they were given some graphs and were asked to walk so as to obtain the same graph by means of the motion sensor. An inspiring reference for the design of this introductory activity was the research on children's use of motions detectors to make sense of Cartesian graphs of position versus time (see, for instance, Nemirovsky, Tierney, & Wright, 1998).

7 Example 1: A Classroom Discussion from a Poll on Argumentation

In the following, we present an excerpt from a grade 7 class discussion starting from a poll belonging to category 2: Polls that ask to compare different answers to a problem. The episode concerns the interpretation of the final part of the graph in Fig. 2. At first, students were asked via a problem worksheet to establish what happens during the last 20 s, and to justify their answers. During the classroom discussion, a poll worksheet was used to focus on the completeness of answers, which may be referred to as FA strategy A (Clarifying and sharing learning intentions and criteria for success). Specifically, the poll required students to identify which is the most complete among three given answers:

Some students of another class wrote these answers. Which of them is the most complete?

(A) During the last 20 s, Tommaso is not walking because we have already said that he has reached the bus stop.
(B) I think that, during the last 20 s, Tommaso is not walking because from the graph it is possible to understand that, in the period between 100 s and 120 s, he is always at the same distance from home, that is 160 m.
(C) I understood that, during the last 20 s, Tommaso is not walking because the line of the graph is horizontal.

Option B represents the most complete answer to the question about what happens to Tommaso during the last 20 s because it refers to the correct interpretation of the graph in terms of time and corresponding distance from home. Option A is the typical

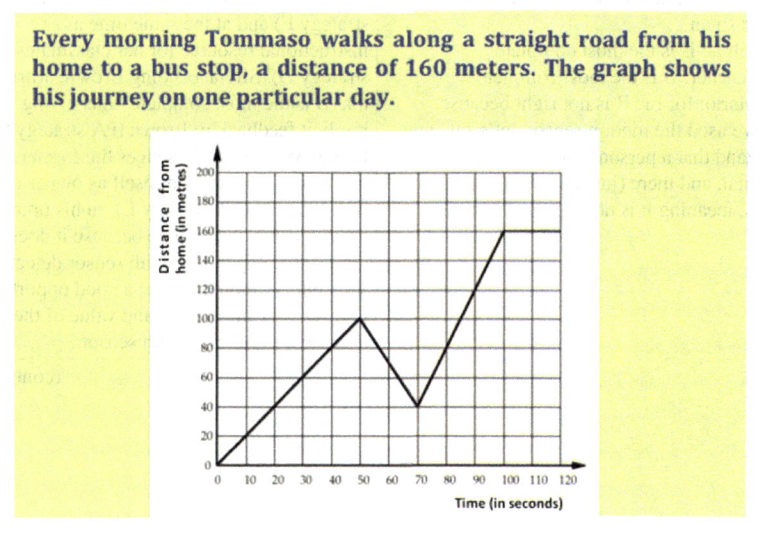

Fig. 2 The time-distance graph of Tommaso's walk

justification provided by a student who is not referring to the graph, but only to their knowledge about "the end of the story" (Tommaso reaches the bus stop). Option C could be the typical answer of those students who prefer to refer to their previous experience with the motion sensor (through which they discovered that a horizontal line represents the fact that an object is not moving), instead of trying to understand how a conclusion could be drawn through a deeper interpretation of the graph.

In order to answer to the poll, students discussed the question in pairs. After all the pairs sent their answers, the teacher displayed the distribution of answers on the IWB: 10% of the students chose option A, 50% chose option B and 40% chose option C. Starting from the display of the results, the discussion took place. The teacher exploited the poll worksheet as a way to engineer effective classroom discussions that elicit evidence of student understanding (FA strategy B). Table 2 presents selected excerpts from the discussion, analysed according to the FaSMEd framework.

Table 2 Excerpts from the class discussion and corresponding analysis

Excerpts from the class discussion	Analysis according to the FaSMEd three-dimensional framework
After brief analysis of A, justifications B and C are compared (353) Teacher: let's look at B and C. Let's hear some explanations of those who chose C, why they chose C and some motivations of those who chose B (354) Brown: we chose B because B specifies also that he (Tommaso) stayed still from 100 to 120 s, while C doesn't say this, saying that they were only 20 s; they could have been 150, 170, 180 and so on… (355) Silvia: B is the most complete (356) Teacher: B is the most complete (357) Mario: for me B is not right because, when we used the motion sensor, let's say, you understand that a person stops when the line is horizontal, and there (justification B) it doesn't say this, meaning it is not the most complete	The teacher encourages the students to discuss the reasons behind the choices of the poll. Her aim is to promote a discussion on the completeness of the two options. This is an instance of FA Strategy A, since the focus is on the requirements that a complete answer must satisfy Suggesting that answer B gives more information on the last trait, Brown activates herself as responsible for her learning (FA strategy E) and at the same time as instructional resource for her classmates (FA strategy D). Silvia, echoing Brown, affirms that B is the most complete, thus giving implicit feedback to Brown (FA strategy C). In line 357 Mario challenges the former evaluation, activating himself as owner of his own learning (FA strategy E): in his opinion, answer B is not complete because it does not refer to the experience with sensor detectors. This intervention provides a good opportunity to discuss again the role and value of the empirical experience with sensors

(continued)

Table 2 (continued)

Excerpts from the class discussion	Analysis according to the FaSMEd three-dimensional framework
(390) Lollo: but if we had not done that activity before… (391) Teacher: the activity with the motion sensor (392) Lollo: we could not have known that if you are still the line is horizontal	Lollo suggests that one cannot refer to the experience with sensors, since the answer should also be intelligible by a reader who did not have such an experience. Lollo turns himself as an instructional resource for his classmates (FA strategy D), in particular giving feedback to Mario (FA strategy C). The teacher reformulates Lollo's intervention so as to involve the other students, turning Lollo into a resource for his classmates (FA strategy D). In this way, she also activates FA Strategy C
(399) Rob: And, anyway, from the graph you can understand why the distance is always the same but the seconds, let's say, go on… (400) Teacher: ok… then, even if we had not had the experience with the motion sensor, that made you understand in an experimental way that if I stay still the line is horizontal, your classmate [Rob] says: "from the graph I can understand it anyway". Why? Rob, could you please repeat it? (401) Rob: because from the graph you can understand that when you don't move, that is to say when there is the horizontal line… (402) Teacher: what does it mean? (403) Rob: the metres remain the same but the seconds go on, let's say	Rob intervenes, stating that in the horizontal trait the distance from home is always the same. This is a shift from an explanation based on the experience with sensors to a theoretical explanation, based on the meaning of the graph. Rob provides other students with feedback to move forward (FA strategy C), turning himself into an instructional resource for his classmates (FA strategy D) The teacher reformulates Rob's intervention, giving all the students feedback that moves them forward (FA strategy C). Reformulation is also a means to activate Rob as a resource for his classmates (FA strategy D)
(413) Teacher: B explains why the line is horizontal, while C just says "the line is horizontal"; B instead explains why the line is horizontal, because the metres remain the same, even if time goes on, isn't it?	As a final intervention, the teacher rephrases the result of the discussion, pointing out what makes answer B more complete than the other options. In this way, she activates FA strategy A

8 Example 2: A Discussion from a Poll on Metacognitive Aspects

In the following, we present an excerpt from a grade 5 class discussion starting with a poll belonging to category 3: Polls on metacognitive aspects. The discussion developed from the results of an instant poll. The instant poll was proposed at the end of the task sequence on time-distance graphs, created on the spot by the teacher (T) and researcher (R) with the aim of boosting a metacognitive reflection on effective ways to tackle graph interpretation tasks. Here is the wording of the poll:

"When interpreting a graph, what is the first thing you look at?"

(a) If the graph starts from the origin
(b) If the graph goes up or down
(c) If the graph has horizontal traits
(d) How many traits compose the graph
(e) How steep is the graph
(f) What is written on the axes.

We may note that, different from the poll in example 1, this poll does not encompass only one correct answer. The subsequent discussion is aimed at making students' strategies visible when approaching a graph and comparing the efficiency of such strategies.

Most students (72%) chose F ("What is written on the axes"); 18% chose A ("If the graph starts from the origin") and 9% chose C ("If the graph has horizontal traits") (see Fig. 3 for the representation of the results that was displayed to the class).

Starting from the display of the results, the discussion took place. The teacher exploited the poll as a way to engineer effective classroom discussions that elicited evidence of student understanding (*FA strategy B*). Table 3 presents selected excerpts from the discussion, analysed according to the FaSMEd framework.

Fig. 3 Results of the instant poll, as displayed on the IWB

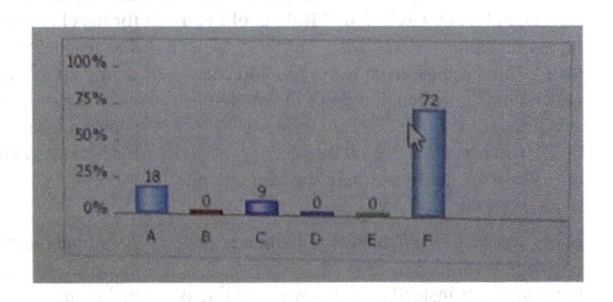

Table 3 Excerpts from the class discussion and corresponding analysis

Excerpts from the class discussion	Analysis according to the FaSMEd three-dimensional framework
1. Researcher: Here we have 72% that answered F 2. Teacher: the axes 3. Researcher: "What is written on the axes". Someone chose A: "If the graph starts from the origin". Someone chose C: "If there are horizontal traits". The other options were not chosen. Some of you said you changed her mind. Would you like to tell me your answer now? (speaking to Sabrina) 4. Sabrina: We chose A, but later we changed our mind. We want to choose F 5. Researcher: So, actually for you it is F?	The functionality of technology is processing and analysing, since the display of data is the starting point for a discussion. The activated FA strategy is B (engineering effective classroom discussions) Immediately after the display of the results, Sabrina and her classmate ask to change their choice: this can mean that they recognise they have answered without deep reflection. Recognising this and asking to amend the answer is an instance of FA strategy E (they make themselves responsible for their own learning)

(continued)

Table 3 (continued)

Excerpts from the class discussion	Analysis according to the FaSMEd three-dimensional framework
6. Researcher: We could start from F. Why do you think the first thing to look at is what is written on the axes? 7. Some students raise their hands 8. Elsa: Because, if you look at what is written on the axes, you can already understand the graph… and you can get some information 9. Researcher: Let's listen to somebody else. Carlo 10. Carlo (he worked in a pair with Elsa): I wanted to say that on the axes it is written what they are, what you have to measure, look at, observe… 11. Researcher: Ok	The initiation of the discussion consists of focusing on the most chosen answer (remember here there is not only one correct answer). The teacher and the researcher choose to focus on answer F also because it is undoubtedly important and efficient to start the work on a graph with analysis of the axis Elsa and Carlo explain to their classmates that, by knowing which variables are represented on the axes, one can get much information on what is represented in the graph. Elsa and Carlo, therefore, turn themselves into resources for their classmates (FA strategy D)
12. Luca: Also on the axes… if it had been the contrary, with here (with gestures, he draws a vertical line) the time and here (with gestures, he draws a horizontal line) the distance, the graph would have changed… (he draws with gestures a possible new graph) 13. Researcher: Did you listen to what Luca said? (she is speaking to the other students) 14. Voices: Yes! 15. Researcher: I guess that somebody did not listen 16. Teacher: He said a very interesting thing 17. Researcher: Would you like to repeat what Luca said? (to Lavinia, who raised her hand) 18. Lavinia: We always have the distance from home (with gestures, she draws a vertical line) and time (with gestures, she draws a horizontal line), but maybe, in order to mislead us… 19. Teacher: Or because it is represented in another way. It could be! Then, it could be written… 20. Lavinia: In another way (with gestures, she draws the vertical and horizontal axes)	Luca points out that inverting the two variables represented on the axes leads to different graphs The researcher, in order to highlight Luca's intervention and to turn Luca into a real resource for the classmates (FA strategy D), carries out the following strategy: she asks if the other pupils listened to what Luca said, and asks another pupil to repeat it. This strategy makes Luca's thinking visible to the classmates Lavinia tries to repeat Luca's idea, activating herself as responsible for her learning (strategy E), but she speaks of "misleading" rather than of a different graph The teacher interrupts Lavinia and clarifies that there could be other possible graphs characterised by different variables represented on the two axes

(continued)

Table 3 (continued)

Excerpts from the class discussion	Analysis according to the FaSMEd three-dimensional framework
21. Researcher: Yes, that time is on the vertical axis and distance on the horizontal one. Luca said that, in that case, the graph changes. Or the same graph is interpreted in a different way. If I go to the blackboard… I was thinking of the impossible graph (in facing one of the preceding problem worksheets of the sequence, the class had worked on an impossible graph: time was represented on the horizontal axis, distance was represented on the vertical axis and the graph contained a vertical trait). Let's draw it… 22. Researcher: Ok. Now I draw only the vertical trait. If the time is here and the distance is here (she draws on the blackboard, see Fig. 4), we said that a vertical trait is impossible, isn't it? 23. Voices: Yes 24. Researcher: And if, instead, I put the time on the vertical axis, as Luca said, and the distance on the horizontal axis, would it be impossible? (she exchanges the variables on the axis in the same graph on the blackboard—see Fig. 5) 25. Voices: No! 26. Student: It is possible!	The researcher recalls to the students an impossible graph that was encountered in a previous problem worksheet. Her intervention is aimed at promoting a collective reflection on the fact that, if the variables change, the graph must be interpreted in another way We highlight the crucial choice of recalling a previous experience, occurred during the task sequence. In this way, the researcher and the teacher may also collect some feedback about the previous activities (strategy C)
27. Researcher: How would it be? Many students raise their hands 28. Researcher: Livio 29. Livio: In my opinion, it is impossible, because he does not move… because… if… (he points to the drawing on the blackboard) 30. Teacher: You can come to the blackboard Livio goes to the blackboard 31. Livio: If this is time (he points to the vertical axis), he spends this amount of time (he points to the vertical trait of the graph, Fig. 6), but he remains still, (he points to the horizontal axis, Fig. 7) in this trait of time (he points again to the vertical trait, Fig. 8)	At first, Livio asserts that the graph is impossible even when variables change; however, afterwards, he formulates a new interpretation Answering the teacher and going to the blackboard, expressing his ideas, Livio acts as responsible for his own learning (FA strategy E) and finally turns himself into a resource for his classmates (FA strategy D) We point out that Livio had not chosen option F when answering the poll. This means that the discussion enabled him to focus his attention on new specific aspects The new question posed by the researcher (concerning the new graph on the blackboard) makes Livio reflect on the importance of analysing the variables on the axes, and understand that the change of variables causes a change in interpretation of the graph

(continued)

Table 3 (continued)

Excerpts from the class discussion	Analysis according to the FaSMEd three-dimensional framework
32. Teacher: At this point … 33. Researcher: He remains still because the distance… 34. Livio: It is zero 35. Researcher: It is not zero 36. Livio: He is still here, he doesn't go there (with gestures he does a horizontal movement)	The aim of the teacher and the researcher is to make Livio realise why the graph on the blackboard represents that a person is still. In this way, they are activating him as a resource for his classmates (FA strategy D). In this phase of the discussion, Livio and the researcher are using the same word ("distance") with different meanings; in particular, Livio seems to interpret the word "distance" in terms of "walked distance" and not "distance from home"
37. Researcher: Let's listen to someone else 38. Teacher: Who agrees with him? Or on the contrary: who doesn't agree? 39. Researcher: Does somebody want to add something to what Livio said? Some students say no. Carlo raises his hand 40. Carlo: I wanted to say that he (Tommaso) spends some time still, at a given distance from home 41. Researcher: You wanted to add this. This idea is very interesting. We must not think that on the horizontal axis there is always the time and on the vertical axis the distance, we must be careful and check these aspects!	Carlo activates himself as a resource for Livio and his classmates (FA strategy D), proposing a correct interpretation of the word "distance" (in term of "distance from home") and clarifying the meaning of what Livio was trying to say (lines 34–36) Finally, the researcher gives feedback that moves the learners forward (FA strategy C), since she highlights the need to carefully analyse the variables represented on the axes as an efficient starting strategy for all activities involving interpretation of graphs

Fig. 4 The first graph drawn on the blackboard

Fig. 5 The new graph on the blackboard

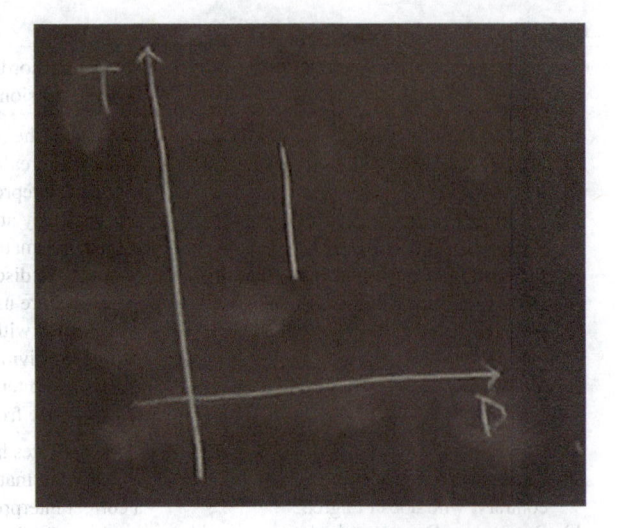

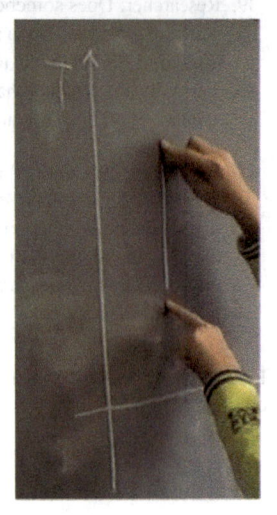

Fig. 6 Livio pointing to the vertical trait of the graph

Fig. 7 Livio pointing to the horizontal axis

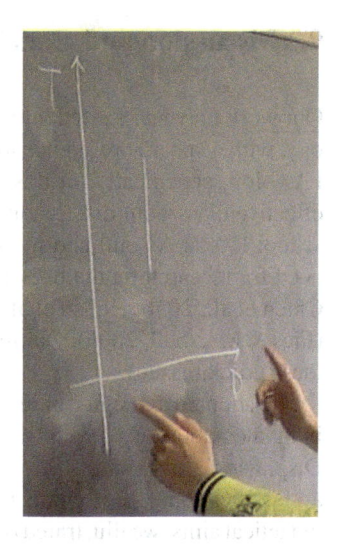

Fig. 8 Livio pointing again to the vertical trait

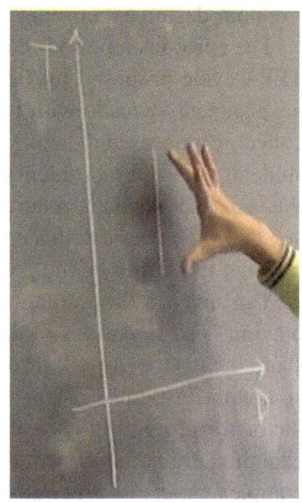

9 Discussion and Conclusions

Our work may be inserted into the stream of research concerning the use of technology, with a specific reference to Connected Classroom Technologies, for promoting FA. More specifically, in this chapter we focused on the use of polls to enhance effective classroom discussions with FA purposes at primary and lower secondary school levels. As outlined in our background paragraph, meta-reviews suggest the need for research on the use of CRS at lower school levels (Kay & Le Sage, 2009; Chien et al., 2016; Hunsu et al., 2016) and on the use of polls by groups of students (Fies & Marshall, 2006). Our research aimed at providing new insights in relation to these crucial issues.

We adopted a design-based approach and studied the use of polls for promoting FA in mathematical lessons through the "Processing and Analysing" and "Sending and Displaying" functionalities of the technology (Aldon et al., 2017). After presenting a classification of polls into four categories in relation to their content and their didactical aims, we illustrated two categories by means of the analysis of two episodes from our design experiments.

The episodes reported in examples 1 and 2 resonate with the main phases of the TEFA cycle proposed by Beatty and Gerace (2009): pose a challenging question; have students wrestle with the question and decide upon a response; use a CRS to collect responses and display a chart; elicit different reasons and justifications from students; develop a student-dominated discussion; provide meta-level comments. We remark a dialectic relationship between group work and poll: students work in small groups to establish a common answer to the poll, and students then discuss all together the results of the poll.

The analysis shows that the "Processing and Analysing" and "Sending and Displaying" functionalities of technology are efficient tools in enabling the teacher to share the results of polls with students and to structure around them a class discussion with FA purposes. This is in line with what research has identified as a crucial factor for promoting class discussion through the use of CRS: asking students to explain and justify their own answers (Chien et al., 2016). In these discussions, a complex variety of FA strategies emerged with the involvement of all the classroom actors (students, peers and teacher). Moreover, the overall data analysis (based on about 450 h of video-recording) allowed us to also highlight different structures of classroom discussions (and corresponding patterns of activated FA strategies) developed from polls. In the following we will present the main structures we identified, referring also to the two examples analysed in this paper.

First of all, we can distinguish between two main ways of initiating classroom discussions according to the percentages of students' answers in the case of polls with only one correct answer (categories 1 and 2).

When percentages of the correct answer and of one or more incorrect ones are quite balanced, the discussion asks students to compare these two (or more) options and to express the motivation for their choice. In this way, it is possible to focus on the mistakes that could lead to the choice of incorrect answers and to make students activate themselves as owners of their own learning (FA strategy E). This is the case shown in example 1, which refers to a poll on argumentation in which the percentage for answer B (the correct one) is 50% and the percentage for answer C (incorrect) is 40%: as we can see from the excerpt, during the discussion the students could recognise their own mistakes and reflect on the reasons for them. In addition, students who chose the correct answer could benefit from the discussion, because they were asked to make their justification explicit; hence, they could develop their awareness about the reasons why they chose a specific option (again activation of FA strategy E). Furthermore, throughout the discussion, students are activated as instructional resources for their classmates (FA strategy D) because they give feedback to each other (FA strategy C) on the reasons why a chosen option is better than another.

When the percentages of students' answers are not balanced, the initiation of the discussion on the poll is different. After displaying the results of the poll, the teacher usually starts the discussion asking those who chose the incorrect answer to explain their choice (phase 1). This strategy is fruitful because it allows the students to focus on the mistakes that led to the choice of incorrect answers, making them activate themselves as owners of their own learning (FA strategy E). In this way, students can receive feedback about the task (FA strategy C) and are, thus, supported in recognising their own mistakes and reflecting on the reasons for them.

Afterwards, the teacher asks those who chose the correct answer (without revealing it is correct), to explain the reasons for their choice (phase 2). In this way, the students are led to focus on the justifications for the given answers and, then, can become more aware of the strategies they applied to solve the task (FA strategy E). As a consequence, the solving strategies are shared and discussed within the class. The students, activated as instructional resources for their classmates (FA strategy D), can then give feedback to each other (FA strategy C). Another important focus of this phase of the discussion is the comparison between the different ways of solving the task, in order to highlight the most efficient and to give feedback about the processing of the task. This kind of meta-level analysis is carried out to make students become aware of the most effective ways of facing specific tasks. In this way, feedback about self-regulation is provided and students could expand the "repertoire" of possible strategies to adopt when again facing similar problems.

During phases 1 and 2 of the discussion, some groups, who faced difficulties in choosing an answer, are asked to share their doubts and difficulties with their classmates. In this way, misconceptions are elicited and, consequently, students could give and receive feedback. Moreover, it often happens that some pairs/groups declare that they changed their mind during the discussion. During these phases of the discussion it is important to enable these students to share the reasons why they changed their mind, in order to activate themselves as owners of their own learning (FA strategy E).

The structure of classroom discussions is different when the discussions refer to a poll in which there is not one correct answer (categories 3 and 4).

Example 2, which refers to a poll on metacognitive aspects (in this case, efficient strategies to address problems containing graphs), illustrates a typical way of initiating the discussion in the case of polls belonging to categories 3 and 4. The discussion is initiated with the analysis of the most chosen option in order to raise meta-level issues. The display of results makes some students revise their initial answer, thus making themselves responsible for their own learning (FA strategy E), or intervene to justify their choice, acting as resources for their classmates (FA strategy D). An example of this activation of FA strategies E and D is in lines 12–15, when Luca expresses his idea (FA strategy E) and the researcher relaunches it, so as to also involve the other students in the reflection and to turn Luca into a resource for his peers (FA strategy D). During the discussion, the teacher and the researcher intervene to give feedback that moves the students' learning forward (FA strategy C). This approach to the activation of classroom discussions is an instantiation of how CRS could be used in order to promote metacognitive strategies in classes, as advocated by research (Hunsu et al., 2016).

Another way of initiating (or developing) the discussion on metacognitive aspects [and, thus, promoting metacognitive strategies in the sense of Husu, Adesope, & Bayly, (2016)] is to focus on the options that where not chosen by students and to ask them to explain why they did not choose these. In this way, other meta-level issues could be faced because students are stimulated to reflect on the reasons why they preferred one option instead of another.

The episode presented in example 2 also represents an example of instant activation of a poll. Other frequent cases when the instant poll is initiated are situations where, during a classroom discussion on a problem worksheet, some incorrect answers emerge and the teacher decides to check whether all the students are aware of the non-correctness of these answers. This use of the poll may be linked to Beatty and Gerace's (2009) instructional purpose of eliciting a misconception. In the case in which this kind of poll is activated, students are asked to express their agreement with the incorrect solution, or to choose between different solutions (this kind of poll belongs to category 1 or 2); the results of the poll are displayed and subsequent discussion is aimed at working together towards a shared correct answer. Students who express the incorrect idea receive feedback (FA strategy C) and are led to become responsible for their own learning (FA strategy E); students who point out that the idea is not correct act also as resources for their classmates (FA strategy D).

Data analysis confirms that the teacher plays a crucial role in structuring the classroom discussion starting from polls to foster the activation of FA strategies. We believe that the different structures that we described, besides contributing to a theoretical reflection on the way polls may foster formative assessment, may also serve as a basis for guiding teacher practice and could be exploited in terms of teacher education. To this regard, we remark that in our design experiments the researcher collaborated with the teachers in developing effective classroom discussion. Further research is needed on the ways of supporting the teachers' autonomous use of this kind of resource.

Acknowledgements Research funded by the European Community's Seventh Framework Programme fp7/2007–2013 under grant agreement No. [612337].

References

Aldon, G., Cusi, A., Morselli, F., Panero, M., & Sabena, C. (2017). Formative assessment and technology: Reflections developed through the collaboration between teachers and researchers. In G. Aldon, F. Hitt, L. Bazzini, & U. Gellert (Eds.), *Mathematics and technology: A CIEAEM source book*. Series 'Advances in mathematics education'. Springer International Publishing.

Ares, N. (2008). Cultural practices in networked classroom learning environments. *Computer-Supported Collaborative Learning, 3*, 301–326.

Bartolini Bussi, M. G. (1998). Verbal interaction in mathematics classroom: A Vygotskian analysis. In H. Steinbring, M. G. Bartolini Bussi, & A. Sierpinska (Eds.), *Language and communication in mathematics classroom* (pp. 65–84). Reston, VA: NCTM.

Beatty, I. D., & Gerace, W. J. (2009). Technology-enhanced formative assessment: A research-based pedagogy for teaching science with classroom response technology. *Journal of Science Education and Technology, 18*, 146–162.

Black, P., & Wiliam, D. (2009). Developing the theory of formative assessment. *Educational Assessment, Evaluation and Accountability, 21*(1), 5–31.

Chien, Y.-T., Chang, Y.-H., & Chang, C.-Y. (2016). Do we click in the right way? A meta-analytic review of clicker-integrated instruction. *Educational Research Review, 17*, 1–18.

Clark-Wilson, A. (2010). Emergent pedagogies and the changing role of the teacher in the TI-Nspire navigator-networked mathematics classroom. *ZDM Mathematics Education, 42*, 747–761.

Collins, A., Brown, J. S., & Newman, S. E. (1989). Cognitive apprenticeship: Teaching the crafts of reading, writing and mathematics! In L. B. Resnick (Ed.), *Knowing, learning, and instruction: Essays in honor of Robert Glaser* (pp. 453–494). Hillsdale, NJ: Lawrence Erlbaum Associates.

Cusi, A., Morselli, F., & Sabena, C. (2017). Promoting formative assessment in a connected classroom environment: Design and implementation of digital resources. *ZDM Mathematics Education, 49*(5), 755–767.

DBRC—The Design Based Research Collective. (2003). Design-based research: An emerging paradigm for educational inquiry. *Educational Researcher, 32*(1), 5–8.

Fies, C., & Marshall, J. (2006). Classroom response systems: A review of the literature. *Journal of Science Education and Technology, 15*(1), 101–109.

Hannula, M. (2011). The structure and dynamics of affect in mathematical thinking and learning. In M. Pytlak, T. Rowland, & E. Swoboda (Eds.), *Proceedings of CERME 7* (pp. 34–60). Rzeszów, Poland: University of Rzeszów and ERME.

Hattie, J. (2009). *Visible learning: A synthesis of over 800 meta-analyses relating to achievement*. London: Routledge.

Hattie, J., & Timperley, H. (2007). The power of feedback. *Review of Educational Research, 77*(1), 81–112.

Hunsu, N. J., Adesope, O., & Bayly, D. J. (2016). A meta-analysis of the effects of audience response systems (clicker-based technologies) on cognition and affect. *Computers & Education, 94*, 102–119.

Irving, K. I. (2006). The impact of educational technology on student achievement: Assessment of and for learning. *Science Educator, 15*(1), 13–20.

Kay, R. H., & Le Sage, A. (2009). Examining the benefits and challenges of using audience response systems: A review of the literature. *Computers & Education, 53*, 819–882.

Nemirovsky, R., Tierney, C., & Wright, T. (1998). Body motion and graphing. *Cognition and Instruction, 16*(2), 119–172.

Roschelle, J., & Pea, R. (2002). A walk on the WILD side. How wireless handhelds may change computer-supported collaborative learning. *International Journal of Cognition and Technology, 1*(1), 145–168.

Schoenfeld, A. H. (1992). Learning to think mathematically: Problem solving, metacognition, and sense-making in mathematics. In D. Grouws (Ed.), *Handbook for research on mathematics teaching and learning* (pp. 334–370). New York: Macmillan.

Vygotsky, L. S. (1978). *Mind in society: The development of higher mental processes*. Cambridge, MA: Harvard University Press.
Wiliam, D., & Thompson, M. (2007). Integrating assessment with instruction: What will it take to make it work? In C. A. Dwyer (Ed.), *The future of assessment: Shaping teaching and learning* (pp. 53–82). Mahwah, NJ: Erlbaum.

Making Good Practice Common Using Computer-Aided Formative Assessment

Shai Olsher

1 Introduction and Theoretical Background

Defining, analyzing, and disseminating good teacher practices in the mathematics classroom are ongoing challenges for the research community (Chazan & Ball, 1999), and are viewed as part of any attempt to improve the quality of mathematics teaching (Cobb & Jackson, 2011). Guided inquiry tasks are open-ended tasks that usually have more than one solution and often require taking into account various dimensions that were not addressed in previous learning, requiring students to go through a problem-solving process. Promoting and evaluating this process presents challenges for teachers, as it requires knowledge of different types of content knowledge for teaching (Ball, Thames, & Phelps, 2008) that come into play when conducting classroom discussions. Such discussions may be facilitated by the teachers' ability to anticipate and monitor the students' responses during inquiry tasks, then select and sequence individual responses in the course of the discussion, finally helping students make mathematical connections (Stein, Engle, Smith, & Hughes, 2008). In the case of computer-based guided inquiry, where students are expected to form conjectures and reason about them, the primary role of the teacher is to promote and organize discussions (Yerushalmy & Elikan, 2010).

One way to enact guided inquiry in a digital environment is by using example eliciting tasks (Yerushalmy, Nagari-Haddif, & Olsher, 2017). Example eliciting tasks are open ended tasks in which the students are required to construct an example that satisfies a set of conditions, and could serve as a way to determine the validity of mathematical statements (Buchbinder & Zaslavsky, 2013). Example based generic arguments (Dreyfus, Nardi, & Leikin, 2012) could be part of student's construction of conjecture, and also assist in assessing students' mathematical reasoning (Zaslavsky & Zodik, 2014).

S. Olsher (✉)
University of Haifa, Haifa, Israel
e-mail: olshers@edu.haifa.ac.il

© Springer Nature Switzerland AG 2019
G. Aldon and J. Trgalová (eds.), *Technology in Mathematics Teaching*,
Mathematics Education in the Digital Era 13,
https://doi.org/10.1007/978-3-030-19741-4_2

Orchestrating the work of students in a technological environment, referred to by Trouche (2004) as instrumental orchestration, while collecting information about students that could be used for formative assessment, presents challenges for teachers (Drijvers, Doorman, Boon, Reed, & Gravemeijer, 2010). For the purposes of evaluation, formative assessment requires teachers to draw on the information collected about the students in the course of lessons and use it as feedback to modify their teaching (Black & Wiliam, 1998). The abundance of analysable data being created when students engage in rich inquiry tasks on a technological platform presents yet another challenge for teachers. Some researchers have suggested using technological platforms to collect and display student answers (Arzarello & Robutti, 2010; Clark-Wilson, 2010), or to conduct polls, the results of which could be easily used in real time (Cusi, Morselli, & Sabena, 2017). Another practice observed by Panero and Aldon (2015) was the combination of automatically collected digital data with traditional paper-and-pencil work, which was used by the teacher in formative assessment in real time. Yet another strategy suggested by Olsher, Yerushalmy, and Chazan (2016) is to offload some of the processing of the data onto a digital platform, automatically categorizing student answers based on mathematical characteristics, and providing access for the teacher to processed data to inform decision making. This method of categorization allows detecting creative, unexpected answers, whose characteristics have not necessarily been predefined, potentially promoting the teachers' ability to address such answers in the classroom.

Although the data are accessible, and methods of this type are being studied, guided inquiry is not a prominent practice in the mathematics classroom. In this chapter, I suggest addressing the challenge by exploring ways to study and promote *good practices* for teachers in technologically rich environments. Such practices are intended to engage students with guided inquiry and use the results to construct guided inquiry tasks that offload some of the teachers' work onto a digital platform, as suggested by Olsher et al. (2016).

2 Design of the Study

This study consists of two parts: (a) an analysis of classroom lessons for categorizing the teachers' practice during a technologically supported guided inquiry lesson; and (b) using this method of categorization to design a new guided inquiry activity. The main research aims were (a) to determine whether it is possible to identify good practices in conducting conjecture-based discussions in the classroom, and (b) whether the categorization of conjectures facilitates good practices by providing predefined categories.

In the first part, I used recorded classroom lessons with high school students to identify the categories for the students' conjectures. Next, I inserted this categorization into a digital platform and tested it within the framework of a university course.

The first part of the study is based on a recorded lesson with 24 9–10 grade students working in pairs. The students worked with a first generation DGE (Geometric Supposer), which was used as a technological platform to elicit conjectures. The recorded lesson was planned to summarize the main theorems concerning the similarity of triangles.

In the second part of the study, I applied the method developed in the first part to design a guided inquiry activity. The activity was designed using DGE diagrams, which were incorporated into the STEP platform (Olsher et al., 2016). The task was carried out with 15 graduate students (replacing the middle school students), as part of a graduate mathematics education course. The task focused on perpendicular bisectors in quadrilaterals, specifically, on the characteristics of quadrilaterals in which perpendicular bisectors meet in a single point. I served as the teacher conducting this inquiry, using the STEP platform to introduce the students to the inquiry activities. I used automatic analysis of student responses in real time during the lecture, according to predefined characteristics, and conducted a discussion based on the automatic categorization of student answers.

3 Methodology

To answer the research question, I initially observed a classroom in which the teacher needs to collect, process, and use information generated by the students conducting an individual inquiry activity using a DGE. Trouche (2004) used the term "instrumental orchestration" to describe didactic configurations and the way in which they are being used in the classroom. He also suggested that these constructs could "give birth to new instrument systems" (ibid., p. 304). This instrumental genesis is based on a two-way interaction that generates the use by the subject (the teacher in this case), as well as the role of the artefact as an instrument (the technological platform), all embedded in a social environment that contextualizes this practice within a community of practice. I used this framework to describe the way in which the teacher processed the students' answers in the lesson I observed and suggested new instrument systems, which I later examined with respect to a different case, with the support of a different technological platform.

For this chapter, I analyzed a recording of a one-hour lesson in a mathematics classroom. I also drew upon the design principles of the STEP platform for use in classrooms equipped with personal digital devices, and upon an assessment activity designed for a mathematics education course. The activity, student responses, and the field notes of the following classroom discussion I conducted are used to examine a possible implementation of the findings as a novel instrument system.

In the next part, I describe the task presented to the students in the original lesson. Next, I analyze the way in which the teacher orchestrated the discussion around the conjectures raised by the students. I analyze the categorization of the conjectures as they are placed on the board (if at all), and the way in which the teacher treated them (acknowledging the difficulty of proving a certain conjecture or specifying the underlying constraints). This orchestration requires a great deal of real-time decision making by the teacher. I therefore introduce a possible implementation of the way in which an automatic analysis tools (e.g., the STEP platform) could offload some of this orchestration, creating new instrument systems and possibly making the process more accessible for all teachers. To demonstrate this implementation, I constructed an activity that facilitates collecting student-generated examples and analyzing them automatically according to various characteristics that fall into predefined categories. Finally, I follow the discussion I conducted using the automatically analyzed answers to advance from a specific characterization to the general case.

3.1 The First Task

Construct an acute triangle ABC (Fig. 1). Draw the altitudes from each of the triangle vertices and mark the feet of the altitudes D, E, and F. Label the intersection point of the altitudes G. Reflect point G over each side of the triangle. What is the relationship between triangle DEF, which is formed by connecting the feet of each altitude, the triangle formed by connecting the image points of G (HIJ), the original triangle, the angles, and the segments? Investigate anything that you can find. Write out formal conjectures as we have been doing in class.

Fig. 1 Sketch of the geometric construction discussed in the classroom

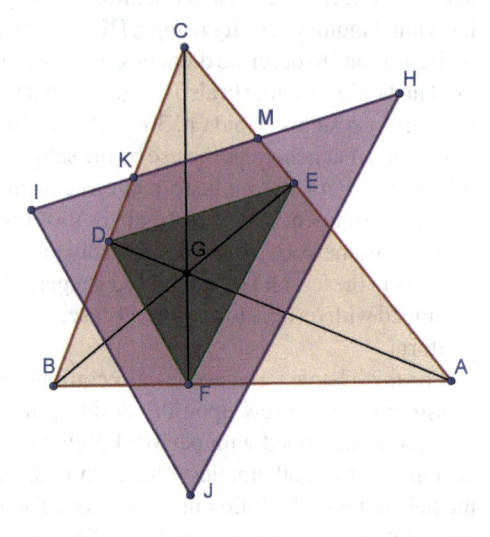

3.2 Data Analysis

During the recorded session, I identified 11 conjectures that were formulated in the classroom (Table 1). After 7 of the 11 conjectures were listed, the teacher initiated a discussion aiming to review the conjectures and the underlying argumentation of each.

I identified four strategies the teacher used to address these conjectures, and I use these as categories of analysis. The first strategy, which can be demonstrated with reference to conjecture A1 (Table 1), was to write the conjecture on the left side of the board, then ask how many of the students agree with it:

Teacher: You think that triangle IHJ is similar to triangle DEF [writes $\triangle IHJ \sim \triangle DEF$ on the board]. Raise your hand if you believe that's true. [All the students raise their hands] Oh. So everyone does. Great.

The second strategy can be demonstrated using conjectures C1 and A3. Following this strategy, instead of writing out the initial conjecture, the teacher refined it himself (turning C3 into B2) or involved the students in doing so, as shown in the following excerpt:

Student 1: Their sides are 2–1.
Teacher: The ratio of their sides is 2–1.
Students: Perimeter.

Table 1 Conjectures raised by students in the order of their appearance on the board

Listed on the left side of the board	Listed on the right side of the board	Conjectures that were raised by students but did not appear on the board
A1. $\triangle IHJ \sim \triangle DEF$	B1. If $\triangle ABC$ is isosceles and acute, then: $\angle ACB$ is geometric mean of $\angle FED$ $\angle FDE$	C1. $\frac{\text{Sides of } \triangle IHJ}{\text{Sides of } \triangle DEF} = 2$
A2. $\frac{\text{Area of } \triangle IHJ}{\text{Area of } \triangle DEF} = 4$	B2. Bisector is the same as altitude. $\overline{BE}$, bisects $\angle DEF$ and $\overline{BE}$ extended bisects $\angle IHJ$	C2. Corresponding sides are parallel
A3. $\frac{\text{Perimeter of } \triangle IHJ}{\text{Perimeter of } \triangle DEF} = 2$	B3. $\angle ACB \cong \angle FED \cong \angle FDE$ ($\triangle ABC$ is isosceles)	C3. $\triangle JHI \sim \triangle DEF$ bisector of H also bisects E
A4. $\overline{IH} \| \overline{DE}, \overline{EF} \| \overline{HJ}, \overline{DF} \| \overline{IJ}$		C4. If ABC is isosceles, it creates two other isosceles triangles

Teacher: The perimeter is 2–1 [writes $\frac{\text{Perimeter of } \triangle IHJ}{\text{Perimeter of } \triangle DEF} = 2$ on the board]. The perimeter of triangle IHJ, to the perimeter of triangle DEF is 2. Which means that the ratio of their sides is also 2–1.

The third strategy, which can be illustrated using conjecture B1, was to write the conjecture with the additional constraints relevant to it on the right side of the board, to the right of the sketch. In this case, the teacher also ascribed ownership for the conjecture and the additional constraints to the students who raised it:

Student 2: If triangle ABC is isosceles, then hmm, the measure of the angle ACB equals either of the two base angles in the two smaller triangles. Because those two smaller triangles are also isosceles.

Teacher: You and Jennifer worked a lot with isosceles triangles, didn't you? [Drawing on the board the sketch shown in Fig. 2] OK, what do you claim?

Student 2: That angle ACB equals, is congruent with angle FED and angle FDE.

Teacher: [writes $\angle ACB \cong \angle FED \cong \angle FDE$ ($\triangle ABC$ is isosceles) on the board].

The fourth strategy applied to general cases represented on the right side of the board. There was one instance of a case like this, conjecture C3, raised and afterwards transformed into conjecture B2, which was finally written on the board. The lesson plan that the teacher prepared for the activity did not include this conjecture.

Table 1 suggests rough categories for the conjectures, as they were addressed by the teacher. The conjectures that appeared on the left side of the board (A1–A4) were those that the teacher expected and went over their justifications in class. The conjectures that did not reach the board (C1–C3) required some rephrasing or generalization to become better defined and represented, and they appeared on the board in their improved form. The conjectures listed on the right side of the board were either case-specific (B1, B3) or more advanced than the content that had been covered in class.

Next, the teacher shared his thoughts and his initial lesson plan with the students. As a well-established authority figure in the classroom, he indicated that he was not

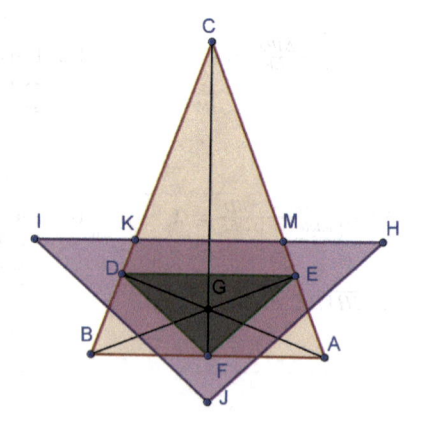

Fig. 2 Sketch of the constrained case students addressed in the classroom

completely prepared for everything that the students suggested. But he was pleased and surprised by student ideas that he did not expect. He asked which conjecture the students thought surprised him, and they pointed to the bisector one (B2). The teacher agreed. He then noted that he would not address all the conjectures, only the ones on the left side because these are the ones that everyone had found, after which he went over the proofs for all of them. Finally, he turned to the right side of the board and categorized the conjectures further: conjectures B1 and B3, referring to the sketch in Fig. 2, he assigned as homework, and conjecture B2 he referred to as a general conjecture, true for any triangle. The teacher said that it may be difficult for the students to prove this, and gave them additional time, offering hints in case they encountered difficulties. The reflection about the surprising aspects in students' conjectures may have served the teacher to categorize them into more and less trivial (expected).

In sum, the teacher categorizes the conjectures based on several dimensions: according to their relevance to the topic at hand (statements that should be initially addressed in the classroom), according to their level of generality (whether or not they addressed an individual case), according to the level of accuracy of the way in which they were shared with the classroom (e.g., statements that had been refined before they were displayed on the board), and according to the complexity to the proof (e.g., whether further elaboration was needed).

3.3 The Second Task

The results of the analysis described in the previous section were used to design a similar guided inquiry learning activity that can be supported by the STEP platform. The goal was to create a task that would require students to submit different answers and to automatically characterize these answers based on predefined categories, to offload some of the teacher's work, enabling instrumental orchestration (Trouche, 2004) and the creation of a new instrument system. Automatic analysis could lead to a meaningful class discussion by enabling the teacher to sequence the responses according to given characteristics, without having to monitor and analyze them during the lesson (Stein et al., 2008). The teacher would also have real-time analytics, showing how many responses with various characteristics were submitted by students.

I chose to design a task based on a topic presented in a study that examined students' conjectures and justifications using a DGE (Marrades & Gutiérrez, 2000). The original task introduced a quadrilateral with 3 fixed points and perpendicular bisectors for all sides, with no angle measurements apparent. The students were asked which conditions needed to be satisfied for all of perpendicular bisectors to meet at a single point. The study followed the students' conjectures and the process of justifying and proving them, using such DGE features as dragging and measuring to explore the dynamic figure.

Given that the STEP platform supports tasks that require multiple student-generated examples, the task was divided into two separate parts. The goal of the first part was for the students to create different examples that meet the condition, without measurements, using dragging only, as shown in Fig. 3. The first part of the task was formulated as follows: "A, B, C, and D form quadrilateral ABCD. They are all dynamic and can be dragged. If possible, create 3 examples that are as different as possible from each other, in which the perpendicular bisectors to the sides of ABCD meet in a single point. If it is not possible, provide an explanation. Is the claim correct [Yes/No]?" Fig. 3 shows a screenshot of the initial state of the dynamic diagram created for this part. The main reason that the exploration did not include measurements was to enable the students to focus, at this initial stage, on the possible variations between the submitted examples, rather than on the common elements, which were addressed in the second part of the task.

The second part of the task asked students to form a conjecture. Students were asked to use the examples from the first part of the task for *pattern exploration* (Mills, 2014) during the activity. During the discussion phase, before forming a conjecture, I used the examples for pattern exploration. Students were asked to form conjectures following their exploration in the first task. These conjectures could later be justified and proven as part of the inquiry activity.

In the preparation stage, I categorized the potential responses to the first part of the task, using the categories derived from the first part of the study to describe the various expected answers of the students. Examples that were labelled "Sum of opposite angles is 180°" were correct answers, containing general examples that the lecturer (I will refer to the instructor in task 2 as the "lecturer," to distinguish him from the "teacher" in task 1) eventually wanted to reach, consistent with the first strategy of the teacher: conjectures that the teacher expected and treated during the class session. Other examples, which served as the planned starting point in the sequence of examples addressed during the discussion, were specific examples that could potentially develop into more general claims. These examples contained additional constraints and therefore provided specific cases. They were consistent with the third strategy: case-specific conjectures. These examples had two labels: "square" and "rectangle." In addition, some characteristics that were not present in the initially suggested categorization were developed. These characteristics described answers that were not correct in general, such as "rhombus" and "parallelogram" (not every example of a rhombus or parallelogram has the required characteristic). Finally, I prepared for some unexpected categories that could emerge during the session.

Students submitted three examples each. Some of the submissions are shown in Fig. 4. In 45 of 51 student submissions, the perpendicular bisectors intersected in a single point.

The discussion was planned to sequence the examples so as to advance gradually from the most specific ones toward a general claim. When discussion of the examples began, the lecturer filtered them according to the predetermined categories, asking students whether the different descriptions fit the phenomenon. The first filter activated was the "square" filter, as shown in Fig. 5. This filter passed all the student examples that were squares.

A, B, C, and D form quadrilateral ABCD. They are all dynamic and can be dragged. If possible, create 3 examples that are as different as possible from each other, in which the perpendicular bisectors to the sides of ABCD meet in a single point. If it is not possible, provide an explanation.

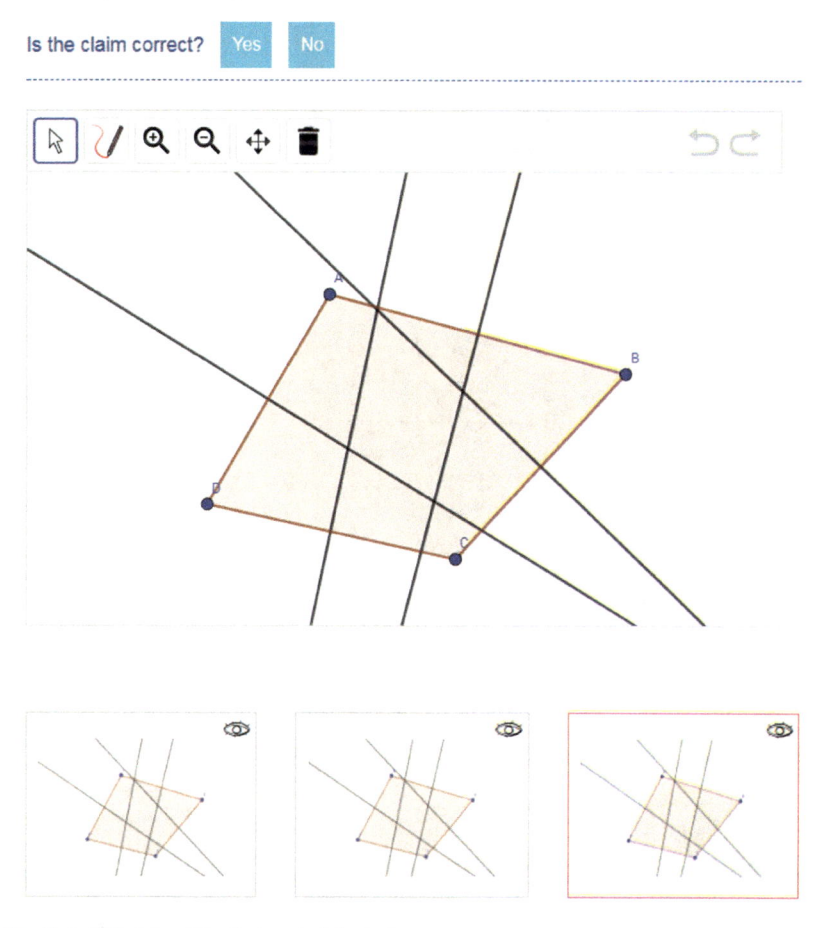

Fig. 3 Initial state of the first part of the task

Figure 5 shows a screenshot of the STEP platform, displaying the available filters at the top, with "Square" selected (on a white background). Students were asked to submit three examples, and the display shows that four students submitted it as their third example, one student submitted it as the first example, and one as the second example. But the examples were not numbered in the students' interface, therefore no meaning should be attached to the order of the examples. The lecturer asked the students to define the characteristics that were necessary for the perpendicular bisectors to meet in a single point, and they all agreed that this happens if the quadrilateral is a square.

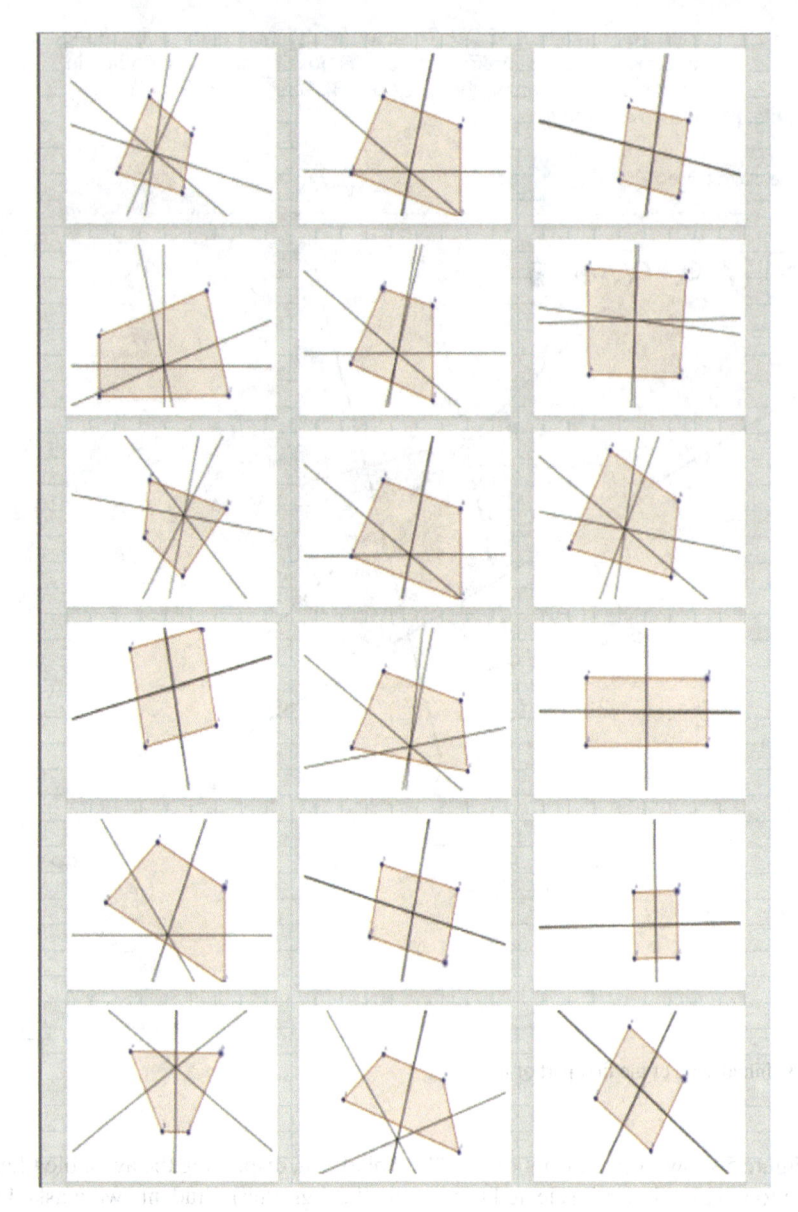

Fig. 4 Part of the complete answer space (one row represents the three answers of one student)

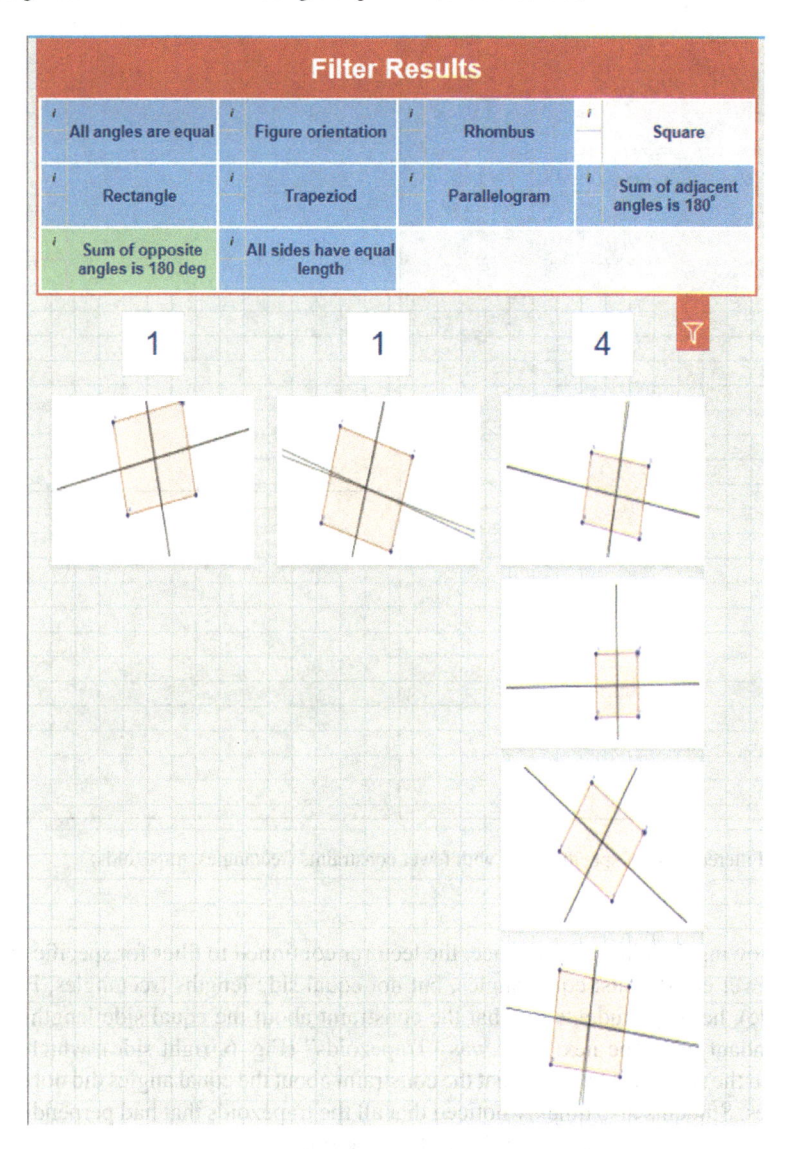

Fig. 5 Filtered view of the case with the most constraints (squares)

Fig. 6 Filtered view of specific cases with fewer constraints (rectangles, trapezoids)

Following the planned sequence, the lecturer continued to filter for specific cases with fewer constraints: equal angles, but not equal side lengths (rectangles, Fig. 6, left side), helping students see that the constraint about the equal side lengths was a redundant one. The next filter was "Trapezoids" (Fig. 6, right side), which now exposed the students to the fact that the constraint about the equal angles did not cover all cases. Students also quickly noticed that all the trapezoids that had perpendicular bisectors meet in one point were isosceles trapezoids.

At this point, one of the students asked whether the kites were part of the quadrilateral family, which had the required characteristic of perpendicular bisectors that meet in one point, as shown in Fig. 7. This characteristic was not one that the lecturer had prepared for, and no automatic filter was available. Nevertheless, the lecturer had time to address this example during the discussion, asking the class to comment about it. This was possible largely because much time was saved during the class session in monitoring, sequencing, and selecting the student examples, most of which was carried out during the design of the task, and performed automatically by the platform.

Fig. 7 Unplanned student example for quadrilaterals (kite)

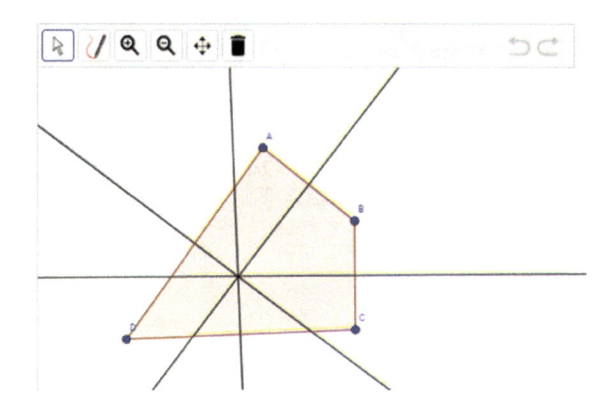

When conducting the discussion in the recorded session, in several cases the teacher asked students to raise their hands if they thought the statement was true. While the second task was being carried, it was not necessary to ask students to raise their hand because this function was also offloaded to the technological platform, which provided the graphic interface shown in Fig. 8. This enables math teachers to see how many examples in the complete student example space have a certain characteristic.

Using the Venn diagram representation makes it possible for teachers to see in real time which examples have more than one characteristic. This representation enables teachers to make informed decisions about the characteristics that are either prominent or missing in the students' answers. In the example shown in Fig. 8, 45 out of 51 answers submitted had the characteristic of *sum of opposite angles is 180°*. Another characteristic that was predefined proved to be present in 15 of the submitted examples. This characteristic, referred to as *figure orientation*, is not directly related to the correctness of the example (Olsher et al., 2016), but rather to a general mathematical characteristic that can occur in correct as well as incorrect examples. In this case, the characteristic is what some researchers call a "prototypical shape" (Rosch & Mervis, 1975), as one of the sides or lines in the geometric shape is either horizontal or vertical. These analytics, provided in real time, enabled the lecturer to address this general characteristic as part of making connections across mathematical topics. This made it possible even for students who got the wrong answer to engage in the discussion in a meaningful way.

4 Discussion

The two cases described discuss the potential added value automatic categorization of student answers could provide for the teacher, and then demonstrate the enactment of such automatic categorization in a guided inquiry session in the classroom.

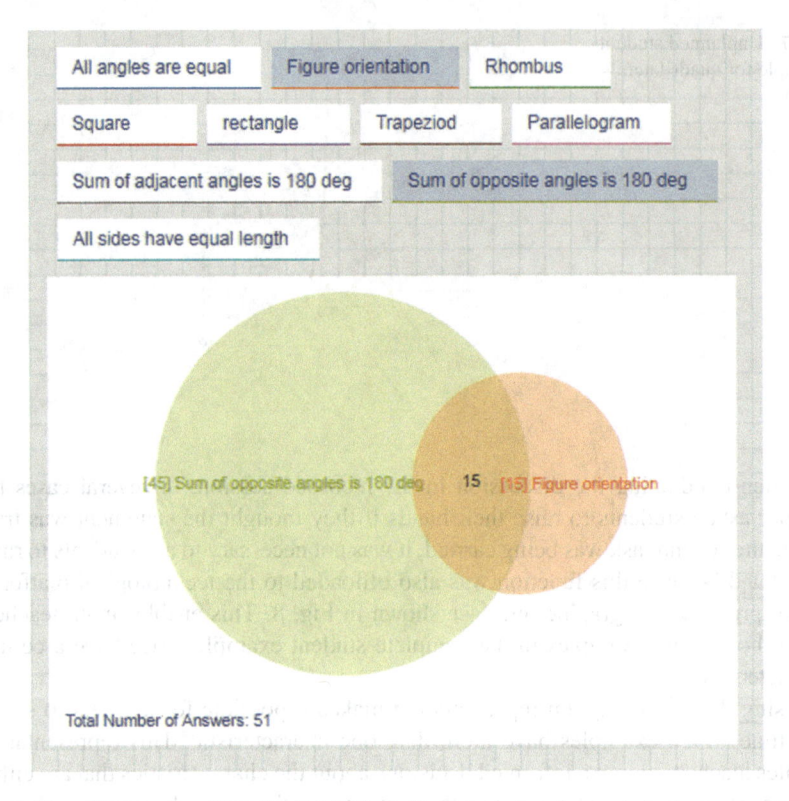

Fig. 8 Venn diagram describing the distribution of two characteristics in student answers

In the first case presented, the categorization performed by the teacher could be mapped using an automatic filtering scheme. The topic of this lesson being similarity, many conjectures that address certain characteristics of similarity are expected to be raised: ratio between sides and areas, relationship between corresponding segments (e.g., parallel segments). Even student mistakes that are common when learning about similarity can be predicted (e.g., mistaking the ratio of segments for the ratio between areas). These relations could be predefined and automatically recognized by STEP or some other platform, collecting relevant data, such as whether a certain relation is addressed by the students, and if yes, by how many. Furthermore, given that present-day DGEs are more flexible than the Geometric Supposer in allowing students to drag pre-constructed figures, and therefore student example spaces may expand, categorization by the mathematical properties stated could become even more important. For example, when filtering student answers it is possible to determine whether they added constraints to the given situation, potentially limiting the generality of their answer, as shown in conjectures B1 and B3. By indicating the expected relations, we are also setting the stage for the unexpected relations to stand out. The teacher could easily address these and automatically determine their correctness. By acknowledging that the platform does not identify the entire space of relations that students raise, we leave room also for acknowledging student creativity, which is an important part of inquiry-based activities, but which may also prevent educators from using automatic assessment platforms. In time, teachers could choose to incorporate these relations into the scheme of detected relations, if they so choose.

In the second case, I illustrated the use of such a platform (STEP) in the design of a mathematics task and its utilization in class. The example follows the design process aimed at classifying student answers according to the categories revealed in the first case, then using them to sequence a classroom discussion based on student-generated examples. This scenario offloaded many aspects of the lecturer's work onto the technological platform. Using the predefined filters and the Venn diagram representation enabled the lecturer to refer to the relevant examples without investing valuable classroom time in identifying them and surveying the class for the level of agreement about them. The lecturer used this time to address unexpected examples and to make cross-topic mathematical connections that were not directly related to the correctness of the answers.

5 Conclusion

Based on the cases described in this paper, I suggest that the automation of categorizing and surveying student answers, with further expansion of the range of categories beyond answer correctness, could serve as a tool for teachers in their instrumental orchestration of the learning environment for technology-based guided inquiry. Expert teachers have the skills and knowledge to filter and categorize student answers during the classroom session even in complex situations of inquiry-based learning. These teachers are able to monitor, select, and sequence student answers and conjectures during the lesson. This ability is not common, however, especially when collecting information from technology-based platforms (Drijvers et al., 2010). As seen in the case presented above, the teacher noted that he might not be able to address in the classroom certain examples that he had not expected. Olsher et al. (2016) suggested the use of automatic filtering of student responses to make information about the characteristics of student answers more accessible for teachers to use in formative assessment, such as the STEP platform, which enables teachers to predefine mathematical properties of student answers. The platform can analyze and categorize the answers automatically to increase the efficiency of assessment.

References

Arzarello, F., & Robutti, O. (2010). Multimodality in multi-representational environments. *ZDM Mathematics Education, 42*(7), 715–731.

Ball, D. L., Thames, M. H., & Phelps, G. (2008). Content knowledge for teaching: What makes it special? *Journal of Teacher Education, 59*(5), 389–407.

Black, P., & Wiliam, D. (1998). Assessment and classroom learning. *Assessment in Education, 5,* 7–74.

Buchbinder, O., & Zaslavsky, O. (2013). A holistic approach for designing tasks that capture and enhance mathematical understanding of a particular topic: The case of the interplay between examples and proof task. In C. Margolinas (Ed.), *Task design in mathematics education: Proceedings of ICMI Study 22* (pp. 25–34). UK: Oxford.

Chazan, D., & Ball, D. (1999). Beyond being told not to tell. *For the Learning of Mathematics, 19,* 2–10.

Clark-Wilson, A. (2010). Emergent pedagogies and the changing role of the teacher in the TI-Nspire Navigator-networked mathematics classroom. *ZDM Mathematics Education, 42*(7), 747–761.

Cobb, P., & Jackson, K. (2011). Towards an empirically grounded theory of action for improving the quality of mathematics teaching at scale. *Mathematics Teacher Education and Development, 13*(1), 6–33.

Cusi, A., Morselli, F., & Sabena, C. (2017). Promoting formative assessment in a connected classroom environment: Design and implementation of digital resources. *ZDM Mathematics Education, 49,* 755–767. https://doi.org/10.1007/s11858-017-0878-0.

Dreyfus, T., Nardi, E., & Leikin, R. (2012). Forms of proof and proving in the classroom. In G. Hanna & M. de Villiers (Eds.), *Proof and proving in mathematics education* (pp. 191–214). New York: Springer.

Drijvers, P., Doorman, M., Boon, P., Reed, H., & Gravemeijer, K. (2010). The teacher and the tool: Instrumental orchestrations in the technology-rich mathematics classroom. *Educational Studies in Mathematics, 75*(2), 213–234.

Marrades, R., & Gutiérrez, Á. (2000). Proofs produced by secondary school students learning geometry in a dynamic computer environment. *Educational Studies in Mathematics, 44*(1), 87–125.

Mills, M. (2014). A framework for example usage in proof presentation. *Journal of Mathematical Behavior, 33,* 106–118.

Olsher, S., Yerushalmy, M., & Chazan, D. (2016). How might the use of technology in formative assessment support changes in mathematics teaching? *For the Learning of Mathematics, 36*(3), 11–18.

Panero, M., & Aldon, G. (2015). How can technology support effectively formative assessment practices? A preliminary study. In N. Amado & S. Carreira (Eds.), *Proceedings of the 12th International Conference on Technology in Mathematics Teaching (ICTMT 12)* (pp. 293–302).

Rosch, E., & Mervis, C. B. (1975). Family resemblance: Studies in the internal structure of categories. *Cognitive Psychology, 7,* 573–605.

Stein, M. K., Engle, R. A., Smith, M. S., & Hughes, E. K. (2008). Orchestrating productive mathematical discussions: Five practices for helping teachers move beyond show and tell. *Mathematical Thinking and Learning, 10*(4), 313–340.

Trouche, L. (2004). Managing the complexity of human/machine interactions in computerized learning environments: Guiding students' command process through instrumental orchestrations. *International Journal of Computers for Mathematical Learning, 9*(3), 281–307.

Yerushalmy, M., & Elikan, S. (2010). Exploring reform ideas of teaching Algebra: Analysis of videotaped episodes and of conversations about them. In R. Leikin & R. Zazkis (Eds.), *Learning through teaching: Developing mathematics teachers' knowledge and expertise in practice* (pp. 191–207). Springer Netherlands.

Yerushalmy, M., Nagari-Haddif, G., & Olsher, S. (2017). Design of tasks for online assessment that supports understanding of students' conceptions. *ZDM Mathematics Education, 49*(5), 701–716.

Zaslavsky, O., & Zodik, I. (2014). Example-generation as indicator and catalyst of mathematical and pedagogical understandings. In Y. Li, E. A. Silver, & S. Li (Eds.), *Transforming mathematics instruction* (pp. 525–546). Cham: Springer International.

Technology Supporting Student Self-Assessment in the Field of Functions—A Design-Based Research Study

Hana Ruchniewicz and Bärbel Barzel

1 Aim of the SAFE Tool

Many digital self-assessment tools are designed similar to, for instance, the online assessment platform mathster from the UK (Fig. 1). These tools generate a set of questions, evaluate the student's answers based on correctness, and then provide the student with feedback in form of the number of correct responses or a score. In addition, items in computer-based assessment tools are predominantly multiple-choice format and single-entry number answers (Stacey & Wiliam, 2013). While the learner works individually in such environments, it is the technology that evaluates the answers and gives feedback. Therefore, the student does not entirely adopt the role of the assessor and the term "self"-assessment refers mostly to the organisation of the assessment for such tools, rather than an evaluation of one's own reasoning and understanding.

To overcome individual misconceptions or fill in gaps to develop one's own learning, it is essential for the student to gain information on his/her own understanding of the specific content (Wiliam & Thompson, 2008). Moreover, a key principal of formative assessment is the active involvement of learners in the formative assessment process (Bernholt, Rönnebeck, Ropohl, Köller, & Parchmann, 2013). Investigating their (mis-)conceptions helps students to develop an awareness for their strengths and weaknesses. In addition, students can discover how to observe and direct their learning processes using metacognitive strategies along with reflection and adopt responsibility for their own learning in the process (Black & Wiliam, 2009; Heritage, 2007). This is why, the greatest challenge for developing the SAFE tool is that the stu-

---

H. Ruchniewicz (✉) · B. Barzel
University of Duisburg-Essen, Essen, Germany
e-mail: hana.ruchniewicz@uni-due.de

B. Barzel
e-mail: baerbel.barzel@uni-due.de

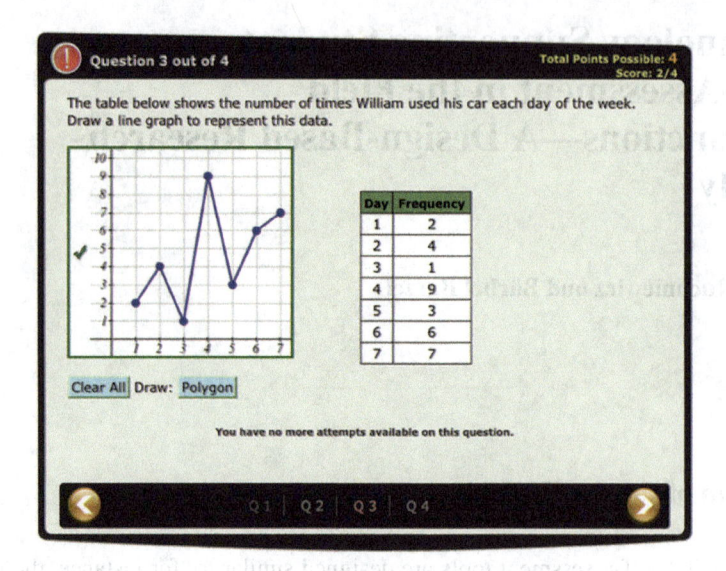

Fig. 1 Example of a common digital self-assessment tool (www.mathster.com)

dent should conduct the assessment, rather than the technology. Hence, a key design feature of our tool is starting with an open assessment task followed by a checklist to guide the student's reflection on his/her own solution after they have completed the open assessment task. The checklist is based on typical misconceptions related to the mathematical content. In this example, the concept of function, specifically the translation from a situational to a graphical representation of a functional relationship. The SAFE tool's development was initiated during the design-based research EU-project "Raising Achievement through Formative Assessment in Science and Mathematics Education" (FaSMEd), which introduced and investigated technology enhanced formative assessment practices (www.fasmed.eu).

2 Theoretical Background

2.1 Conceptualising Formative Assessment

Formative assessment (FA) is "the process used by teachers and students to recognize and respond to student learning in order to enhance that learning, during the learning." (Bell & Cowie, 2001, p. 540). It results in the active adaptation of classroom practices to fit students' needs by continuously gathering, interpreting and using evidence about ongoing learning processes (Black & Wiliam, 2009). The required data can be elicited and exploited during the different phases of these processes. Wiliam and Thompson (2008) refer to Ramaprasad (1983) and focus on three central steps in teaching and

	Where the learner is going	Where the learner is right now	How to get there
Teacher	1 Clarifying learning intentions and criteria for success	2 Engineering effective classroom discussions and other learning tasks that elicit evidence of student understanding	3 Providing feedback that moves learners forward
Peer	Understanding and sharing learning intentions and criteria for success	4 Activating students as instructional resources for one another	
Learner	Understanding learning intentions and criteria for success	5 Activating students as the owners of their own learning	

Fig. 2 Key strategies of formative assessment (Wiliam & Thompson, 2008)

learning, namely establishing: where the learners are, where the learners are going and how they might get there. The authors state that FA can be conceptualised in five key strategies (Fig. 2). These strategies enable teachers, peers and students to close the gap between the students' current understanding and the intended learning goals.

While Wiliam and Thompson (2008) take into account central steps of the learning process and the agents (teacher, peers and learners) who act in the classroom, their framework regards mainly the teacher to be responsible for the process of FA. It is the teacher who creates learning environments to investigate students' understanding (strategy 2), who gives feedback (strategy 3), who activates students as resources for one another (strategy 4) and as owners of their own learning (strategy 5). This framework was refined in the FaSMEd project to provide greater emphasis on the responsibility of all three agents in each of the steps and key strategies of FA. Further, the framework was extended to allow the characterisation and analysis of technology enhanced FA processes in three dimensions: agent/s, FA strategies and functionalities of technology (Fig. 3).

In the FaSMEd framework, the "agent/s" dimension specifies who is assessing: the student, peer/s, or the teacher. Ideally, all agents would be involved in FA as the "assessment activity can help learning if it provides information that teachers and their students can use as feedback in assessing themselves and one another [...]" (Black, Harrison, Lee, Marshall, & Wiliam, 2004, p. 10). Moreover, an active involvement of students by peer and self-assessment is stated as a key aspect of FA. It includes opportunities for learners to recognize, reflect upon and react to their own/their peers' work. This helps them to use metacognitive strategies, interact with multiple approaches to reach a solution and adapt responsibility for their own learning process (Black & Wiliam, 2009; Sadler, 1989).

The "FA strategies" dimension of the FaSMEd framework refers to the five key strategies of FA, described by Wiliam and Thompson (2008), but understands them in a broader sense by acknowledging that all agents can be responsible for FA. For example, while the teacher engineers classroom discussions and learning tasks that elicit evidence of student understanding (strategy 2), a student can also elicit

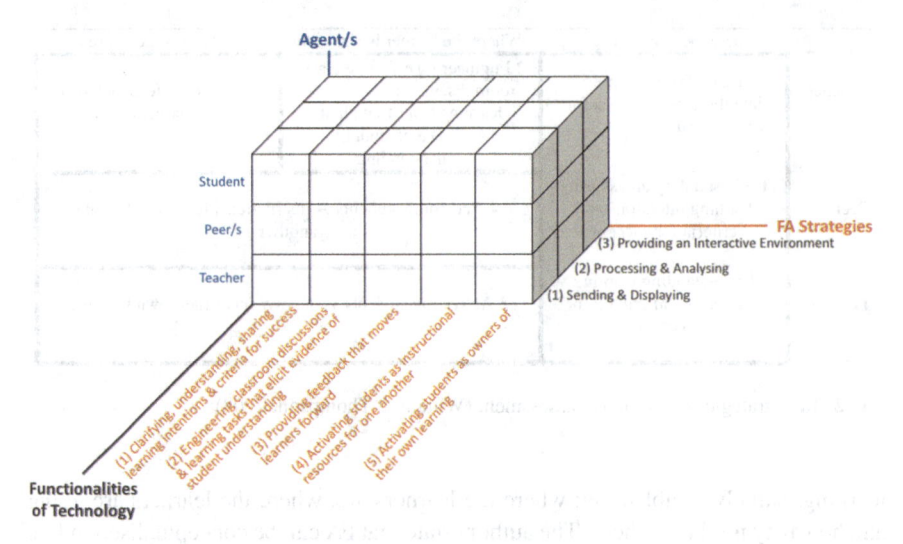

Fig. 3 The FaSMEd framework. Adapted from Aldon, Cusi, Morselli, Panero, and Sabena (2017)

evidence on his/her own understanding by working and reflecting on such assessment tasks. Furthermore, peers could provide effective feedback (strategy 3) or a student might control his/her own learning process using metacognitive activities (strategy 5) without being activated by a teacher.

To specify the different functionalities that technology can resume in FA processes, FaSMEd introduced a third dimension to the framework: "functionalities of technology". Based on the different FA approaches explored in the project, we distinguish three categories:

(1) *Sending and Displaying*, which includes all technologies that support communication by enabling an easy exchange of files and data. For example, the teacher sending questions to individual students' devices or displaying one student's screen to discuss his/her work with the whole class.
(2) *Processing and Analysing* considers technology converting collected data. This includes software that generates feedback and results to an operation or applications which create statistical diagrams of a whole class' solution, for example after a poll.
(3) *Providing an Interactive Environment* refers to technology that enables students to work in a digital environment that lets them explore mathematical or scientific contents interactively. This category includes, for example, shared worksheets, Geogebra files, graph plotting tools, spread sheets or dynamic representations (www.fasmed.eu).

2.2 The Mathematical Content: Functions

During the development of a self-assessment tool, its mathematical content needs careful consideration. Bennett (2011) states that "to realise maximum benefit from formative assessment, new development should focus on conceptualising well-specified approaches […] rooted within specific content domains" (p. 5). Thus, a content analysis needs to evaluate, for example, which competencies or skills students need to master, what a successful performance entails and which conceptual difficulties might occur. This 'a priori' content analysis revealed three aspects relating to functions relevant for the SAFE tool's development: different mental models that students need to acquire for a comprehensive understanding, translating between mathematical representations and known misconceptions.

There are different theories to describe the meaning that a person links to a mathematical content. Tall and Vinner (1981) propagate "the term *concept image* to describe the total cognitive structure that is associated with the concept, which includes all the mental pictures and associated properties and processes" (p. 152). While their integral approach includes a person's pre- and misconceptions and, thus, helps to explain students' behaviour and difficulties when engaging in a mathematical activity, it does not include a normative dimension, which helps to identify adequate interpretations of the content as a guideline for teaching or designing teaching tools (Klinger, 2018). This is why the German tradition of subject-matter didactics specifies the idea of mental models in the concept of *Grundvorstellungen* (GVs). GVs "characterize mathematical concepts or procedures and their possible interpretations in real-life" (vom Hofe & Blum, 2016, p. 230) thereby, identifying the different approaches to a content that makes it accessible for students. GVs describe, which mental models learners have to construct in order to use a mathematical object for describing real-life situations. In this sense, GVs act as mediators between mathematics, reality and the learners' own conceptions. Hence, they include a normative as well as descriptive aspect (vom Hofe & Blum, 2016).

There are three GVs that require consideration when using the graph of a function to describe a given situation: *mapping*, *covariation* and *object*. In a static and local view, a function maps one value of an independent quantity to exactly one value of a dependent quantity. Thus, the graph of a function can be seen as a collection of points that originate from uniquely mapping values of one quantity to another. In a more dynamic view, a function describes how two quantities systematically change in relation to each other. Considering a functional relation with this focus allows a graph to embody the simultaneous variation of two quantities. Finally, a function can be seen as a mathematical object. Then, the graph is viewed as a whole from a global perspective (Vollrath, 1989).

Similar aspects are highlighted by Dubinsky and Harel (1992) when describing an individual's construction of the function concept. They identify four constitutive stages. At first, a student has a *prefunctional* understanding that is not useful for solving function related tasks. On the level of an *action conception*, a learner can perform mental or physical manipulations of functions. For example, an x-value can be inserted into a numerical representation of a function to calculate the corresponding y-value by following a sequence of commands given by the equation. Thus, the *action conception* is static in the way that a student on this level will regard one step of the calculation at a time. If the learner can view a function as a dynamic transformation of depending quantities that will produce the same transformed quantity whenever the same original quantity is used, it is viewed with a *process conception*. Then, the function is understood as a complete process. Finally, a student can reach an *object conception* that lets him/her understand functions as entities that can be acted upon and transformed. The cognitive mechanism of abstracting a function into a new mathematical object is referred to as "encapsulation" stressing that students with this perception of a function reach a higher level of understanding (Dubinsky & Harel, 1992, p. 4). This concretization of a process towards an object is also described as "reification" by Sfard (1991, p. 19), who focuses on the process-object-duality of mathematical concepts. While the *object conception* of Dubinsky and Harel (1992) can be treated as equivalent to the *object GV*, the *mapping* and *covariation GV* appear to be relevant for both the *action* and the *process* conception. In contrast to Dubinsky and Harel (1992), who regard their conceptions to be consecutive, the GV theory postulates that all three GVs need to be build evenly for a full understanding (Vollrath, 1989). This is most comparable with DeMarois and Tall (1996), who add a fifth *proceptual conception* as the highest level of understanding the concept of functions. By referring to Gray and Tall (1994), the authors define "procept" to be the merger of a process, a concept and a shared symbol that can evoke either one of these perceptions. Thus, it is indicated that students at this stage are able to flexibly move between viewing a function as a process or an object. All these theoretical aspects, especially the APO (action-process-object) as well as the GV theory (mapping, covariation, object) help to grasp the concept of function and give concrete indications for designing the SAFE tool. As the tool is not meant to serve the individual student's initial learning process, but rather to support the assessment of already learned contents that a student is able to retrieve, all of the aspects and mental images in the theories above have to be regarded.

Besides constructing relevant mental images of a function, a comprehensive understanding of the concept requires students to be able to translate between different forms of representations of a function (Duval, 1999). Functional relations appear in a range of semiotic representations. Learners can encounter functions, for instance as situational descriptions, numerical tables or Cartesian graphs. Each of these emphasizes different characteristics of the represented function. Thus, transforming one form into another makes other properties of the same mathematical object explicit (Duval, 1999). Further, Duval (1999) stresses that mathematical objects are only accessible through their semiotic representations. Therefore, each mathematical activity can be described as a transformation of representations. Duval (1999) differs between treatments, meaning the manipulation within the same semiotic system, and conversions, meaning the change of one representational register to another while preserving the meaning of the initial representation. The author identifies conversions between different registers to be the "threshold of mathematical comprehension for learners […]" (Duval, 2006, p. 128) and concludes that "only students who can perform register change do not confuse a mathematical object with its representation and they can transfer their mathematical knowledge to other contexts different from the one of learning" (Duval, 1999, p. 10). Hence, asking students to draw a graph based on a given situation means assessing a key aspect of their understanding of the concept of functions.

As students' mistakes can mirror their conceptual difficulties, typical misconceptions in the field of functions were considered for the development of the SAFE tool. For instance, Clement (1985) states that many students falsely treat the graph of a function as a literal picture of the underlying situation. They use an iconic interpretation of the whole graph or one of its specific features instead of viewing it as an abstract representation of the described functional relation (Clement, 1985). To overcome this error, known as the 'graph-as-a-picture mistake', students need opportunities to consider graphs symbolically. Thus, a task might ask learners to interpret a graph point by point or to describe the change of the dependent quantity for certain intervals. Another example of a typical cognitive issue when graphing functions is the 'swap of axes' labels. This mistake can arise when students name the axes intuitively without regarding mathematical conventions (Busch, 2015). Hadjidemetriou and Williams (2002) even speak of the "pupils' tendency to reverse the x and the y co-ordinates and their inability to adjust their knowledge in unfamiliar situations" (p. 4). In order to correctly label the axes for a given situation, learners need to understand the functional relation between two quantities from its description and apply the convention to record the independent quantity on the x-axis and the dependent quantity on the y-axis of a Cartesian coordinate system (Busch, 2015).

These are examples of some of the findings on typical misconceptions that were used in the design of the SAFE tool that both anticipate certain student difficulties and provide hints to foster the desired competencies. In addition, the design is based on the integration of technology. Digital media not only offer new opportunities for assessment but allow visualizations and activities that help students to overcome misconceptions in the field of functions.

2.3 *Potential of Digital Technology to Support Students' Self-Assessment in the Field of Functions*

Digital technologies have the potential to support student's self-assessment by changing this process due to, for example, new types of tasks, feedback, representations or even assessed skills. Furthermore, they offer numerous affordances for learning the function concept by, e.g. providing dynamic visualizations (Drijvers et al., 2016). In order to evaluate, which role digital tools can play in this context, it is essential to consider their potential but also regard possible risks. Some of the most important arguments for using digital technology to support students' self-assessment in the field of functions are:

- *Fast availability of representations*:
  The fast availability of functional representations leaves time to examine the functional relationships, that are visualized, to generate examples, or check one's hypotheses. However, the great amount of representations and the speed of their availability might lead students to feel overwhelmed. This might hinder them from reflecting on their actions (Barzel, Hußmann, & Leuders, 2005). Cavanagh and Mitchelmore (2000, p. 118), for instance, identified the students' tendency "to accept whatever was displayed in the initial window without question" as a main error, which occurs when upper secondary students were asked to interpret graphs on calculator screens. They missed to reflect the visualization or relate it to the inserted algebraic equation (Cavanagh & Mitchelmore, 2000).
- *Multiple representations*:
  Multiple representations of the same function can be displayed quickly. As each type of representation focuses on different aspects of the function, this can help students to build connections and grasp the underlying concept (Duval, 2006).
- *Dynamic representations*:
  In static media, such as pen-and-paper, variations of a mathematical object have to be perceived and interpreted as well as projected upon its representations by the user. Digital media offer the potential of using dynamic representations. With these students are able to explore changes—for example in the quantities of a functional relationship—directly (Kaput, 1992). Therefore, dynamic visualizations support the covariation GV, as they simplify the identification of changes in a function's values.
- *Interactivity and linked representations*:
  Technology allows students "not simply to display representations but especially to allow for actions on those representations" (Ferrara, Pratt, & Robutti, 2006, p. 242). Furthermore, representations can be linked to each other, so that learners are able to investigate functional relationships by changing one representation and observing the direct effect these changes have on another. This immediate feedback encourages students to translate between different functional representations. However, one must be concerned that the technological speed of students' manipulations of functional representations entail a risk of learners missing to reflect on their actions (Zbiek, Heid, Blume, & Dick, 2007).
- *Effecting student actions*:
  The design of a technological learning or assessment environment influences the students' actions. For instance, a digital tool might already provide learners with a coordinate system when asked to draw a graph, so that they do not need to focus on choosing correct axes as well as their labels or scale. Kaput (1992, p. 526) refers to a tool's "constraint-support structure" stating that "whether a feature is regarded as one or the other does not depend inherently on the material itself, but on the relation between the user's intentions and those of the designer of the material and the contexts for its use."

Technology Supporting Student Self-Assessment in the Field ... 57

In our study, these arguments for using technology to support students' conception of the function concept are used to guide the design of the digital self-assessment tool (SAFE tool).

3 Design of the Digital Self-Assessment Tool

The structure of the SAFE tool draws on a set of self-assessment materials originating from the KOSIMA (German acronym for: contexts for meaningful mathematics lessons) project (Barzel, Prediger, Leuders, & Hußmann, 2011). The tool comprises of five parts: *Test, Check, Info, Practice* and *Expand*. These are connected in a hyperlink structure and labelled with different symbols to support easy learner orientation regarding the SAFE tool's use (Fig. 4).

As explained in the beginning of this chapter, the aim is to create a tool that allows students to conduct the assessment process themselves. This is why, the design intends to create a balance between providing enough information as well as autonomy for the learners. The SAFE tool does not provide students with any direct feedback in order to help them to adopt the role of the assessor. However, it supports learners by providing a hyperlink structure to guide their assessment, dynamic graphing windows to explore their thinking, sample solutions and assessment criteria to reflect their own work, and explanations as well as tasks to engage them in further learning activities.

The initial step of the self-assessment process is for the student to identify the learning goal. It is specified and made transparent in our tool by the question: "Can I sketch a graph based on a given situation?", which appears on the top of the first screen (Fig. 5a). The learner is provided with the *Test* task (labelled with a magnifying glass icon). This *Test* presents the story of a boy's bike ride and asks the student to build a graph that shows how the boy's speed changes as a function of the time. To solve the *Test*, the learner can label the axes by selecting an option from drop-down menus, and build a graph out of moveable and adjustable graph segments. These are dragged into the graphing window and can be arranged in any order that the student chooses. In addition, the slope of the single segments can be altered by the user. Therefore, we use an interactive representation in the *Test* task that helps students to investigate their understanding of drawing a graph based on a given situation, while constraining some of their actions with regards to the given coordinate system and list of axes labels to choose from.

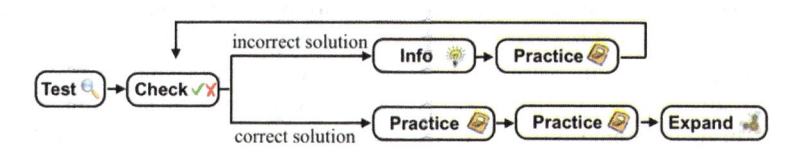

Fig. 4 Hyperlink structure of the SAFE tool

Fig. 5 **a** *Test* and **b** *Check* of the SAFE tool

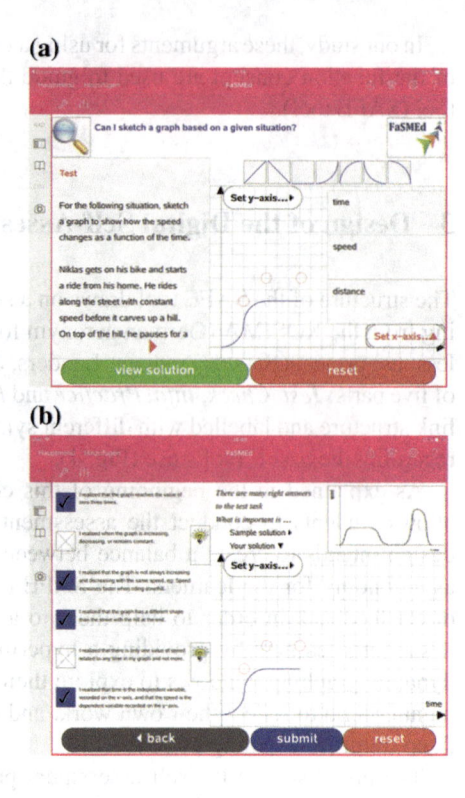

After submitting a graph, a sample solution and *Check* are presented to help evaluate one's individual answer (Fig. 5b). The *Check* (labelled with two check marks) presents the student with six statements regarding important aspects of the functional relation at hand alongside common mistakes that could arise when solving the *Test* task. For example, one of the *Check*-points addresses the graph's slope: "I realized when the graph is increasing, decreasing or remaining constant.", or another represents the graph-as-a-picture mistake: "I realized that the graph does not look like the street and the hill." The learner decides for each statement, if it is true for his/her solution, in which case it is marked off. For this diagnostic step, the student's screen not only presents the *Check*-list, but his/her answer as well as a sample solution to make a comparison easy. Thus, the *Check* helps the learner to self-assess his/her solution by presenting criteria for successfully solving the *Test* and therefore constraining his/her assessment in order to reduce its complexity. Furthermore, by using a multiple representation of the student's as well as the sample solution, the learner is encouraged to reflect his/her own solution in comparison to the sample solution and *Check*-points. Additionally, the *Check* serves as a directory through the tool's hyperlink structure (Fig. 4). This way, the student is encouraged to take further steps to move his/her learning forward, while the technology offers a direct link to the chosen contents due to its hyperlinks.

If an error is identified by the learner, he/she can choose to work on the *Info* and *Practice* task corresponding with the *Check*-point's statement. The *Info* (labelled by a lightbulb) entails a general explanation that is intended to repeat basic classroom contents to overcome the certain mistake. Moreover, the general explanation is followed by a concretization using the time-speed context of the *Test* as an example. In addition, an illustration is included to ensure a visual help and to encourage the learner to change between the two semiotic representations: verbal description as well as Cartesian graph. When the student finishes reading the *Info* section, he/she moves to the *Practice* task (labelled with an exercise book icon). The *Practice* lets the student test his/her understanding of the repeated content. There are four different task types implemented in the SAFE tool that vary in the way that the learner presents his/her answer: graphing, open answer, selection and matching. For the graphing tasks, the student builds a graph with the interactive graphing window as described for the *Test* task. If a *Practice* task requires the student to give reasons for, and verbalize his/her thinking, the student is asked to insert a solution via an open answer text box. Other tasks are solved by selecting situations that fit certain criteria from a list. Finally, for the matching tasks, the learner identifies which situation belongs to which graph shown to them. This is another dynamic task type as moving the graphs from one situation to another is easy via the iPads touchscreen. Depending on the contents, the *Practice* tasks may combine these task formats or be comprised of only one of them. After solving a *Practice* task, the user can go back to the *Check* and work on the next statement.

If the sketched graph is stated as correct or if the learner has worked on all *Info* units and *Practice* tasks for his/her identified mistakes, two further *Practice* tasks and one *Expand* task (labelled with a gearwheels icon) with a more complex context are provided. In this case, the *Expand* task asks the student to draw two different graphs for the same situation.

The SAFE tool aims to challenge the student to reflect on his/her own solutions and reasoning. This is why, besides offering a *Check*-list, it presents sample solutions for all tasks. It is the learner who decides whether their own answer is correct by comparing it to the sample solutions. No direct feedback is provided by the technology as we want the students to adopt the responsibility for their own assessment process.

4 Methodology

Besides generating a well-grounded tool for formative self-assessment, the study aims to examine the following research questions:

When students work with the SAFE tool:

(1) which formative assessment strategies do they use?
(2) which functionalities does the technology have within the student's FA processes?

To answer these questions, a design-based research approach is used that connects the conception and evaluation of the SAFE tool. Design-based research is a "formative approach to research, in which a product or process is envisaged, designed, developed, and refined through cycles of enactment, observation, analysis, and redesign, with systematic feedback from end users" (Swan, 2014, p. 148). Thus, different versions of the SAFE tool were designed, evaluated by analysing case studies and re-designed leading into the next cycle of implementation and evaluation. Two different forms of case studies are applied: class trials and student interviews. The purpose of the class trials is to evaluate the effectiveness of the tool's implementation by exploring whether: self-assessment is possible using the SAFE tool, the structure is clear, and any technical issues are identified. Class trials are conducted during a lesson where students work on the digital self-assessment tool individually or in pairs. Data is collected in the form of the researcher's notes on the lesson and a classroom discussion about the students' experiences with the tool. In addition, task-based interviews with individual students aim for a more detailed understanding of the learners' FA processes. This is why, students are asked to "think out loud" during their work with the SAFE tool and interviewers are instructed to only intervene the students' self-assessment to remind them to verbalise their thoughts or to help with technical issues. At the end, reflecting questions about the students' experience with the tool are asked. The interviews are videoed and transcribed to serve as the main data pool for qualitative analyses. These lead to the reconstruction of FA processes using the FaSMEd framework (Fig. 3).

In each cycle of development, the investigation of the research questions using the FaSMEd framework (Fig. 3) informs the re-design of the SAFE tool. On this account, several development cycles took place in the study since 2014. A first pen-and-paper version of the tool was designed to pilot the SAFE tool's structure as well as its tasks, checklist items and formulations. It was evaluated through interviews with eleven grade eight students from two different secondary schools in Germany.

Following the tool's redevelopment, two digital prototypes were created using different technologies: JACK and TI-Nspire Navigator. JACK is a server-based system for online assessment developed by the Ruhr Institute for Software Technology at the University of Duisburg-Essen. While the software has several useful options, such as being able to generate automatic feedback based on student answers, create statistical overviews of submitted solutions and insert tasks with variable contents, the JACK prototype proved to be unfit for implementation of our tool. There were three main reasons the JACK prototype was unsuitable for our tool: Firstly, the SAFE tool's hyperlink structure could only be implemented in a restricted way. It was not possible to display the entire *Check*-list at once, but only single *Check*-points. This could make orientation within the tool's structure and deciding which parts to work on hard for students. Secondly, the software has a limited number of task types that are mainly in form of multiple choice or open answer formats while the SAFE tool includes, for example graphing or matching tasks that could not be implemented. Thirdly, JACK requires an internet connection, but most schools in Germany do not have access to wireless internet in their classrooms, which would limit its potential use. The second digital prototype was programmed in Lua script using the TI-Nspire Navigator software, which enabled the tool's hyperlink structure to be realized, allowed offline access and provided a choice of using the tool on a computer or iPad. Moreover, the options for implementing open tasks were greater and dynamic visualisations could be inserted. Hence, the tool's design was implemented only for TI-Nspire Navigator. The subsequent classroom trial of the digital tool run on iPads involving 18 grade ten students led to further redevelopments.

The finished TI-Nspire Navigator version was trialled in two grade ten classrooms at two further secondary schools and associated student interviews (one per class) were recorded. Additionally, another set of student interviews with two second semester university students were held. The wide range of data in different age groups and schools resulted in a thorough evaluation of the tool's potentials and constraints. As it is intended to assess and repeat basic mathematical competencies, its use is not limited to one specific group of learners. First experiences with the tool show that students in all of the tested class levels (grades 8, 10 and university) had similar issues concerning mathematical understanding as well as technical problems. This article focuses on the two single student interviews recorded in grade ten.

5 Results

This chapter reports on two students' work with the SAFE tool. Their FA processes are presented and analysed using the FaSMEd framework (Fig. 3). Thus, we describe and interpret their actions and comments in the interviews with regards to the three dimensions of the framework: agent/s, FA strategies and role of technology. The identified categories in these dimensions are specified in brackets after the explanation of why they are found in the students' self-assessment processes. For each self-assessment process, we can then highlight several cuboids in the representation of the FaSMEd framework to visualize their reconstruction (see Figs. 9 and 12). Both learners (S1 and S2) are female and sixteen years old, but visit different secondary schools. Their interviews were chosen for the analysis because they both trialled the same digital version of the tool (TI-Nspire Navigator) and produced similar answers for the *Test* task. This suggests that their FA processes could easily be comparable. Nevertheless, their reconstructed FA processes differ from each other showing that their self-assessment is not influenced by their initial task solution. Both students start with the *Test* task (Fig. 6).

For the following situation, sketch a graph to show how the speed changes as function of the time.

Niklas gets on his bike and starts a ride from his home. He rides along the street with constant speed before it carves up a hill. On top of the hill, he pauses for a few minutes to enjoy the view. After that he drives back down and stops at the bottom of the hill.

Fig. 6 *Test* task of the SAFE tool

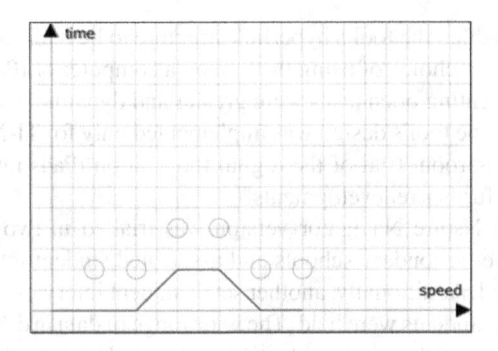

Fig. 7 S1's *Test* solution

5.1 Reconstruction of S1's FA Process Regarding the Axes Labels

S1 built her graph (Fig. 7) by dragging moveable graph segments into the graphing window and selecting labels for both axes from drop-down menus.

As she (student) solved the assessment task, she demonstrates her understanding of sketching graphs of given situations (strategy 2) while the tool provides an interactive learning environment due to its interactive graphing window (functionality 3). After reading the sample solution out loud, S1 moved to the *Check* and was silent for a while. The interviewer asked what she was thinking about. The student mentioned being unsure about which *Check*-list items to mark off because she "saw in the sample solution that there was another graph and this was missing in [her] own solution." With the "other graph" she means the last part of the sample graph, that increases at first and then decreases again representing the bike riding downhill and stopping at the bottom the hill, which she indicated by gesturing its shape on the screen with her finger. It can be concluded that the *Check* stimulated S1 to assess her answer by comparing her own graph to the sample solution. By reflecting on her answer, S1 (student) uses a metacognitive strategy and, thus, adopts some responsibility for her own learning process (strategy 5). The tool displays the information she needs for the diagnostic step in form of the sample solution and *Check*-list (functionality 1). Furthermore, the student decided to evaluate the last statement in the *Check*. It reads *"I realized that the time is the independent variable recorded on the x-axis and that the speed is the dependent variable recorded on the y-axis."* S1 stated that this was not true for her graph, which means that she understands a criterion to successfully solve the *Test* (strategy 1). What is more, she reflects on her solution by comparing it to the *Check*-point statement (strategy 5) and formulates a self-feedback (strategy 3): "The speed and time were wrong because there [she points to the x-axis] needs to be the time and there [she points to the y-axis] the speed. I did not realize this." Here, the technology is once more functioning as a display of information in the form of the *Check*-list item (functionality 1).

At that point S1 decided the next step in her learning (strategy 5) when she read the associated *Info* about the independent quantity being recorded on the x-axis and the dependent quantity being recorded on the y-axis. As she read through the *Info* quickly and didn't express her thoughts on it, we can't make assumptions about her learning progress or whether she understood the contents of the additional help provided (functionality 1). After the interviewer reminded her of the possibility to do an exercise related to her mistake, S1 worked on the linked *Practice*. This helped her to elicit evidence about her understanding of the independent and dependent quantity of a functional relation (strategy 2). The tool provides the task and sample solution (functionality 1). The task presented the learner with ten different situations describing the functional relation between two quantities. For each one, the learner was asked to assign labels to the axes of a coordinate system (given that he/she imagined drawing a graph based on the situation in the next step). The labels were chosen from a number of given quantities: temperature, distance, speed, time, pressure, concentration, money, and weight. S1 solved six out of ten items correctly. While she seemed to have no difficulties with situations in which time appeared as the independent quantity, she struggled to label the y-axis when time was being dependent on another quantity. For example, in the situation *"In a prepaid contract for cell phones, the time left to make calls depends on the balance (prepaid)."* S1 chose "time" as the label for the x-axis and "money" as the label for the y-axis. However, she explained "if you have a prepaid phone, you can only make calls as long as you have money." Therefore, she grasped the relation in the real-life context but couldn't use this knowledge when asked to represent it in form of a graph. Moreover, the student repeated this mistake of 'swapping the axes' even in situations that didn't include time as a quantity. For instance, S1 selected "distance" as the label for the x-axis and "speed" for the y-axis in the situation *"Tim's running speed determines the distance he can travel within half an hour."* Nonetheless, she explained correctly that "the speed specifies how far he can run." A possible explanation for her repeating mistake could be her approach to the task. S1 selected a label for the y-axis first before going on to the x-axis. This could mean that she does not fully understand the conventions of drawing a Cartesian coordinate system. However, her mistake could also originate from a deeper misunderstanding as Hadjidemetriou and Williams (2002) speak of the "pupils' tendency to reverse the x and the y co-ordinates and their inability to adjust their knowledge in unfamiliar situations" (p. 4). This would show a need for further interventions. However, S1 was able to identify two out of her four mistakes by comparing her answers to the sample solution (strategy 5) before she returned to the *Check* and marked off the respective *Check*-point statement.

In summary, S1's work with the SAFE tool concerning the naming of the axes can be depicted as shown in Fig. 8. She solves a diagnostic task, identifies a mistake by understanding criteria for success, reflecting on her answer and comparing it to a sample solution and displayed statement. She gives herself feedback and decides to take further steps in her learning by revising information on her error and practicing. Though she is not able to fully overcome her mistake, the SAFE tool supports S1 to think about her work on a metacognitive level and adopt responsibility for her learning.

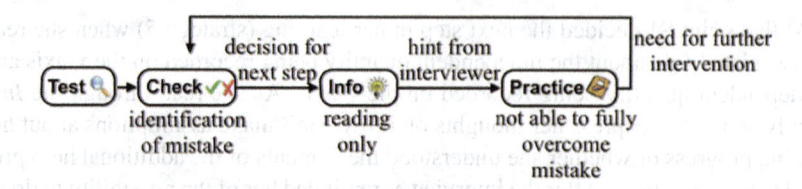

Fig. 8 Structure of S1's FA process

Thus, S1 uses four FA strategies, while the tool's functionality can be labelled as displaying information or, in case of the *Test* task, providing an interactive environment. S1's formative assessment process can be characterised using the FaSMEd framework as shown in Fig. 9. Each of the highlighted cuboids stresses how the dimensions of the framework interact in parts of the reconstructed assessment process. For example, while working on the *Test* task, we identified S1 (the student) to be the active agent of the assessment as she is eliciting evidence of her understanding of sketching graphs (strategy 2) while the digital tool serves as an interactive environment (functionality 3). Thus, for S1's work on the *Test*, we can highlight the second cuboid in the last row on the student level. In another step of S1's self-assessment process, she worked on the *Check*. Here, S1 (student) could be identified as the agent, who is actively using three different FA strategies. She understands a criterion for success, namely labelling the coordinate axes correctly (strategy 1), she is giving herself feedback by identifying a mistake in her solution and correcting it (strategy 3) and she is regulating her own learning process by reflecting on her own *Test* solution on a metacognitive level (strategy 5). Meanwhile, the SAFE tool

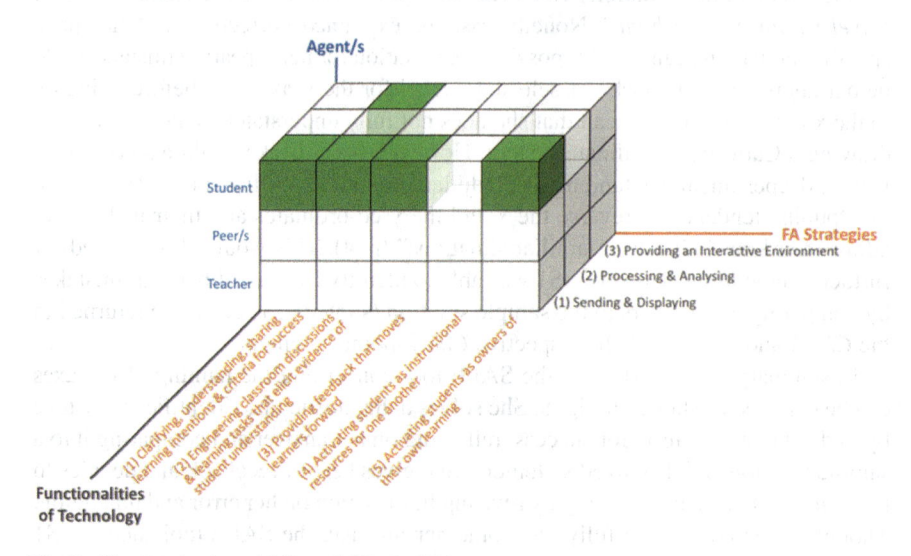

Fig. 9 Characterisation of S1's and S2's 2nd FA process

displays the information she is using in form of the sample solution and checkpoints (functionality 1). Thus, for S1's work with the *Check*, we can highlight three cuboids in the FaSMEd framework: the cuboids regarding FA strategies 1, 3 and 5 in the first row on the student level. These two examples show how the representation in Fig. 9 is constructed. The representation makes it possible to characterise and compare the student's reconstructed formative self-assessment process to others. However, it does not demonstrate whether a FA strategy was used more than one time during a self-assessment process. Therefore, it is possible for two different self-assessment processes to have the same characterisation in regards to the FaSMEd framework (Fig. 9).

5.2 Reconstruction of S2's 1st FA Process Regarding the Axes Labels

S2 also sketched a graph (Fig. 10) to solve the *Test* and elicit evidence of her understanding (strategy 2) using the tool's interactive graphing window (functionality 3).

In the *Check*, she didn't mark off the statement concerning time being the independent and speed being the dependent quantity. Thus, S2 identifies what she considered to be an error in her answer based on the displayed *Check* statement (functionality 1). Even though she labelled the axes correctly, S2 decided to read the *Info* concerning her alleged mistake and is, thus, adopting responsibility for her learning (strategy 5). When reading the *Info*, she realized: "Oh, that is correct as well because I did it in the same way." She not only states a self-feedback (strategy 3), but also compares the displayed information (functionality 1) to her own *Test* answer and reflects on her assessment (strategy 5). Then S2 went back to the *Check* and marked off the statement correcting the error in her previous assessment autonomously. In conclusion, S2 identifies a correct aspect about her work, which means she now understands a criterion for success (strategy 1).

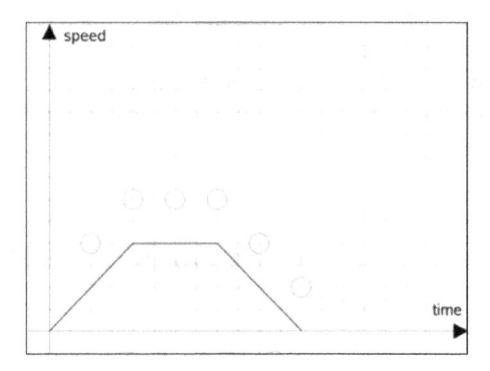

Fig. 10 S2's *Test* solution

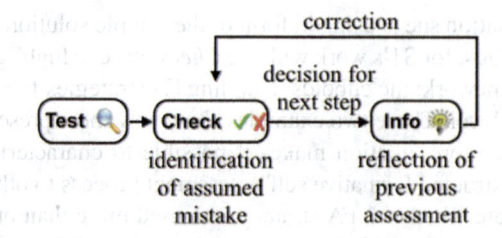

Fig. 11 Structure of S2's 1st FA process

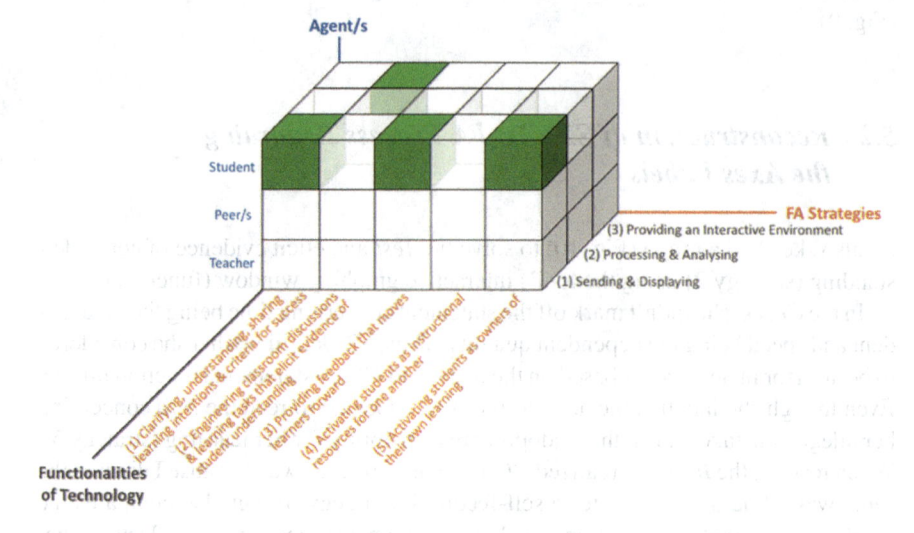

Fig. 12 Characterisation of S2's 1st FA process

In summary, S2's work with the SAFE tool concerning the naming of the axes can be illustrated as in Fig. 11. She works on a diagnostic task, identifies an assumed mistake and decides to gather more information on it. Then, S2 identifies an error in her previous self-assessment by comparing her *Test* solution to the displayed *Info*. Finally, she corrects her assessment.

The analysis shows that in this process, she uses four different FA strategies, while the tool functions mainly as a display of information and for the *Test* provides an interactive environment (Fig. 12).

5.3 Reconstruction of S2's 2nd FA Process Regarding the Graph Reaching Zero

Another process of formative self-assessment can be reconstructed by looking at S2's interview regarding this *Check*-point: "*I realized that the graph reaches the value of*

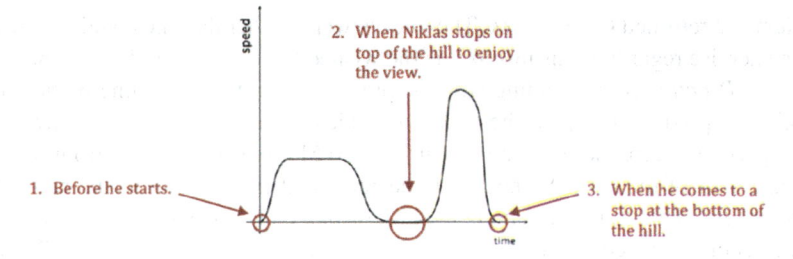

Technology Supporting Student Self-Assessment in the Field ...

Fig. 13 Illustration displayed in *Info 1* of the SAFE tool

zero three times." During the *Check*, S2 (student) didn't mark off this statement as she gestured to the points in which her graph reaches the value of zero and stated: "But it is only two times for mine." Here, S2 identifies a mistake in her *Test* solution by reflecting on her graph (strategy 5), understands a criterion for a correct solution (strategy 1) and formulates a self-feedback (strategy 3). The tool displays the necessary information (functionality 1). Afterwards, the student clicked on the 'lightbulb button' for this *Check*-point and was forwarded to the associated *Info*. Thus, S2 adopts responsibility for her learning process by deciding to seek assistance regarding her mistake (strategy 5). When she read the *Info* text, the learner remarked that she understood this then. After the interviewer asked her to elaborate her thoughts further, S2 explained with the help of the illustration displayed in the *Info* (Fig. 13):

> I did not do it like this. I did it so that Niklas rides along the street [she points to the first increasing part of the graph] and then here [she points to the first graph segment that remains constant] he rides along the hill and then he stops, but I did it so that he goes back again [she points to the first decreasing part of the graph]. I did not do it with the second zero, when Niklas stands on top of the hill, he has no speed anymore.

Therefore, the student reflects on her previous reasoning (strategy 5) in comparison to the explanations and visualizations displayed in the *Info* unit (functionality 1). We can conclude that her original rationale indicates a graph-as-a-picture mistake as she associates: (a) the increasing graph segment with riding along the street and uphill, (b) the constant graph segment with riding along the top of the hill and (c) the decreasing graph segment with going back or riding downhill. Due to her reflection, S2 gains the insight that while standing still, the bike "has no speed". By addressing the "second zero", she indicates that she is connecting the speed reaching the value of zero with the graph touching the x-axis. Therefore, S2 seems able to correct this local aspect of her graph-as-a-picture mistake. Nevertheless, she did not mention her other misinterpretations of the graph's slope. This could mean that the learner has not overcome her graph-as-a-picture mistake globally. However, this cannot be expected from S2 at this stage within her work with the SAFE tool because the *Info* only addresses the matter of when the graph reaches the value of zero. The student's focus is brought on only one criterion for a correct graph.

Next, S2 returned to the *Check*. The interviewer told her that she could also work on an exercise regarding this mistake, if she wished to do so. Then, the student went on to the *Practice* corresponding with the previous *Info*, thus, deciding on her next learning step (strategy 5). In the task, the student is asked to imagine drawing a time-speed graph for the storyline of a girl named Marie walking home from school. For each part of the story, S2 needed to decide whether the graph would reach the value of zero. Therefore, by solving the task displayed by the tool (functionality 1), S2 can elicit evidence on her understanding of the previous *Info* (strategy 2). She solved this task correctly and was only unsure about the fourth part of Marie's walk: "*After Jana says goodbye, Marie goes on more quickly.*" S2 explained that she was not certain about the meaning of 'saying goodbye' in this situation as she thought that the girls in the story would have to stop in order to do so, but then the text stated that Marie goes on more quickly. After the interviewer clarified that in the described part of the situation, the 'goodbye' has already happened in the past, S2 correctly responded that the speed would not reach the value of zero for this part of the story. After looking at the sample solution, the student acknowledged that she solved the task correctly. Therefore, she reflects on and evaluates her previous work (strategy 5) while the tool displays the necessary information in form of the sample solution (functionality 1). In this *Practice*, S2 proves that she understood from the *Info* that the speed reaches the value of zero whenever someone stands still and does not move in the described context. As she is pointing to the associated sections of the graph in her reflection of the *Info*, we can assume that she is also able to connect the speed reaching the value of zero to the graph reaching the value of zero. However, S2 does not mention the graph while solving the *Practice* task. Furthermore, the interview does not reveal whether she realizes that the reason for the graph reaching the value of zero whenever the speed reaches the value of zero is that the speed is the dependent quantity in the described functional relation.

In summary, S2's work with the SAFE tool regarding the issue of when a graph reaches the value of zero can be depicted as shown in Fig. 14. After solving a diagnostic task and identifying a mistake as well as generating self-feedback in the *Check*, S2 decides to gather more information on her mistake and reflects on her previous reasoning. By solving the *Practice* task correctly, the student shows that she was able to overcome her previous graph-as-a-picture mistake at least locally and in the given context. What is more, she autonomously checked her *Practice* answer by comparing it to the sample solution and is, thus, reflecting her work metacognitively to move forward in her learning. The analysis shows that we could identify four FA strategies and two functionalities of the technology in S2's second FA process. Because these are consistent with the ones described in S1's example, we can again use Fig. 9 to characterise the student's FA process.

5.4 Examples Revealing Constraints of the SAFE Tool

While the previous examples have reconstructed self-assessment processes, in which the students were able to identify their own mistakes or correct aspects about their work using the SAFE tool, the student interviews also revealed that the learners were not able to recognize all of their mistakes. In contrast to S2, S1 did not mention the *Check*-point concerning the graph reaching the value of zero three times. She did not proceed to the corresponding *Info* or *Practice* in the course of her interview even though this could be expected based on her *Test* solution (Fig. 7). This could indicate that S1 was not able to identify her mistake, but it could also mean that she misunderstands the functionality of the *Check*. She seems to think that she is only allowed to tick off one statement in the *Check* rather than all that are correct for her individual *Test* solution. This becomes clear in her interview as she only marks off one *Check*-point and states: "I am considering which *Check*-point I should pick." Another interesting case of the students not being able to identify their mistakes appears in the interviews when we look at their self-assessment regarding the graph-as-a-picture mistake. Both students' *Test* solutions (Figs. 7 and 10) indicate that they drew the hill that the bike is driving over as the time-speed graph. Nevertheless, neither S1 nor S2 marked off the corresponding *Check*-point: "*I realized that the graph has a different shape than the street with the hill at the end.*" However, both students seem to have different reasons for drawing their graphs and for not choosing this *Check*-point. S1 explains her reasoning for the hill-shaped part of her *Test* graph (Fig. 7) by associating driving uphill with a linear increasing graph segment, standing still with a constant graph segment and the decreasing graph segment with "going back down". In the *Check*, she did not address the statement in question. This could be because S1 did not notice her mistake or because her graph-as-a-picture mistake is only local while the *Check* statement addresses a global aspect of the graph. The student did not interpret the constant graph segment as a literal picture of the situation, thus, showing only local graph-as-picture mistakes for riding up- and downhill. In contrast, S2 did show a global graph-as-a-picture mistake as she explained her *Test* solution (Fig. 10) to show that "he goes uphill [...] and then I think he rides straight on because he is on top of the hill [...] and then back down again." However, while S2 reflects on the *Check*-point, she does not seem to connect a sketch of the described situation to the real-life shape of a street and hill. She stated: "[...] the shape is –

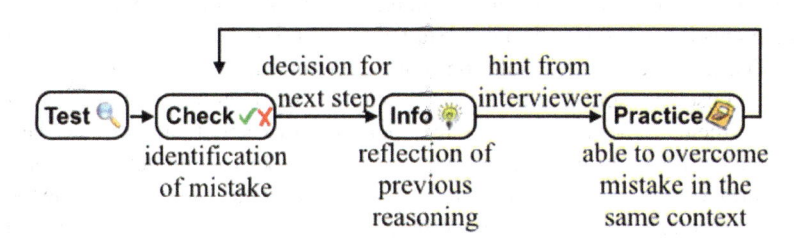

Fig. 14 Structure of S2's 2nd FA process

when I do it normally, on the street, it is much different than here on the graph." Thus, she misunderstands the *Check* statement and is not able to identify her mistake, even though the SAFE tool prompts her to reflect on her solution with regards to the overall shape of the graph compared to the storyline of the bike ride.

Finally, we can find examples in the student interviews in which the students falsely marked off *Check*-points that are not adequate based on their *Test* solutions. Both students thought that they realized when the graph is increasing, decreasing or remaining constant. Therefore, the students were making mistakes in their self-assessment even though they had seen a sample solution with explanations previous to the *Check*.

6 Conclusions and Further Steps

The analysis of the two cases shows that the SAFE tool does have the potential to support students' formative self-assessment concerning their ability to draw a graph based on a given situation. We were able to reconstruct three different FA processes within the student interviews that showed the following characteristics. Firstly, it is the user, who holds the responsibility to identify mistakes and decide on next steps in the learning process. Secondly, the tool stimulates learners to actively use four different key strategies of formative assessment: the clarification and understanding of criteria for success, eliciting evidence on student understanding, formulating feedback and being activated as the owners of one's own learning.

However, the case studies highlight some constraints of the SAFE tool, which, in the cyclic process of the study, lead to a redesign that is currently being programmed. In the interviews, it became clear that students are uncertain about assessing themselves as they did not meet the expected assessment for all *Check*-points and mentioned that they expect validation from either the teacher or the technology. This is why the redesign focuses on improving students' comprehension of the learning goal, namely the change of representation from situation to graph, and simplifying the learners' self-evaluation. Hence, the static picture of the *Test's* sample solution will be replaced with a dynamic simulation of the described bike ride linked to the sample graph. Additionally, students will be able to view a changing qualitative speed-o-meter as they play the simulation to grasp the change of speed during the simulated bike ride (Fig. 15a). Furthermore, the *Check* will contain a positive and negative check mark for each point. Students will not only have to tick the statements that are true for their solution, but select for each *Check*-point whether it is true or false for their graph (Fig. 15b). This design feature might support their self-assessment further as they are encouraged to think about each of the Check's statements and have to make a visible decision in order to complete their assessment. Furthermore, the procedure of the *Check* becomes more intuitive for the tool's user to avoid misunderstandings concerning the execution of this central step of the self-assessment like described for S1. To counter both students' problems with the third *Check*-point concerning the graph-as-a-picture mistake, it was rephrased into:

Fig. 15 **a** Bike ride simulation as *Test* sample solution and **b** *Check* of the SAFE tool's redesign

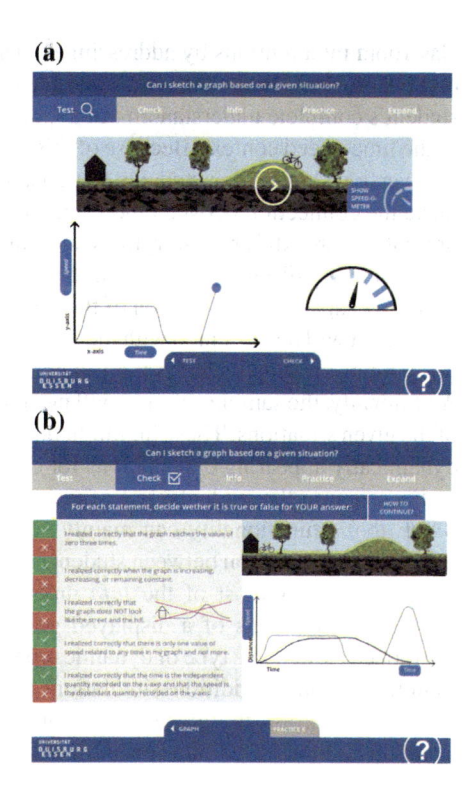

"*I realized correctly that the graph does NOT look like the street and the hill.*" The new formulation leaves out the words "different form", which got misinterpreted by S2. What is more, a crossed-out sketch of the described route of the bicycle was added to the *Check*-point to make the meaning of the statement clearer by using multiple representations—verbal and pictorial (Fig. 15b). Additionally, the sample solutions of the *Practice* tasks will be displayed more simultaneously with the user's answer for easier comparison. Finally, a help button labelled by the question mark on the bottom right side of the screen will be introduced (Fig. 15a, b). It will link to a video tutorial that leads through the SAFE tool's structure and use. This will enable learners to seek technical assistance independently.

Besides these changes intended to simplify the tool's use and improve students' comprehension of graphing the given situation in the *Test*, several other alterations are motivated by the previous analysis. The students' interview statements and S1's case, in which she was unable to fully overcome her mistake, revealed that it will not be possible for all students working with the SAFE tool to (re)learn the change of representation from situation to graph on their own. Further interventions not included in the tool might be necessary. Therefore, the newest redesign will save the individual student's work. A teacher page will be included that makes it possible to review students' solutions and enable more effective planning of post-assessment classroom interventions by addressing students' needs directly. Moreover, S2's second FA process revealed that the *Practice* task described is not suitable to elicit the student's complete understanding of a graph reaching the value of zero independent of the time-speed context. Because of this, the *Practice* will be redesigned in the new tool version. It aims for students to transfer their knowledge to other contexts and make the connection between the value of the graph and the dependent quantity by presenting three different storylines and functional relations: (a) speed as a function of the time, (b) distance to a destination as a function of the distance from a start and (c) filling capacity as a function of the time. In contrast to the previous task version, the student will be presented with the entire storyline at once and asked to answer how many times the graph reaches the value of zero as well as provide his/her reasoning. Additionally, the sample solution will present the user with graphical representations of the given situations. These highlight the addressed aspect of the functions by using dynamically appearing explanations for each time that a graph touches the x-axis.

Focusing on the students' use of technology, the two cases demonstrate that the SAFE tool's functionality can mainly be described as displaying information. To increase the interaction between students and tool, the redesign will include dynamic visualisations for most of the *Info* units. These will enable students to click on highlighted segments of a displayed graph to open and read an explanation. As mentioned earlier, this type of dynamic representations is also used for one *Practice*'s sample solution. In addition, simulations as described for the *Test*'s sample solution, that allow to grasp the given context and make connections between the real-life situation and the graph of a function, will be used in some of the *Practice* tasks as well.

Finally, the interviews show that more detailed analyses are necessary to gain a deeper understanding of students' formative self-assessment processes. Working with the SAFE tool did not help learners to overcome all of their mistakes and, based on their *Test* solutions, choose all of the expected *Check*-points. However, it encouraged them to reflect on their own solutions on a metacognitive level, which seems to be key for their success in doing self-assessment. Thus, a category system for a qualitative content analysis of the interviews is currently being developed. It focuses on three main categories regarding the students': metacognitive activities, tool activities and content-related activities. The aim is to observe which metacognitive activities are prompted through which design aspects of the SAFE tool and how this can help the students' conceptual understanding of the content of functions.

References

Aldon, G., Cusi, A., Morselli, F., Panero, M., & Sabena, C. (2017). Formative assessment and technology: Reflections developed through the collaboration between teachers and researchers. In G. Aldon, F. Hitt, L. Bazzini, U. Gellert (Eds.), *Mathematics and technology: A C.I.E.A.E.M. sourcebook* (pp. 551–578). Springer International Publishing.

Barzel, B., Hußmann, S., & Leuders, T. (Eds.). (2005). *Computer, Internet & Co. im Mathematikunterricht*. Berlin: Cornelsen Scriptor.

Barzel, B., Prediger, S., Leuders, T., & Hußmann, S. (2011). Kontexte und Kernprozesse: Aspekte eines theoriegeleiteten und praxiserprobten Schulbuchkonzepts [Contexts and core processes—Aspects of a school book concept guided by theory and practice]. *Beiträge zum Mathematikunterricht*, 71–74.

Bell, B., & Cowie, B. (2001). The characteristics of formative assessment in science education. *Science Education, 85*(5), 536–553.

Bennett, R. E. (2011). Formative assessment: A critical review. *Assessment in Education: Principles, Policy & Practice, 18*(1), 5–25.

Bernholt, S., Rönnebeck, S., Ropohl, M., Köller, O., & Parchmann, I. (2013). *Report on current state of the art in formative and summative assessment in IBE in STM: Report from the FP7 project ASSIST-ME* (Deliverable 2.4). Kiel: IPN—Leibnitz-Institut für die Pädagogik der Naturwissenschaften und Mathematik.

Black, P., Harrison, C., Lee, C., Marshall, B., & Wiliam, D. (2004). Working inside the black box: Assessment for learning in the classroom. *Phi Delta Kappan, 86*(1), 8–21.

Black, P., & Wiliam, D. (2009). Developing the theory of formative assessment. *Educational Assessment, Evaluation and Accountability, 21*(1), 5–31.

Busch, J. (2015). Graphen "laufen" eigene Wege. Was bei Funktionen schiefgehen kann [Graphs "run" their own ways. What can fail with functions]. *Mathematik lehren, 191,* 30–32.

Cavanagh, M., & Mitchelmore, M. (2000). Graphics calculators in mathematics learning: Studies of student and teacher understanding. In M. Thomas (Ed.), *Proceedings of the International Conference on Technology in Mathematics Education* (pp. 112–119). Auckland: The University of Auckland and Auckland University of Technology.

Clement, J. (1985). Misconceptions in graphing. In L. Streefland (Ed.), *Proceedings of the 9th Conference of the International Group for the Psychology of Mathematics Education* (Vol. 1, pp. 369–375). Utrecht: PME.

DeMarois, P., & Tall, D. (1996). Facets and layers of the function concept. In L. Puig & A. Gutierrez (Eds.), *Proceedings of the 20th Conference of the International Group for the Psychology of Mathematics Education* (Vol. 2, pp. 297–304). Valencia: PME.

Drijvers, P., Ball, L., Barzel, B., Heid, M. K., Cao, Y., & Maschietto, M. (2016). *Uses of technology in lower secondary mathematics education: A concise topical survey.* London: Springer Open.

Dubinsky, E., & Harel, G. (1992). The nature of the process conception of function. In G. Harel & E. Dubinsky (Eds.), *The concept of function: Aspects of epistemology and pedagogy* (pp. 85–106). Washington: Mathematical Association of America.

Duval, R. (1999). Representation, Vision and Visualization: Cognitive Functions in Mathematical Thinking. Basic Issues for Learning. Retrieved from http://files.eric.ed.gov/fulltext/ED466379.pdf.

Duval, R. (2006). A cognitive analysis of problems of comprehension in a learning of mathematics. *Educational Studies in Mathematics, 61*(1), 103–131.

Ferrara, F., Pratt, D., & Robutti, O. (2006). The role and uses of technologies for the teaching of algebra and calculus. In A. Gutièrrez & P. Boero (Eds.), *Handbook of research on the psychology of mathematics education: Past, present and future* (pp. 237–273). Rotterdam: Sense.

Gray, E., & Tall, D. (1994). Duality, ambiguity, and flexibility: A "proceptual" view of simple arithmetic. *Journal for Research in Mathematics Education, 25*(2), 116–140.

Hadjidemetriou, C., & Williams, J. (2002). Children's graphical conceptions. *Research in Mathematics Education, 4*(1), 69–87.

Heritage, M. (2007). Formative assessment: What do teachers need to know and do? *Phi Delta Kappan, 89*(2), 140–145.

Kaput, J. J. (1992). Technology and mathematics education. In D. A. Grouws (Ed.), *Handbook of research on mathematics teaching and learning* (pp. 515–556). New York, NY: Macmillan.

Klinger, M. (2018). *Funktionales Denken beim Übergang von der Funktionenlehre zur Analysis: Entwicklung eines Testinstruments und empirische Befunde aus der gymnasialen Oberstufe* [Functional thinking in the transition from function theory to calculus: Development of a test and empirical findings from upper secondary students]. Wiesbaden: Springer Spektrum.

Ramaprasad, A. (1983). On the definition of feedback. *Systems Research and Behavioral Science, 28*(1), 4–13.

Sadler, D. R. (1989). Formative assessment and the design of instructional systems. *Instructional Science, 18*(2), 119–144.

Sfard, A. (1991). On the dual nature of mathematical conceptions: Reflections on processes and objects as different sides of the same coin. *Educational Studies in Mathematics, 22*(1), 1–36.

Stacey, K., & Wiliam, D. (2013). Technology and assessment in mathematics. In M. A. K. Clements, A. J. Bishop, C. Keitel, J. Klipatrick, & F. K. S. Leung (Eds.), *Third international handbook of mathematics education* (pp. 721–752). New York: Springer.

Swan, M. (2014). Design research in mathematics education. In *Encyclopedia of mathematics education* (pp. 148–152). Springer Netherlands.

Tall, D., & Vinner, S. (1981). Concept image and concept definition in mathematics with particular reference to limits and continuity. *Educational Studies in Mathematics, 12*(2), 151–169.

Vollrath, H.-J. (1989). Funktionales Denken [Functional thinking]. *Journal for Didactics of Mathematics, 10*(1), 3–37.

vom Hofe, R., & Blum, W. (2016). "Grundvorstellungen" as a category of subject-matter didactics. *Journal for Didactics of Mathematics, 37*(1), 225–254.

Wiliam, D., & Thompson, M. (2008). Integrating assessment with learning: What will it take to make it work? In C. A. Dwyer (Ed.), *The future of assessment: Shaping teaching and learning* (pp. 53–82). Mahwah, NJ: Erlbaum.

Zbiek, R. M., Heid, M. K., Blume, G. W., & Dick, T. P. (2007). Research on technology in mathematics education: A perspective of constructs. In F. K. Lester (Ed.), *Second handbook of research on mathematics teaching and learning* (pp. 1169–1207). Charlotte, NC: Information Age.

Students' Self-Awareness of Their Mathematical Thinking: Can Self-Assessment Be Supported Through CAS-Integrated Learning Apps on Smartphones?

Bärbel Barzel, Lynda Ball and Marcel Klinger

1 Introduction

A range of digital technology to support mathematics learning is freely available on the internet. Students can use these technologies in an informal way to replace the procedural requirements of mathematics and to get support when doing homework, for example when solving an equation. Examples of online technologies include *Wolfram alpha*, offering students immediate access to many mathematical concepts and procedures; *Math 42* and *Photomath*, two computer algebra apps, offering graphical capabilities and step-by-step-solutions for school-relevant algebra procedures; or the large range of mathematical tutorials, such as tutorial systems (e.g., *Cognitive Tutor Algebra I*, see Pane, Griffin, McCaffrey, Karam, 2014). Based on the number of downloads for online support systems (e.g., *Math 42* on the Google Play store has in excess of 100,000 downloads) it can be conjectured that many students look for online support to do and learn mathematics outside the classroom. This is done in an informal, self-regulated way, often through accessing apps to support performance of mathematical procedures. The authority for mathematics learning no longer rests solely with the teacher, nor the textbook. However, curriculum and topics studied in class are still determined by the teacher, hence students are likely to be using digital technologies to support their learning in the context of those topics and the curriculum being studied in class.

B. Barzel · M. Klinger (✉)
University of Duisburg-Essen, Essen, Germany
e-mail: marcel.klinger@uni-due.de

B. Barzel
e-mail: baerbel.barzel@uni-due.de

L. Ball
MGSE, University of Melbourne, Melbourne, Australia
e-mail: lball@unimelb.edu.au

© Springer Nature Switzerland AG 2019
G. Aldon and J. Trgalová (eds.), *Technology in Mathematics Teaching*,
Mathematics Education in the Digital Era 13,
https://doi.org/10.1007/978-3-030-19741-4_4

In this chapter, we analyze some learning apps (available for smartphones) with an integrated computer algebra system (CAS) that may offer support when learning how to solve equations. Our aim is to provide an indication of the potential, strengths and weaknesses of the approaches presented to students by these CAS-apps, through consideration of the approaches for one area of mathematics. Our investigation is focused on the solution of quadratic equations (our example for reference which will be used later: $2x^2 + 5x - 3 = 0$). We regard the knowledge about informal ways of learning as an important consideration for teachers, designers and researchers as individual learning with apps cannot only be used to refresh previously taught knowledge, or to deepen students' knowledge, it also has the potential to assist in the learning of new mathematical knowledge.

In the context of solving quadratic equations, use of apps in an informal way to learn how to solve not only touches on learning issues in the field of algebra, but also aspects of students' self-regulation and the use of technology. These different aspects are discussed in the theoretical background and are used to guide our methodological approach to analyze different CAS-apps. We finish with a discussion of how to capitalize on the potentialities of these apps, as well as overcome their weaknesses, followed by a discussion of the potential role of these apps in the future design of teaching.

2 Theoretical Background

2.1 The Challenge of Learning How to Solve a Quadratic Equation

The aim of any teaching of mathematics at school is to develop students' understanding of mathematical concepts and procedures and help them to develop flexibility in performing these procedures. For example, learning to solve an equation requires more than memorization of a procedural routine. Stacey (2011) speaks of a "multi-year journey" to learn to write and solve algebraic equations, as the capabilities to learn about equations are diverse. Students should understand that algebraic equations play different roles: They can describe mathematical connections in a general way to cover more than one situation (such as geometric formulas), but they can also serve as an instrument to model and solve problems—either purely mathematical problems or real-world problems (Barzel & Holzäpfel, 2017). In addition, students must develop symbol sense (Arcavi, 2005) or algebraic insight (Pierce & Stacey, 2002, 2004) to get a feeling for the inner structure of an equation and the types of solutions to expect. This is not only important to model a given situation and to find an appropriate equation to represent the situation, but also in deciding on an appropriate way to solve the equation and to judge whether a solution is correct and realistic. All in all, procedural capabilities when doing mathematics are still important, but they must be developed through flexible and productive exercises in order to learn and understand the mathematical structure behind it (Winter, 1984). An example of a flexible and productive exercise in the context of solving a quadratic equation is to determine the value of p in $x^2 + px + 2 = 0$, so that the graph of the function only cuts the x-axis once. One important goal when learning procedures is to develop number sense (Dehaene, 1997) as well as symbol sense (Arcavi, 2005) to be prepared for a flexible and adaptive transformation of the procedures to other contexts and situations, which is not only important for success at school but also an important feature in a lot of professions at work.

When solving quadratic equations, it is important for students to develop a range of methods for solving, developing an understanding that there can be more than one method for solving, thus enabling consideration of efficiency of various approaches. Depending on the parameters of an equation having the form $ax^2 + bx + c = 0$ there are different approaches that students can use to solve:

- Graphic approach (looking for zeroes of a quadratic function's graph)
- Numeric approach ("guess, check and improve")
- Algebraic approach:
  - Do the same to both sides, when b = 0, then: $ax^2 + c = 0$, then $ax^2 = -c$ and then taking the root
  - Using 'zero-product' or the 'null factor law' (e.g. $(x - 2)(x + 3) = 0$), so x − 2 = 0 or x + 3 = 0
  - Quadratic formula $x = \frac{-b \pm \sqrt{b^2 - 4ac}}{2a}$ or an equivalent algebraic form.

Block (2015, p. 391) describes *flexible algebraic action* as the ability to choose an appropriate method for solution according to the specificity of the task. To give an example: A student must be aware that using the quadratic formula for the equation $(x - 2)(x + 3) = 0$ or $x^2 = 9$ is not an efficient method. Block (2015) investigated students' work when solving different types of quadratic equations and found, like de Lima and Tall (2006), that most of the students solved quadratic equations by trial-and-error or using the quadratic formula—mostly without success. Students in the study often only focused on one element of the equation and connected it with use of one specific method to find a solution; they did not identify that multiple approaches could be used to solve quadratic equations. One example is where a zero on one side of the equation was interpreted as a signal to use the quadratic formula; students considered this independently from the structure of the expression on the other side of the equation. Clearly use of the quadratic formula is not always an efficient approach, for example, when a quadratic is given in factorized form, so identifying a quadratic equation in the form $f(x) = 0$ as a prompt to use the quadratic formula is not always efficient. Another example was where students observed brackets and focused only on this aspect when considering an approach for solving; many of these students interpreted existence of brackets as a sign to immediately expand the expression within the brackets, which is not always helpful when dealing with quadratics.

To become capable of flexible algebraic actions for solving an equation, it is important to learn the different methods in a generic way, while reflecting on the potential and strengths of the different methods, as well as their weaknesses and possible pitfalls, according to the specific structure of a quadratic equation. The ability to use apps to explore quadratic equations may help students to develop such flexibility.

2.2 Informal Learning by Using Smartphone Apps

The big challenge when designing teaching and learning processes is to find the right balance between reaching the normative goals of a mathematics curriculum and each learner's current state of learning. Vygotsky calls this the Zone of Proximal Development (ZPD), *"the distance between the actual developmental level as determined by independent problem solving and the level of potential development as determined through problem solving under adult guidance or in collaboration with more capable peers."* (Vygotsky, 1978, p. 86). Following Vygotsky, there is a challenge faced by the teacher or tutor, namely to find the hardest tasks the learner can do to go further in their individual development; the teacher must find tasks within and not outside each student's ZPD, which can be challenging in a mixed-ability classroom. But what happens when tasks during the "formal learning" (Werquin, 2007) in the classroom are too challenging for a student and thus outside his or her ZPD? Then an additional scaffolding with hints, examples and instructions (Brown, Ellery & Campione, 1998), either personal or digital, is needed as guidance; this could occur either in class or outside the formal institutional structures (i.e. through apps, etc.). The recognition of non-formal learning and informal learning outcomes is of increasing interest for policy makers (see OECD-framework on non-formal and informal learning; Werquin, 2007) as it enables people at any stage of their learning and professionalization process to overcome individual gaps in performance, which is important for any field of work (i.e. education, industry, economy or production). Non-formal learning is usually organized (i.e., in institutes for private tutoring) whereas informal learning is never organized and is usually initiated by the learners themselves, for example when doing homework (Werquin, 2007).

When defining the ZPD, Vygotsky conceptualized the role of a guiding adult or capable peer providing experiences at a student's ZPD. Nowadays one should widen this definition, since tutorial systems, which are easy to acquire (e.g. as smartphone apps) offer an increasing range of informal learning opportunities for students at any age to overcome individual difficulties, thus providing students with access to different learning paths (Nattland & Kerres, 2009; Beal, Arroyo, Cohen, & Woolf, 2010). These systems offer hints, examples, instructions or dialogues and enable students to self-select guidance in order to develop their own mathematical knowledge. Tutorial systems which offer these possibilities, together with adaptive features such as immediate feedback, prompts and assessment tasks allowing individualized instruction, are referred to as *Intelligent Tutorial Systems* or *ITS* (Ma, Adesope, Nesbit, & Liu, 2014; Nattland & Kerres, 2009; Hillmayr, Reinhold, Ziernwald, & Reiss, 2017). There is considerable research indicating that ITS can be effective for learning in a large variety of domains (e.g. adult learning: Woolf, 2009; Physics: Graesser, McNamara & VanLehn, 2005; VanLehn et al., 2005; Chemistry: Walsh, Moss, Johnson, Holder, & Madura, 2002; Arithmetics/fractions: Beal et al., 2010, Algebra: Ritter, Anderson, Koedinger, & Corbett, 2007).

Ma et al. (2014) performed a meta-analysis including 107 studies with comparative pre-post-test-designs about use of ITS in diverse subject areas. This analysis resulted in generally positive test outcomes for ITS when compared to teacher-led, large-group instruction, non-ITS computer-based instruction, and textbooks or workbooks. However, no significant differences were found when comparing learning with ITS and learning from individualized human tutoring or from small-group instruction. These results were mainly independent from the specific subject area, the level of education and the way ITS was used (e.g. as the principle means of instruction or as an aid to do homework).

Given these positive effects reported by Ma et al., the high popularity of current (intelligent) tutorial systems available as smartphone apps (e.g. apps for learning languages, mathematics, programming, or even for acquiring a driving license) is not at all astonishing. For example, an independent research association in Germany continuously collects data from 1200 young people via phone-interview about their use of media. In 2017, 24% of the 12–19-year-old girls and 19% of the 12–19-year-old boys stated that they use tutorials as additional support to supplement their learning at school on a daily basis. In comparison, explanatory videos on the internet are used by 10% of the girls and 17% of the boys at least once a day, with 87% of these accessed via smartphones (Feierabend, Plankenhorn, & Rathgeb, 2017). This suggests that many school students are accessing online information, often via smartphones, to supplement their studies at school.

Apps like *Photomath*, with some capabilities of a computer algebra system, are freely available and offer a wide variety of functionalities and features (Klinger, 2019). If not introduced in formal learning settings by the teacher, they seem to be accessed by students as an important part of every-day informal learning for tasks such as homework (Webel & Otten, 2016). For example, Photomath is reviewed by more than 500 thousand users and rated 4.5 out of 5 stars on average; this demonstrates the popularity of this and similar apps. For this reason, our investigation will mainly focus on apps offering support when doing mathematics and which include a CAS which is capable of solving quadratic equations.

3 Reflection, Self-Regulation and Self-Assessment

Schoenfeld (2014) addressed the importance of students' reflecting on their own learning to be able to deepen both their understanding and the ability to transfer cognition from one situation to another. It is an important aim of education to enable students to use mathematics in a flexible way, especially when preparing students for their future professional life in industry, economy and production, where they should reason and evaluate solutions and where procedural work will likely be delegated more and more to machines and software.

Not only mathematical capabilities are important for students' future life—even more important for life-long learning is the development of self-regulation skills. In a modern democratic world with quick changes of jobs and challenges, the essentials for success are individual flexibility, learning on the go, logical thinking and the ability to adapt available knowledge to new situations and problems. Bandura (1986) highlights three metacognitive aspects for self-regulation in his theory of self-regulation: At the beginning, it is a *self-observation* when reflecting on one's own performance. This is followed by *self-judgement* when comparing one's own performance with the performance of others or with given standards. At this point students develop self-awareness about the status of their own performance, as they have criteria to judge their performance against. Last, Bandura calls the reaction to this judgement the *self-response*, when the individual takes responsibility for steps to overcome their own problems and gaps in their learning. This could be a point where a student takes advantage of technology, such as an app, to support his or her own learning.

A student who realizes gaps when doing homework and who thinks about how to overcome these gaps to enhance his or her own learning follows these steps of self-regulation and therefore is within the process of *formative assessment* (Bell & Cowie, 2001). Wiliam and Thompson (2008) describe formative assessment strategies from the perspective of a teacher, where "Activate students as owner of their own learning" is given as the fifth formative assessment strategy and the highest standard.

Wiliam and Thompson (2008) highlight that the teacher is not the only actor in formative assessment, it can also be the peer or the student him- or herself. The question of the actor, who is the person responsible for the formative assessment, was also an important issue in the FaSMEd framework, a collaborative project on formative assessment in mathematics and science education (www.fasmed.eu; Aldon, Cusi, Morselli, Panero, & Sabena, 2017). In this framework the role of technology when doing formative assessment is categorized according to three types of interaction: (1) *Sending and Displaying* when technology is used to support communication among the agents, (2) *Processing and Analyzing* where technology converts collected data (e.g. of a learner), including feedback and results, to an operation and finally (3) *Providing an Interactive Environment* refers to technology that enables students to work in a digital environment that enables them to explore mathematical or scientific contents interactively and collaboratively. For investigating CAS-integrated smartphone apps only the second and third category of technical support is relevant, since these apps are not intended to enable opportunity for communication among different students. All the apps deliver correct results immediately for a given problem but may react differently to common mistakes (category 2) and some of them may even offer interactive support by a (virtual) tutor or another learner (category 3).

4 Leading Questions

The following two questions guided our investigation of the different CAS-apps on smartphones:

1. Do CAS-integrated learning apps support students to be self-aware of their own learning of mathematics?
2. To what extent do different CAS-integrated learning apps support understanding of how to solve an equation?

5 Methodology

5.1 Presentation of the Sample

In this investigation, we focus only on smartphone apps, which are either freely available or available for a low price and which can be used for informal learning at secondary level.

For our sample, we chose apps that are either available in the Google Play- or Apple App Store and which had a relatively good position according to download counts. The apps we consider are *Math 42*, *Photomath*, *GeoGebra*, *Mathway*, *Cymath*, and *Socratic*. All these app comprise a CAS which solves given equations, such as our model case $2x^2 + 5x - 3 = 0$. Every app offers a mathematical keyboard for problem input, whereas some of them offer the possibility of character recognition of printed or written letters as an additional input method.

Although some of these apps offer a range of different functionalities and features we concentrate on solving equations of second order, since this is a challenging task for a lot of students. Furthermore, it offers a variety of possible approaches and visualizations (see above). Last, the focus on one single problem which all apps are capable of solving guarantees a high comparability.

5.2 Instrument According to the Theoretical Aspects

We followed our investigation of the apps by addressing three main criteria which consist of (but are not limited to) several subcategories.

- Surface-structure of the app: Which topics does the app cover and what kind of properties does the app offer to enable ease of use?
    - *Topics*: Which mathematical topics are covered by the apps?
    - *Type of input*: Does the app offer a keyboard and/or character recognition? Is it possible to manipulate a problem once an input has been entered?

- *Format of possible support*: Does the app offer explanation videos, pop-ups, references to other services, e.g. YouTube, etc.?
- Learning issues: What characteristics does the app offer to learn how to solve an equation?
  - *Cognition*: What features does the app offer to understand the topic (such as aspects of concept images or connections to previous knowledge)?
  - *Characterization of possible support*: Does the app offer support, multiple explanations of procedures or information on related concepts?
  - *Language*: Is the language easy to understand and suitable for students or coined by mathematical terminology and a technical style?
  - *Multiple representations*: Does the app offer different types of representations?
  - *Multiple approaches*: Does the app offer multiple approaches for solving quadratic equations (see above)?
- Agents of formative assessment: Does the app offer opportunities for learners to become the owner of their own learning?
  - *Metacognition*: Does the app initiate activities of reflecting and organizing a student's own learning?
  - *Registration*: Does the app offer the possibility to sign up for an individual and personalizable account (e.g. to save previous equations)?

6 Results

The apps we selected are *Math 42*, *Photomath*, *GeoGebra*, *Mathway*, *Cymath*, and *Socratic* (see methodology above). These apps are all easily available for students, easy to access and based on a computer algebra system (CAS). *GeoGebra* is the broadest app as it is a smartphone version of a computer software, which combines features of a geometry package, a CAS, a spread sheet and a calculator. *GeoGebra* is a powerful app, where the user is free to choose from a variety of features in nearly all areas of mathematics. Like working with a calculator the user must follow the menu structure and commands given by the tool, for example for solving an equation: "solve $(2x^2 + 5x - 3 = 0, x)$". The other apps are different. Although they are based on a CAS, one cannot call them a CAS in a classical sense as they offer reduced capability and focus on supporting students doing algebraic procedures. Therefore, the user has only to put in the equation, either by typing in the equation or by photographing it. The apps transfer the photo (or a handwritten equation) to the correct format of an equation required to use the algebraic features of the app. *Cymath*, *Mathway*, *Photomath*, *Socratic* allow both types of input, *Math 42* only allows the input via keyboard. *Photomath* and *Math 42*, in addition to algebraical features, also offers graphical features and the ability to plot a function which can be interpreted as another representation for solving an equation. *Photomath* also includes probability features and *Socratic* works as a type of search engine as it offers immediate links to other math apps.

An analysis of the selected apps indicates that none of the apps is meant to introduce the mathematical concepts needed to solve an equation (except Math 42 which offers a compendium of mathematical notions). *GeoGebra* could also be used as a CAS to introduce the topic—but only when it is framed in this way through the tasks used or given by a teacher in a classroom. The other apps assume that the topic is not totally new to the user and just addresses students not having developed a full understanding in class, yet. These apps focus on the pure mathematical procedure without any connected aspects to develop a concept image or an overall idea of the procedure. There is no hint why this procedure is needed or meaningful, such as finding a zero or finding the intersection point between two functions, which could either be a pure mathematical question or a question connected to a real-world problem. None of the apps offers a chance to connect the procedure with previous knowledge related to quadratic equations, which could be in the field of solving linear equations or in the field of quadratic functions.

A quite positive feature of the apps is that they do not just provide the result of solving a given equation, instead they all offer further information about how to solve the equation. In this sense these apps always offer the possibility to let a black box become a white box (Buchberger, 1990; Heugl, Klinger, & Lechner, 1996). A calculation can be displayed step-by-step and in some of the apps even be unfolded to see more details.

Apart from *GeoGebra*, the apps tend to develop students' procedural capacities by providing considerable explanations for students through offering verbal descriptions of the single steps of solution or by including sample solutions. The remaining problem here is that all these descriptions are really reduced to just describe the single steps of the solution process (see Fig. 1); there is no description of what the whole process is about and what is the aim of the algebraic transformations.

Regarding the language of the explanations we must mention that descriptions are mostly technical and use mathematical terminology (such as factor out, isolating x, etc.), without providing any reasoning to help students to develop their ability to choose and use these procedures in other problems. This may not be optimal for students who are accessing an app to supplement their understanding. The apps do not offer a "language of learning" (Ehret, 2017) but more a mathematical language which is essentially designed to condense information effectively in a few words, which can be a major pitfall for learners (Leuders & Prediger, 2016) who need reasoning behind procedures. In one of the apps (*Socratic*) the use of mathematical symbols was not always correct and did not follow pen-and-paper conventions, which would be necessary to support students' learning. For example, when $f(x) = 3x$ is entered, the app shows $fx = 3x$, which is not the standard notation for a function. The app then deals with $fx = 3x$ as a function, even though the notation is not correct.

Some of the apps (*Photomath*, *Math 42*, and of course *GeoGebra*) also provide students with different representations of quadratic functions in addition to an algebraic, in particular a graphical representation. This fact is very positive, but it may be a pitfall for students as there is no mention of the connection between the graph-

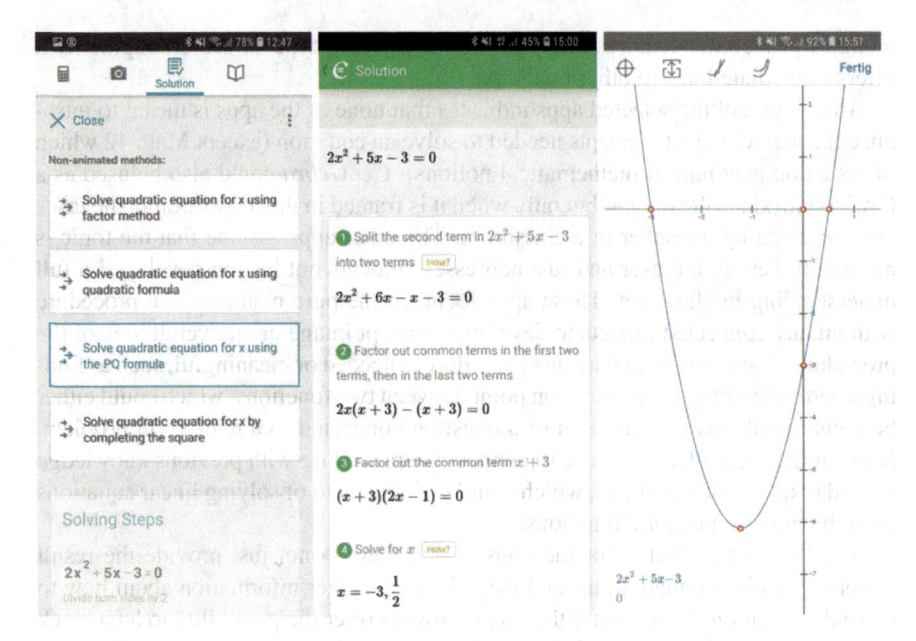

Fig. 1 Choosing a method for solution in *Photomath* (left), solving with factoring method and "How?"-Buttons in *Cymath* (middle), unexplained graphical approach in *Math 42* (right)

ical representation and the algebraic solution. It remains an open challenge to the students to build this bridge themselves. Hence the affordance to make connections across representations or change between them to foster understanding is not realized explicitly through the design of the technology and instead is reliant on the student.

We appreciate that some of the apps (*Photomath*, *Math 42*, *Mathway*) offer a range of approaches, showing different solution methods. The highest level from a didactical point of view for this aspect is the app *Mathway* where students must first decide on a method to use, prior to being provided with the result of solving an equation (see Fig. 1). Other apps just give the result immediately, so that the student must not decide on a method, however some of these apps provide the opportunity for a student to choose an alternative method of solution to the method privileged by the app. Of course, just offering methods does not help in developing the ability to choose appropriate ones. It is a pity that none of the apps provide support, ideas or reasons for choosing the specific solution method. But it is exactly this type of information about the dependency between the structure of the equation and the solution method which would be good support in helping students to develop insight about a procedure (for example "do the same to both sides when you have $x^2 = 9$", or zero product/null factor law instead of expanding brackets in $(x-2)(x+3) = 0$). The apps explored did not initiate students' reflection on the methods used or efficiency or appropriateness of a given method.

The overall question about whether these types of apps foster metacognition has to be answered in a twofold way—according to the surface-structure or the ability to foster reflection on learning.

Regarding surface-structure, the pure existence and availability of these apps gives students the chance to informally foster their learning. It is already a big step of metacognition when a student looks for support individually and informally and then uses it.

The apps support these kinds of metacognitive activities by enabling easy access to mathematical information on a platform familiar to students from their private social life; this is a big benefit. Some of the apps allow students to sign up for an individual account, where the history of activities is saved, thus enabling students to revisit their work at a later date and providing the ability to reflect on, and build on, prior work. This is the case for *Mathway*, *Photomath* and *Math 42*, where students can register and where they can start to work where they have ended the previous time of use. This can be especially beneficial if the app offers training by providing a sample solution, such as in *Math 42*.

Beside these surface-structure elements allowing metacognition to organize your individual work, the other side of metacognition in the sense of reflecting the "inner cognitive part" of the student's own mathematical learning is not explicitly initiated in these apps. Of course, Rittle-Johnson, Schneider, and Star (2015) have indicated, that there is a bidirectional relation between conceptual und procedural knowledge and maybe some students will develop aspects of understanding through considering sample solutions and by following routines. But this understanding is reduced to these routines, students may also learn by this more than one way to solve an equation and they may even understand that a solution means that you have found values for x that makes an equation true. But this understanding of the routines may not result in students being able to apply the understanding in non-routine contexts. Understanding the whole procedure would cover much more than just the routines, such as realizing the characteristic pattern of the equation and choosing an appropriate solution method according to the specific structure of the equation to solve it effectively.

Of course, any user of a technology is free to reflect, or not reflect, on their own work. There may be users of these apps who look back and check what they have done and try to find patterns so that they are able to transfer the ideas to other examples. To get an impression on this, we read some of the manifold reviews of the apps. Looking at the third comment in Fig. 2 shows that this person criticizes the type of explanations and tries to get more information by paying a fee. But even then, the person reflects on the fact that he or she still finds the explanations confusing, with steps of reasoning missing. This person is already active and reflective, as they are using an app to support their work, but they regret the lack of features offered by the app. In a certain sense this person is already the active agent, the "owner of their own learning". This student wants an app that supports his or her learning, rather than just provides answers.

Within the apps investigated it did not appear that the apps were designed to initiate intense reflection by the student on the topic. In this "inner" sense the students using

fantabulous fangirl March 8, 2018
★★★★★

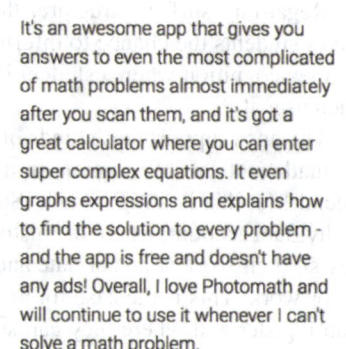

It's an awesome app that gives you answers to even the most complicated of math problems almost immediately after you scan them, and it's got a great calculator where you can enter super complex equations. It even graphs expressions and explains how to find the solution to every problem - and the app is free and doesn't have any ads! Overall, I love Photomath and will continue to use it whenever I can't solve a math problem.

March 10, 2018
★★★★★

Photo math really has helped me in many ways, including my school work which can be difficult at times. Photomath is easy to use and explains step by step in depth the problems I struggle with in just one quick scan of a problem. I really enjoy this helpful app and use daily.

Very Helpful but Sometimes the Steps get Confusing   Sep 27, 2017
★★★☆☆   Bobbyantor

I love this app it helps me solve the equation by telling me the answer. I did of course get stuck on some problems and did not know how to solve some of them so I payed the for monthly fee. It was good at first but then it was confusing on how they told me to do each step. I clicked on "show more", but they would even skip some steps in there. I was confused on how they got from one point to the other. Maybe if they had arrows point to the new solution instructing us, so that it would be easier to understand. Other than that is really good.

If you are just looking for the awnser to calculus problems this is the app for you.

If they were to be more specific on their explaining process I would give them five stars.

Fig. 2 Reviews of *Photomath* on the American Google Play store (upper row), review of *Mathway* on the British Apple App store (lower row)

these apps are more passive and satisfied with the reduced information provided which purely focuses on the steps of the procedure. This can be seen by looking at the other two reviews in Fig. 2. They exemplify most of the reviews on these apps and also highlight that students find apps such as this providing effective support when doing homework. It is a shame, that homework seems only to involve procedural routines and skill training.

7 Conclusion and Outlook

In this study, we have investigated six smartphone apps which are essentially based on a computer algebra system and mainly free of charge. *GeoGebra* has to be mentioned separately as it is a smartphone version of a whole software package including a CAS with a variety of features and the typical command structure of a CAS. All the other apps only focus on how to deal with algebraic procedures.

We have looked at these apps with the aim to capitalize the potentialities and weaknesses of these apps when students are using these to assist in learning how to solve a quadratic equation and we also wanted to get an impression on how self-awareness and metacognition may be fostered when students work with these apps. We followed three perspectives to gain insight into the content and structure of the apps. We looked at the surface-structure of the apps (covered topics, type of input), learning issues (e.g. cognition, kind of support, representations, approaches) and tried to explore metacognition, in particular how the apps allow the user to be an active agent of the own learning.

We conclude our investigation in two claims:

1. *Easy to use apps may offer a benefit for individual learning.*
2. *Current CAS-apps provide a range of features, but they are not yet in line to enhance understanding.*

i. Regarding the large numbers of downloads and reviews, these apps are used by a lot of students independently from their learning in class. Apps are used by students at home, or outside class, in an informal way and to inform their learning. The discussions about assessment do not often take account of this kind of formative assessment performed by a student, so that they can personally determine what they need to learn next and then use technology to facilitate this learning. These actions of metacognition, based on self-reflection of their own learning, in particular the choice to access online support for informal learning of mathematics, is of big benefit but is still an area that requires further investigation and research.

ii. The investigated apps offer some remarkable features which are far beyond just delivering a result when solving an equation. You can find:

- Different representations beside the algebraic solution (such as a table or a graph)
- Explanations (for the single steps of transformation)
- Different approaches for finding the solution to a quadratic equation (such as square addition, quadratic formula, factorizing)
- Links to further information (e.g. other apps, websites, YouTube)
- Registration allowing an individual account, which can enable the student to have a history of their prior work
  Although we really appreciate these features, we see that further development is necessary to enhance the opportunities for deeper understanding of the mathematics behind some procedures:

- There must be a link between the algebraic and the other representations when solving an equation (e.g. finding a zero as one task to solve an equation can be highlighted in the graph).
- Explanations are not only necessary to explain single steps of a transformation but must also cover the overall idea of solving an equation and help students to develop their ability to find an appropriate method according to a specific equation.
- Apps should not be reduced to one type of task ("Solve the equation!") but should offer a range of algebraic activities to deepen learning and understanding of algebra (e.g. tasks where one has to just find the appropriate method of solution to given tasks, word-problems, multiple choice problems, etc.).
- Apps should allow for more assessment with powerful items, including typical misconceptions, for use as formative assessment by the teacher (such as SMART, see http://www.smartvic.com/; Stacey, Steinle, Gvozdenko, & Price, 2013) or by the students themselves (see Ruchniewicz & Barzel, Chap. 3 in this book)

Teachers need to be aware that following or performing routines in algebra is no longer challenging for students when CAS-apps are so easily and freely available. More than ever it is necessary that algebra classrooms focus on development of algebraic concepts and skills and involve rich tasks (Klinger & Schüler-Meyer, 2019). This is not at all a new claim—since the eighties a lot of researchers have brought this up (e.g. Küchemann, 1981; Pierce & Stacey, 2002, 2004; Arcavi, 2005; Nydegger, 2018). Their aim was, and still is, to enrich algebra classrooms so that they lead to a deep understanding of algebra with the specific symbols (such as variables) and their power to model different types of situations and problems either mathematic problems or real-world problems. Students should develop "algebraic insight" to be able to use algebra in flexible ways (Pierce & Stacey, 2004). For decades, there has been a worldwide discussion about enriching algebra classrooms by using CAS (on calculators or on computers) and a lot of studies showed the value (Heid, 1988; Barzel, 2007) but till today a lot of countries do not allow CAS or teachers do not use it, even if they can. One problem teachers mention is that integrating CAS would prevent students from learning algebraic procedures by hand (Thurm, Klinger, Barzel, & Rögler, 2017; Klinger, Thurm, Itsios, & Peters-Dasdemir, 2018). But quite often the knowledge about manifold tasks that enrich the algebra classroom is missing. Nowadays, when many students have access to a smartphone with CAS-apps this discussion becomes a farce (Klinger & Schüler-Meyer, 2019). It is no longer time to discuss whether for example CAS should be integrated in teaching—as it is already in students' hands and easily available on a smartphone (Webel & Otten, 2016). The current situation highlights that the algebra classroom can no longer focus on pure routines and skill training, as these features can be carried out using a machine, and we want algebra classrooms to focus on deep understanding of algebra.

This study has only focused on the apps themselves, not on the usage of them. But we certainly need further research to understand how students learn with the existing apps and what they learn from them. The knowledge about typical learning trajectories of students using these apps could give valuable hints for the further design of apps. When apps are a joint adventure of mathematical educators together with software companies to develop useful and good apps to learn algebra in a broad sense and in a deep and manifold way there is an opportunity to revitalize algebra learning in school.

References

Aldon, G., Cusi, A., Morselli, F., Panero, M., & Sabena, C. (2017). Formative assessment and technology: Reflections developed through the collaboration between teachers and researchers. In G. Aldon, F. Hitt, L. Bazzini, & U. Gellert (Eds.), *Mathematics and technology: A C.I.E.A.E.M. sourcebook* (pp. 551–578). Cham: Springer.
Arcavi, A. (2005). Developing and using symbol sense in mathematics. *For the Learning of Mathematics, 25*(2), 42–47.
Bandura, A. (1986). *Social foundations of thought and action: A social cognitive theory.* Englewood Cliffs: Prentice-Hall.
Barzel, B. (2007). New technology? New ways of teaching—No time left for that! *International Journal for Technology in Mathematics Education, 14*(2), 77–86.
Barzel, B., & Holzäpfel, L. (2017). Gleichungen verstehen. *mathematik lehren, 169,* 2–7.
Beal, C., Arroyo, I., Cohen, P., & Woolf, B. (2010). Evaluation of animal watch: An intelligent tutoring system for arithmetic and fractions. *Journal of Interactive Online Learning, 9*(1), 64–77.
Bell, B., & Cowie, B. (2001). The characteristics of formative assessment in science education. *Science Education, 85*(5), 536–553.
Block, J. (2015). Flexible algebraic action on quadratic equations. In K. Krainer & N. Vondrová (Eds.), *Proceedings of the Ninth Congress of the European Society for Research in Mathematics Education* (pp. 391–397). Prague: ERME.
Brown, A. L., Ellery, S., & Campione, J. (1998). Creating zones of proximal development electronically. In J. Greeno & S. Goldman (Eds.), *Thinking practices: A symposium in mathematics and science education* (pp. 341–368). Hillsdale: Lawrence Erlbaum.
Buchberger, B. (1990). Should students learn integration rules? *ACM SIGSAM Bulletin, 24*(1), 10–17.
de Lima, R. N., & Tall, D. (2006). The concept of equations: What have students met before? In J. Novotná, H. Moraová, M. Krátká, & N. Stehlíková (Eds.). *Proceedings of the 30th Conference of the International Group for the Psychology of Mathematics Education* (Vol. 4, pp. 233–240). Prague: PME.
Dehaene, S. (1997). *The number sense: How the mind creates mathematics.* Oxford: Oxford University Press.
Ehret, C. (2017). *Mathematisches Schreiben: Modellierung einer fachbezogenen Prozesskompetenz.* Wiesbaden: Springer Spektrum.
Feierabend, S., Plankenhorn, T., & Rathgeb, T. (2017). *JIM-Studie 2017: Jugend, Information, (Multi-) Media – Basisuntersuchung zum Medienumgang 12- bis 19-Jähriger.* Stuttgart: Medienpädagogischer Forschungsverbund Südwest.
Graesser, A. C., McNamara, D. S., & VanLehn, K. (2005). Scaffolding deep comprehension strategies through Point & Query, AutoTutor and iSTART. *Educational Psychologist, 40*(4), 225–234.
Heid, M. K. (1988). Resequencing skills and concepts in applied calculus using the computer as a tool. *Journal for Research in Mathematics Education, 19,* 3–25.
Heugl, H., Klinger, W., & Lechner, J. (1996). *Mathematikunterricht mit Computeralgebra-Systemen: Ein didaktisches Lehrbuch mit Erfahrungen aus dem österreichischen DERIVE-Projekt.* Bonn: Addison-Wesley.

Hillmayr, D., Reinhold, F., Ziernwald, L., & Reiss, K. (2017). *Digitale Medien im mathematisch-naturwissenschaftlichen Unterricht der Sekundarstufe: Einsatzmöglichkeiten, Umsetzung und Wirksamkeit.* Münster: Waxmann.

Klinger, M. (2019). "Besser als der Lehrer!" – Potenziale CAS-basierter Smartphone-Apps aus didaktischer und Lernenden-Perspektive. In G. Pinkernell, & F. Schacht (Eds.), *Digitalisierung fachbezogen gestalten: Herbsttagung vom 28. bis 29. September 2018 an der Universität Duisburg-Essen* (pp. 69–85). Hildesheim: Franzbecker.

Klinger, M., & Schüler-Meyer, A. (2019). Wenn die App rechnet: Smartphone-basierte Computer-Algebra-Apps brauchen eine geeignete Aufgabenkultur. *mathematik lehren*, 215.

Klinger, M., Thurm, D., Itsios, C., & Peters-Dasdemir, J. (2018). Technology-related beliefs and the mathematics classroom: Development of a measurement instrument for pre-service and in-service teachers. In B. Rott, G. Törner, J. Peters-Dasdemir, A. Möller, & S. Safrudiannur (Eds.), *Views and beliefs in mathematics education: The role of beliefs in the classroom* (pp. 233–244). Cham: Springer.

Küchemann, D. (1981). Algebra. In K. Hart (Ed.), *Children's understanding of mathematics: 11-16* (pp. 102–119). London: Murray.

Leuders, T., & Prediger, S. (2016). *Flexibel differenzieren und fokussiert fördern im Mathematikunterricht.* Berlin: Cornelsen.

Ma, W., Adesope, O., Nesbit, J., & Liu, Q. (2014). Intelligent tutoring systems and learning outcomes: A meta-analysis. *Journal of Educational Psychology, 106*(4), 901–918.

Nattland, A., & Kerres, M. (2009). Computerbasierte Methoden im Unterricht. In K.-H. Arnold, U. Sandfuchs, & J. Wiechmann (Eds.), *Handbuch Unterricht* (pp. 317–324). Bad Heilbrunn: Klinkhardt.

Nydegger, A. (2018). *Algebraisieren von Sachsituationen: Wechselwirkungen zwischen relationaler und operationaler Denk- und Sichtweise.* Wiesbaden: Springer Spektrum.

Pane, J. F., Griffin, B. A., McCaffrey, D. F., & Karam, R. (2014). Effectiveness of cognitive tutor algebra I at scale. *Educational Evaluation and Policy Analysis, 36*(2), 127–144.

Pierce, R., & Stacey, K. (2002). Algebraic insight: The algebra needed to use computer algebra systems. *Mathematics Teacher, 95*(8), 622–627.

Pierce, R., & Stacey, K. (2004). Monitoring progress in algebra in a CAS active context: Symbol sense, algebraic insight and algebraic expectation. *International Journal for Technology in Mathematics Education, 11*(1), 3–12.

Ritter, S., Anderson, J. R., Koedinger, K., & Corbett, A. (2007). Cognitive Tutor: Applied research in mathematics education. *Psychonomic Bulletin & Review, 14*(2), 249–255.

Rittle-Johnson, B., Schneider, M., & Star, J. R. (2015). Not a one-way street: Bidirectional relations between procedural and conceptual knowledge of mathematics. *Educational Psychology Review, 27*(4), 587–597.

Schoenfeld, A. (2014). What makes for powerful classrooms, and how can we support teachers in creating them? A story of research and practice, productively intertwined. *Educational Researcher, 43*(8), 404–412.

Stacey, K. (2011). Eine Reise über die Jahrgänge: Vom Rechenausdruck zum Lösen von Gleichungen. *mathematik lehren, 169*, 6–12.

Stacey, K., Steinle, V., Gvozdenko, E., & Price, B. (2013). SMART online formative assessments for teaching mathematics. *Curriculum & Leadership Journal, 11*(20).

Thurm, D., Klinger, M., Barzel, B., & Rögler, P. (2017). Überzeugungen zum Technologieeinsatz im Mathematikunterricht: Entwicklung eines Messinstruments für Lehramtsstudierende und Lehrkräfte. *mathematica didactica, 40*(1), 19–35.

VanLehn, K., Lynch, C., Schulze, K., Shapiro, J. A., Shelby, R., Taylor, L., et al. (2005). The Andes physics tutoring system: Five years of evaluations. In G. McCalla, C. K. Looi, B. Bredeweg, & J. Breuker (Eds.), *Artificial intelligence in education* (pp. 678–685). Amsterdam: IOS Press.

Vygotsky, L. S. (1978). *Mind in society: The development of higher psychology processes.* Cambridge: Harvard University Press.

Walsh, M., Moss, C. M., Johnson, B. G., Holder, D. A., & Madura, J. D. (2002). Quantitative impact of a cognitive modeling intelligent tutoring system on student performance in balancing chemical equations. *Chemical Educator, 7,* 379–383.

Webel, C., & Otten, S. (2016). Teaching in a world with Photomath. *Mathematics Teacher, 109*(5), 368–373.

Werquin, P. (2007). Moving mountains: Will qualifications systems promote lifelong learning? *European Journal of Education, 42*(4), 459–484.

Wiliam, D., & Thompson, M. (2008). Integrating assessment with learning: What will it take to make it work? In C. A. Dwyer (Ed.), *The future of assessment: Shaping teaching and learning* (pp. 53–82). Mahwah: Lawrence Erlbaum.

Winter, H. (1984). Begriff und Bedeutung des Übens im Mathematikunterricht. *mathematik lehren, 2,* 4–16.

Woolf, B. P. (2009). *Building intelligent interactive tutors: Student-centered strategies for revolutionizing e-learning.* Burlington: Morgan Kaufman.

Part II
Innovative Technologies and Approaches to Mathematics Education: Old and New Challenges

Michèle Artigue
LDAR (EA4434), Université Paris-Diderot, UA, UCP, UPC, URN.

This section of the book is devoted to innovative technologies and approaches to mathematics education. There is no doubt that since the emergence of calculators and computers, these technologies have been seen as a lever to make evolve mathematics education practices and support innovation in that area, as made clear, for instance, by the first ICMI Study devoted to the theme (Churchhouse, 1986) or the pioneering vision and work by Seymour Papert (Papert, 1980). This is still the case today, in a technological world that has dramatically changed since then. A decade ago, ICMI launched a second study on the theme (Hoyles & Lagrange, 2010) that contrasted the technological evolution having taken place within two decades with the resistant difficulties that mathematics education met at up-scaling productive use of these innovative technologies, even those present on the educational scene for decades, making clear that important challenges have not been successfully taken up. Since the publication of the second ICMI Study, technological innovation as well as the diverse ways through which digital technologies increasingly influence our private, social and professional life have been amazing. In many countries, children play with tablets from their early ages, and the generalization of tactile interfaces has changed our technological gestures and the concept itself of direct manipulation; the development and reducing cost of technologies such as augmented or virtual reality make it possible to seriously envisage their educational use; new concepts such as those of MOOC and flipped classroom have become familiar to us in a few years; teachers' professional work is substantially impacted by the dramatic increase of educational resources accessible online everywhere, at any moment, as shown by the development of the documentational approach to didactics (Gueudet, Pepin, & Trouche, 2012), and by the new forms of interaction and mediation offered by technological evolution.

Research in the area has thus to face two different challenges. On the one hand, it must explore the potential for learning and teaching mathematics offered by technologies that have recently entered the educational sphere or could enter it in the future and inspire their development; on the other hand, it must find innovative and more productive ways of using technologies that have entered this scene decades ago, and go on evolving, as is the case for instance for the DGS Cabri-geometry. This DGS, indeed, has moved from a microworld for the teaching and learning of 2D and then 3D geometry to a complex system also offering an online kit of didactic tools that teachers and teacher educators can use to create resources based on this microworld, and fully developed didactic resources that users can combine and adapt to their personal context and views (Laborde, 2018).

The three chapters constituting this section in fact reflect this double concern. The first one, entitled *Dynamic mathematical figures with immersive spatial displays: The case of HandWaver*, co-authored by Justin Dimmel and Camden Bock, reports on the design and development of HandWaver, a gesture-based mathematical making environment for use with immersive, room-scale virtual reality, at the IMRE Laboratory of the University of Maine, with a beta version just released at the time of ICTMT13, in Spring 2017. The authors begin with an historical introduction on how human have been making inscriptions since the pre-historical times, presenting immersive spatial display technologies such as HandWaver as "technologies where *space itself* is the canvas for making inscriptions", and "proto-versions of the means for producing, viewing, and sharing three-dimensional inscriptions [...] that, over time, will mark the beginning of an evolution of how we communicate with each other" and "will transform how people generate, disseminate, and interact with knowledge." Their design and development work is framed by the following research question: "How do gesture-based interactions with virtual objects help students home their spatial reasoning skills?" In the HandWaver environment, more precisely, users construct and explore figures directly with their hands, using three main different types of gestures: pinching, stretching, and spinning. The article carefully explains the design rationale and choices, making visible the influence on these both of economic and ergonomic reasons (for instance those governing the choice of gesture-tracking hardware) and of established scientific knowledge. Affordances of HandWaver are thus, for instance, connected to the potential that this environment offers in terms of *dimensional deconstruction* according to Duval, of engaging the learner in geometrical work at different scales, from the micro to the macro-space, combining different perspectives on mathematical objects (insider/outsider), not to mention the accumulated body of knowledge on the role of gestures and embodiment in mathematics learning. There is no doubt at reading this chapter, that HandWaver should allow its users to experience a geometry of movement really innovative when compared with that offered by the dynamic geometry environments we are used to, or our physical activity in the real space. What could they learn from such experiences? At the moment, answers are highly hypothetical. The authors report about a few studies carried out with teachers, for instance a study with science teachers investigating how these used non-measuring virtual tools to make and test mathematical claims about the volume of pyramids, with promising results, but this remains somewhat anecdotic. The authors also point out that issues of instructional implementation such as the following: "How do practicing and presevice teachers imagine incorporating spatial display technologies into their teaching? What support do they need? What barriers do they anticipate?" are fully open. However, this chapter made me share their ambition to see the development of HandWaver and research associated with its use contributing to "ensure that research-based ideas about the nature of productive mathematical activity are represented in this next generation of virtual learning environments."

The second chapter of the section entitled *Design and evaluation of digital resources for the development of creative mathematical thinking: A case of teaching the concept of locus* is co-authored by Mohamed El-Demerdash, Jana Trgalová, Oliver Labs, and Christian Mercat. It reports on research carried out in the framework of the MC Squared European project (MC2 in the following), the aim of which was to study the processes of collaborative design of innovative resources called *c-books* intended to enhance mathematics creativity in students. The chapter first discusses the idea of creativity at the core of this project. The authors point out the diversity of existing approaches regarding creativity in general and mathematical creativity in particular, and the dual process/product vision underlying this diversity. Referring to a distinction introduced in the literature between little-c or ordinary creativity accessible to anyone of us, and big-C Creativity that very few individuals are able to achieve, they justify their choice of adopting the "little-c" creativity paradigm in MC2, leading to define mathematical creativity (CMT) as "an intellectual activity generating new mathematical ideas or responses in a non-routine mathematical situation" and to consider that such creativity "can be developed in students through appropriate learning situations."

The resources created in MC2 pursue this goal. As explained by the authors, they are based on a C-book authoring environment especially developed for the project "including diverse dynamic widgets, an authorable data analytics engine and a tool supporting the collaborative design of resources called c-books". The c-books themselves "consist of pages than can embed text, several interoperable interactive widgets, links towards external online resources, videos, etc.," and "more than 60 c-books have been created over the three years of the project." The study reported regards the design of the experimental geometry c-book devoted to the teaching and learning of geometric locus of points. The importance of the topic is stressed, considering both the historical development of this area and its applications beyond the sole mathematics field, and the reader is also reminded that educational research has already explored the potential of digital technologies, especially DGS, for its teaching and learning. In fact, the c-book invites students "to experiment geometric loci generated by intersection points of special lines of a triangle while one of its vertices moves along a line parallel to the opposite side". This seems a prioria simple and not so innovative situation but, as shown by the authors, implemented in the c-book, it offers a rich context for "exploring, conjecturing, experimenting, and proving". The authors describe, in fact, in a detailed way, the three sections of the c-book and the different tasks proposed in these to students, making clear its innovative character and the support a priori offered to creative mathematical thinking, especially thanks to the original interoperability achieved between widgets.

What about students' actual use of this c-book? Unfortunately, this issue was not part of MC2. However, we are presented with the interesting results of two a priori evaluations of the c-book, the first one by three researchers involved in the project, according to the four cognitive components of CMT (fluency, flexibility, originality, and evaluation) and the social and affective aspects that c-books are expected to integrate and promote, the second one by four secondary teachers. As was the case for the first chapter, programs of research need now to be developed in order to understand how teachers and students could make meaningful and productive use of these innovative resources at the cutting edge of technological affordances and scientific knowledge, and to inspire didactic action.

The third and last chapter of this section entitled *WIMS: An interactive exercise software 20 years old and still at the top*, authored by Magdalena Kobylanski, is rather different from the two previous ones, and this for many reasons. To mention just some of them: the technology it relies on is an old technology, 20 years old, as made clear by the title; WIMS is an open-source e-learning platform hosting a huge number of online interactive exercises covering diverse disciplines, not just mathematics, and addressing all levels of schooling; one cannot say that WIMS has developed in close connection with mathematics education research, despite the fact the WIMS community has progressively established productive links with the didactic community. The chapter is divided into two main parts. The first part describes the history of WIMS and how an international community of developers and users has progressively grown around it. The author insists on a fundamental feature that, for a long time, has differentiated it from most other bases of exercises: its random feature allowing the generation of "great and almost infinite variations" of the same type of task (for instance, computing the sum of two numbers or two algebraic expressions), and flexibility in the expression of correct answers, thanks to a CAS-based system of evaluation of students' answers. This historical part also shows how the long term and collaborative development of WIMS has generated an impressive growth in scope, in possibilities offered to users, both students and teachers, up to the current sophisticated and original e-learning system that the author presents in a very detailed way.

The second part of the chapter investigates the advantages that WIMS can provide to students' learning. It takes first the form of a reflective discourse based on the four steps of learning proposed by de la Garanderie, and a diversity of other factors likely to influence learning processes such as motivation, working memory, attention, self-regulation, etc. Characteristics of WIMS are matched with this vision of learning and factors. However, strangely for an environment used in so many contexts and during so many years, no empirical evidence coming from WIMS research is used to support this discourse. The reflection is then complemented by some observations coming from an experimentation carried out at the University Paris-Est with students in their first university year, and some regular critics made to WIMS are also addressed, but this second part of the chapter contrasts with the richness of the first one. WIMS is an old technology, but certainly a technology whose development and productive use should benefit from more systematic connection with mathematics education research.

More globally, this chapter raises the issue of balance in research interests between the old and the new, and also between technologies designed to promote inquiry-based pedagogies versus technologies more likely to support consolidation activities.

References

Churchhouse, R. F. (Ed.). (1986). *The influence of computers and informatics on mathematics and its teaching. ICMI study series*. Cambridge: Cambridge University Press.

Gueudet, G., Pepin, B., & Trouche, L. (2012). *From text to 'Lived resources': Mathematics curriculum material and teacher development*. New York: Springer.

Hoyles, C., & Lagrange J. B. (Eds.). (2010). *Mathematics education and technology—Rethinking the terrain. The 17th ICMI study*. New York :Springer.

Laborde, C. (2018). Part de la didactique des mathématiques dans les choix de conception de Cabri. In J. B. Lagrange & M. Abboud-Blanchard (Eds.), *Environnements numériques pour l'apprentissage, l'enseignement et la formation: perspectives didactiques sur la conception et le développement. Cahier du LDAR n°10* (pp. 33-44). Paris: Université Paris-Diderot. http://www.irem.univ-paris-diderot.fr/up/IPS18001.pdf.

Papert, S. (1980). *Mindstorms*. New York: Basic Books.

Dynamic Mathematical Figures with Immersive Spatial Displays: The Case of Handwaver

Justin Dimmel and Camden Bock

1 Introduction

Humans have been making inscriptions for tens of thousands of years (Senner, 1991). Even as the technology of inscribing has evolved—from paintings on the walls of caves, to cuneiform scratched on papyrus, to digital images displayed on screens—fundamental qualities of the activity of inscribing have remain unchanged. The primary purpose of making inscriptions is to communicate—to leave records of what occurred, to convey messages across time and space (Harris, 1986; Senner, 1991). And for all of history, inscriptions have been achieved by using a marking-tool (e.g. a stylus, a quill, sound, a touch pad) to modify a surface (e.g., clay, scrolls, vinyl, a screen). Inscriptions have presupposed a substrate that is inscribed. But with the advent of virtual and augmented reality technologies, this supposition is no longer binding.

Virtual and augmented reality head-mounted displays (such as the HTC Vive or the Microsoft Hololens) are examples of what we refer to as *immersive spatial display technologies* (i.e., spatial displays): Technologies where *space itself* is the canvas for making inscriptions.[1] We use *spatial displays* as a generic, umbrella term to refer to virtual reality, augmented reality, and related media (e.g., holographic

---

[1] We recognize that spatial inscriptions are made possible by 0 and 1 s inscribed on computer disks, but such inscriptions, from the user's perspective, are secondary to the three-dimensional inscriptions one can interact with in virtual spaces.

J. Dimmel (✉) · C. Bock
Maine IMRE Lab, University of Maine, Orono, USA
e-mail: justin.dimmel@maine.edu

C. Bock
e-mail: camden.bock@maine.edu

© Springer Nature Switzerland AG 2019
G. Aldon and J. Trgalová (eds.), *Technology in Mathematics Teaching*,
Mathematics Education in the Digital Era 13,
https://doi.org/10.1007/978-3-030-19741-4_5

projection, volumetric video[2]) that facilitate the creation, storage, and transmission of three-dimensional inscriptions. We use *spatial displays* in recognition of the rapidly changing technological landscape and in an effort to capture the defining feature of these new modes for displaying and interacting with information. In fifteen or even 1,500 years, there will be different portals that facilitate access to what we now refer to as virtual- or augmented-reality worlds, but what will relate these as-yet-unknown technologies to the technologies that are currently the state-of-the-art is the potential for making inscriptions in space.

With the commercial availability of spatial display technologies, we are at the dawn of a new era of human communication. When our ancestors developed writing, they adapted our capacity for making inscriptions into a technology that could encode our very thoughts in artifacts that endure for millennia (Harris, 1986; Senner, 1991). As a result, the invention of writing "profoundly affected the way people came to think and to argue...it brought in its wake a restructuring of human mental processes" (Harris, 1986, p. 24). Writing made possible more complex forms of discourse, relieved human mental faculties of the burden of storing knowledge that was once known only orally, and made it possible for cultures to project their influence throughout the world and into the distant future—evident in the enduring legacies of ancient people from across the globe. Written language has been described as both the glue that binds civilization together and also as the principal driver of its progress (Gnanadesikan, 2011; Harris, 1986).

We offer this brief review of the history of writing and the central role that writing has played in advancing the ongoing project of generating, preserving, and disseminating humanity's knowledge to provide a framework for understanding the significance of the emergence of immersive spatial display technologies. The spatial display technologies available in the early 21st century are proto versions of the means for producing, viewing, and sharing three-dimensional inscriptions. They are our cave paintings, that, over time, will mark the beginning of an evolution of how we communicate with each other.

Historically, the advent of new modes for preserving and sharing information has had far-reaching impacts: writing made it possible to preserve speech, moveable type made it possible to mass produce writing, and the internet has made it possible to widely and instantly share any writing that is produced. The emergence of spatial display technologies is potentially generative because human communication is not only literal or verbal, it is also gestural (McNeil, 2008). The accumulating research on the relationships between gestures, communication, and cognition indicates that how we gesture affects how we think (Alibali & Nathan, 2012). Spatial display technologies can track, recognize, and link gestures to the production of virtual figures. Spatial displays thus have the potential to do for gestures what writing did for speaking. In the near future, these technologies will transform how people generate, disseminate, and interact with knowledge.

---

[2]*Volumetric video* describes capturing real-world objects from multiple perspectives so that complete three dimensional models of those objects can be rendered in virtual or augmented reality environments (Ebner, Feldmann, Renault, & Schreer, 2017).

Figure 1 shows how gesture-tracking technology can be integrated with an immersive virtual world to create a gesture-based user interface. The panel in the lower-left corner of the figure shows how the sensor tracks hand movements that trigger the creation of spatial inscriptions (there are two images because the sensor uses left and right cameras): the foreground view shows how a user's arms, wrists, joints, and digits are virtually reconstructed from the information tracked by the sensor; the background view shows the user's actual hands and arms. The main panel shows how users in the environment see the virtual versions of their hands. In these panels, the user has moved his hands in space to create a line segment.

How mathematical knowledge is represented and communicated could be especially influenced by the affordances of spatial displays. Mathematical experts could be recorded in volumetric video while generating a mathematical proof, and the gestures that the speaker makes could be recorded as spatial inscriptions. The different stages of the expert's gestures (McNeil, 2008), such as when a particular gesture begins or ends, could be tracked not only visually in the video, but spatially, with the immersive environment logging the speed and (x, y, z) positions of the expert's hands as they move. Eventually, as the accuracy and reliability of spatial display technologies improves, gesture-based languages could emerge that allow people to create mathematical figures in spatial displays from movements of their hands.

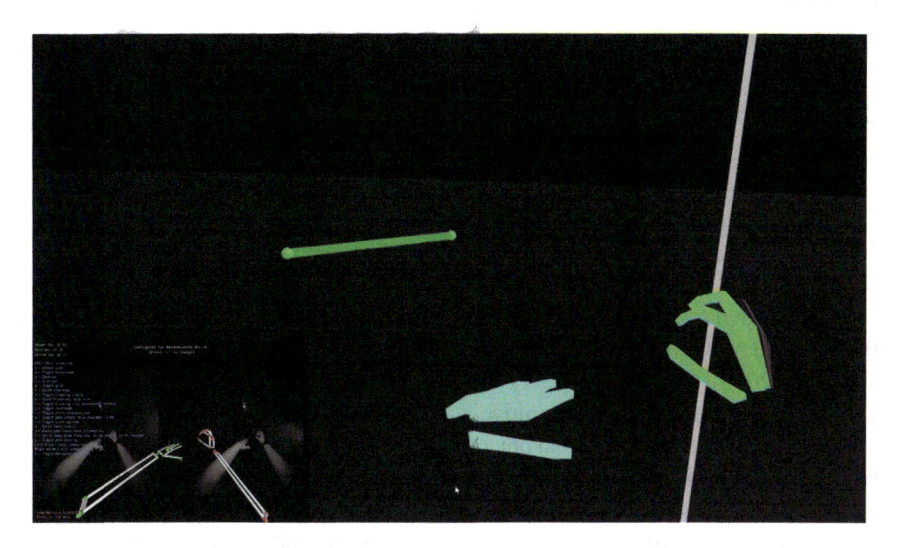

Fig. 1 Two views of gesture-based actions in an immersive virtual world

2 Research Question

Dynamic geometry environments introduced new modes of mathematical sense-making, such as the ability to investigate invariant properties of figures by continuous transformations (Hollebrands, 2007). Immersive spatial display technologies, combined with gesture-based interfaces, have the potential to advance how visualizations are used for the teaching and learning of mathematics. Our design and development work is thus framed by a broad research question: How do gesture-based interactions with virtual objects help students hone their spatial reasoning skills? In an initial effort to explore this question, the Immersive Mathematics in Rendered Environments (IMRE) Lab at the University of Maine has developed *HandWaver*, a gesture-based virtual mathematical making environment. We say the environment is gesture-based because users construct or explore figures in the environment through pinching, stretching, or spinning gestures that they make directly with their hands—no gloves or controllers necessary. We describe the environment as a mathematical making environment to emphasize that the purpose of the environment is to provide a spatial canvas where users can make mathematical figures. In this chapter, we report on the design and development of *HandWaver* and consider priorities for research and development if immersive spatial display technologies are to become integrated into mathematics classrooms.

3 Design Rationale

Our primary goal in developing *HandWaver* was to create an environment where learners could use their hands to act directly on mathematical objects, without the need to mediate their intuitions through equations, symbol systems, keyboards, or mouse clicks (Sinclair, 2014). We designed *HandWaver* around natural movements of a user's hands—i.e., pinching, stretching, and spinning gestures—to foreground the connection between diagrams and gestures (de Freitas & Sinclair, 2012; Chen & Herbst, 2013). Gestural interfaces (Zuckerman & Gal-Oz, 2013)—where objects can be manipulated in natural, intuitive ways by movements of one's hands—allow a degree of direct access to virtual objects that have been shown to facilitate learning (Abrahamson & Sánchez-García, 2016) while minimizing cognitive barriers (Barrett, Stull, Hsu, & Hegarty, 2015; Sinclair & Bruce, 2015). Spatial displays are naturally suited to gesture-based user interfaces because they have affordances for translating multimodal cues—e.g., head or hand movements—into mathematical operations, such as projecting a plane figure into three dimensions by pulling it up into space.

The gesture-based interface employed in *HandWaver* is constrained by the reliability and field-of-view of commercially available gesture-tracking sensors.[3] Given

---

[3] We have found that the *Leap Motion* sensor is both reasonably priced and functional for our purposes. This sensor can be mounted to the front of an HTC Vive headset and provides a sixty-degree field of view for tracking gestures.

these constraints, we have designed the environment around three types of gestures: *pinching, stretching,* and *spinning*. As gesture tracking hardware improves and as gesture-recognition software becomes more discerning, it will be possible to develop more nuanced gesture-based inputs that could facilitate more natural modes of constructing figures in space with hand movements, such as being able to bend, twist, cut, or glue figures together.

Affordances of Spatial Display Technologies for Representing Mathematics. The commercial availability of spatial display technologies raises the question of what these technologies allow us to do that can't already be done with projectors, calculators, tablets, phones, smart boards, or other readily available technologies. What are the *affordances* (Collins, 2010) of spatial displays that warrant using them to design and develop environments for representing mathematical ideas? One affordance is that spatial displays allow direct access to virtual spatial figures. By direct access, we refer to being in the same shared space as a three-dimensional inscription. Users have direct visual access to spatial figures in that such figures are viewed as spatial figures rather than as planar projections of spatial figures. They also can grasp, move, or otherwise manipulate those figures with their hands. The access to objects available with spatial displays combines the tangibility of real-world things with the malleability of dynamic figures.

Direct access to virtual spatial figures can facilitate *dimensional deconstruction*—the process of resolving geometric figures into components, rather than seeing them as whole, fixed shapes (Duval, 2014). Examples of dimensional deconstruction would include unfolding a cube into a planar net of six squares, slicing solids with planes to create two-dimensional cross-sections, or analyzing a polyhedron by investigating the relationships among its faces, edges, and vertices. We conceptualize dimensional deconstruction broadly to also include *generating* higher-dimensional figures from lower-dimensional primitives—e.g., extruding plane figures into prisms and revolving points, curves, or surfaces around axes to create curves, surfaces, and solids. The *stretching* and *spinning* gestures (described below) for constructing spatial figures in *HandWaver* are instantiations of dimensional deconstruction.

A second affordance is that spatial display technologies provide a setting where geometric figures can be constructed and explored at different scales than what is possible when geometric figures are constrained by relatively small two-dimensional screens. There is simply more room in immersive spatial displays for making inscriptions. Issues of scale—and in particular, how we learn to model spaces from the micro- to the macro-scale—are fundamental to spatial reasoning (Herbst, Fujita, Halverscheid, & Weiss, 2017). In *On Proof and Progress in Mathematics,* Thurston (1994) observes:

> An interesting phenomenon in spatial thinking is that scale makes a big difference. We can think about little objects in our hands, or we can think of bigger human-sized structures that we scan, or we can think of spatial structures that encompass us and that we move around in. We tend to think more effectively with spatial imagery on a larger scale: it's as if our brains take larger things more seriously and can devote more resources to them. (p. 165)

The familiar figures of plane geometry are usually encountered as small objects in textbooks and notebooks where the direction of the viewer's gaze is normal to the display pane. This is the mathematical analog of the third-person perspective in storytelling: The mathematical observer is external to the figure, can see it in its entirety, and knows all of the information that is given about the figure. Choices about size, angle of view, or orientation of a diagram influence how the diagram is read by a viewer (Dimmel & Herbst, 2015), and, in turn, how it will be deciphered as a mathematical text (O'Halloran, 2005). Immersive spatial displays allow representations of geometric figures to be explored at larger scales and from different perspectives than what is possible with small, two-dimensional diagrams (see Fig. 2).

Fig. 2 Scaling and exploring a dodecahedron

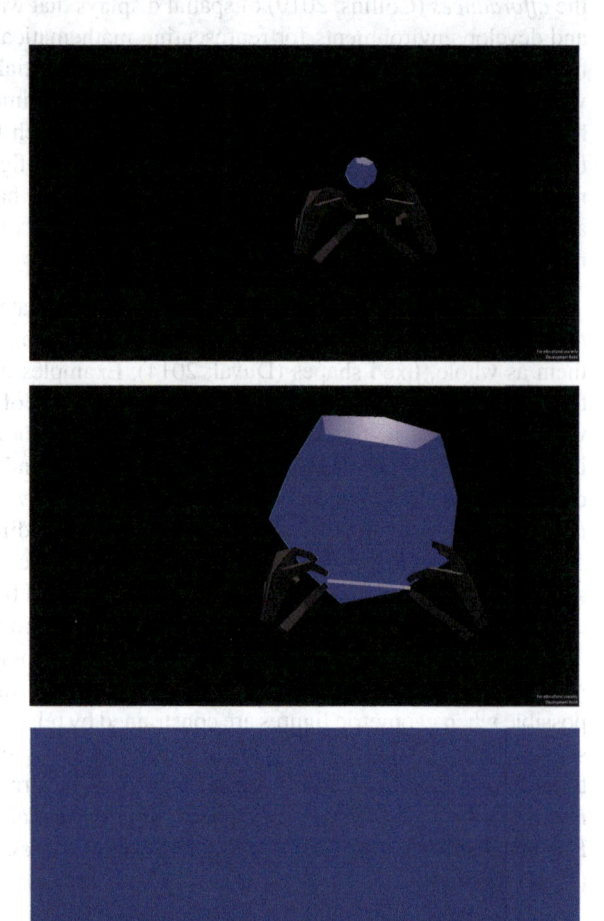

Spatial displays can facilitate movements between figures of different scale. For example, the top panel in Fig. 2 shows a dodecahedron that fits in the palm of a human hand. The second panel shows a user enlarging that figure to be the size of a large ball. And finally, the bottom panel shows that the figure has become large enough for a person to step inside of it. Moving between small-scale and large scale versions of a figure could help learners observe qualities that are otherwise difficult to see, such as how a polyhedron looks when viewed from inside compared to how it looks when viewed from outside. New perspectives on figures that are available in spatial displays could help learners grasp how ideas like congruence can be deduced from observable qualities like symmetry or regularity.

A third affordance is that spatial displays make available an additional spatial dimension for representing and exploring geometric figures. Humans are skilled at projecting higher-dimensional figures onto lower-dimensional canvases, as evidenced by the techniques of perspective drawing. Still, the ability to see and reason about higher-dimensional figures from their lower-dimensional projections is more developed in some than in others. The possibilities of exploring figures with spatial displays could help people refine their spatial reasoning skills—e.g., transforming two-dimensional projections of three-dimensional figures by moving them in space, or visualizing folding a net of a hypercube in three-dimensions (as opposed to on a flat screen). Just as writing freed our mental faculties from the demands of memorization—and, as a consequence, restructured our thinking (Harris, 1989)—spatial displays could reduce the mental cost of visualizing higher-dimensional figures.

Spatial displays now make it possible to observe and directly interact with mathematical objects that once only existed theoretically, such as dynamic, three-dimensional projections of four-dimensional figures. Already, spatial displays are being used by mathematicians to model and explore mathematical structures in new ways (Hart, Hawksley, Matsumoto, & Segerman, 2017a, 2017b). In the decades to come, students of mathematics at all levels of schooling will be able to project higher dimensional figures into dynamic, three-dimensional models that are realized as spatial inscriptions. For these future students of mathematics, three-dimensional projections of four-dimensional figures could be as mundane as two-dimensional drawings of three dimensional figures are for us now. Routine access to dynamic, spatial representations of higher dimensional figures will no doubt shape the mathematical intuitions and imaginations of future mathematicians.

Designing Virtual Environments for Transformative Learning and Teaching. The affordances described above suggest possibilities for how spatial displays could give rise to new methods of investigating mathematical figures. Spatial displays could support virtual mathematics laboratory experiences (Bock & Dimmel, 2017) where students could use virtual tools—designed to facilitate dimensional deconstruction, explorations of objects at different scales, or constructions of spatial figures that would not otherwise be possible—to explore the spatial properties of different figures. Our work to design a virtual environment where such experiences will be possible is guided by the *high level conjecture* (Sandoval, 2014) that *immersive spatial display technologies have affordances for representing information that can trans-*

form mathematics education. We report on the initial stages of the development of *HandWaver* below.

4 Overview of Handwaver

HandWaver is intended for use with immersive, room-scale virtual reality head-mounted displays, such as the HTC Vive. The environment is open source and available for download at: www.handwaver.org. *HandWaver* (Bock & Dimmel, 2017) allows users to construct zero, one, two, or three dimensional geometric figures through iterations of gesture-based operators. The name of the environment is an attempt to reposition *hand waving*—a term used to criticize mathematics that is insufficiently rigorous—as a means for doing mathematical work.

Hardware. Room-scale virtual reality refers to positional and perspectival immersion in a physical space—the activity space—in which a virtual environment is projected via a stereoscopic head-mounted display (HMD). As a user navigates a real physical space, that user's position within the space is tracked via sensors mounted to the HMD. The positional data tracked by the sensors updates the user's virtual position in a rendered environment, giving the user six-degrees of freedom of movement: three degrees of angular movement (linked to the user's head position) and three-degrees of spatial movement (i.e., up/down, left/right, front/back). By contrast, stationary virtual reality experiences, such as those that are currently available via phone-based virtual reality viewers (e.g., Google Carboard, Gear VR by Samsung), provide the user with three-degrees of freedom from a fixed vantage point that does not vary as one's position within the physical world changes.[4]

Room-scale virtual reality hardware is more expensive, less portable, and requires more space than stationary virtual reality viewers. But with current technology, room-scale virtual reality also enables more varied interactions with virtual objects, such as the ability to pick objects up and transform them with one's hands, or to approach objects in space from different perspectives. We developed *HandWaver* for use with room-scale virtual reality to capitalize on the possibilities of these more varied interactions.

Advances in hardware have made significant improvements in performance and cost for room-scale virtual reality HMDs. Advances in consumer graphics processing units (GPUs) have expanded access to the processing power required to drive HMDs to consumer workstations. Early generations of virtual reality triggered motion sickness, vertigo, or other disorienting sensations in users (Hettinger & Riccio, 1992). The combination of improvements in graphics processing and in the HMDs themselves has minimized previous issues with motion sickness. While it is still the

---

[4]With phone-based virtual reality viewers, it is possible to view a virtual world from different vantage points that can be accessed by teleportation or to experience visually immersive simulated movements (e.g., roller coaster rides), but what happens in the virtual world does not depend on a user's position in the physical world.

case that spatial displays can be disorienting for some users, the current generation of consumer-grade HMDs deliver consistent positional tracking and reliable frame rates[5] for rendering images that are a vast improvement over what was available even five years ago.

The HTC Vive and Oculus Touch HMDs both support room-scale virtual reality experiences at costs that are comparable to other classroom technology (e.g., Interactive White Boards). We chose the HTC Vive for it's larger activity space (a 4 m by 8 m rectangular area), early room-scale availability, and its support for local multiplayer in a shared activity space—i.e., two different users could be in the same physical space and have their positions tracked and rendered in virtual worlds. The support for local multiplayer was an important consideration because when designing instructional activities for use with spatial displays, we anticipate that it will be useful—if not essential—for more than one user to be immersed in a virtual world at a time.

A final hardware component for exploring *HandWaver* is a *Leap Motion* sensor that is mounted to the front of the HTC Vive HMD. This is the infrared sensor that was described above (see Fig. 1). The *Leap Motion* tracks the positions of a user's hands in space and makes it possible to define a geometry of movement that is based on hand gestures.

Geometry of Movement (i): *Stretching*. There are two types of movements users in *HandWaver* can employ to construct figures. The first movement is a *stretching* gesture, as if one were pulling something apart. Figure 3 shows iterations of *stretch*: a point is stretched into a line segment; the line segment is stretched into a plane figure; the plane figure is stretched into a prism.

The action of *stretching* a lower-dimensional figure to create a higher-dimensional figure is grounded in the idea that n-dimensional figures consist of adjoined ($n - 1$)-dimensional figures: line segments are adjoined points, plane figures are adjoined line segments, and solids are adjoined plane figures. *Stretching* thus acts on objects to effect a kind of multiplication (Davis, 2015), whereby lower-dimensional figures are transformed into higher-dimensional figures.

Extruding polygons to solids via stretching is one mode of constructing three-dimensional figures in space in *HandWaver*. Users can also extrude polygons to pyramids by using a pyramid generator tool (bottom two panels, Fig. 4). Users select the tool from a virtual shelf by pinching its icon with their thumb and index finger. Once they have gripped the tool, users can move it anywhere in the virtual space by moving their hand, as they would if they were holding and moving a real thing. The tool works by touching a point on the interior of the polygon and then pulling this point up into space. The pulled point becomes the apex of the pyramid. The action of generating a pyramid by pulling a point of a polygon up into space is grounded in the idea that a pyramid is a stack of similar polygonal slices, each of whose area uniformly decreases as it gets closer to the apex of the pyramid. The

---

[5] A frame rate of 90 frames per second is necessary to ensure that the virtual worlds we view through spatial displays are real enough for our visual system.

Fig. 3 Different cases of the *stretch* operator: a point is stretched into a line segment, the segment is stretched into a plane figure, and the plane figure is stretched into a solid

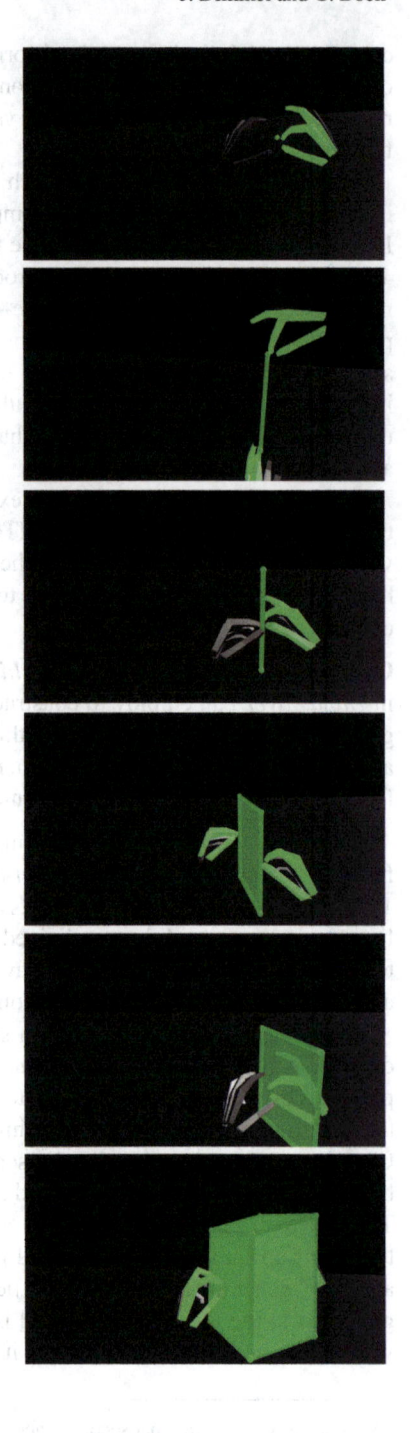

Fig. 4 Pulling a line segment into a triangle and a triangle into a pyramid

pyramid generator tool also allows users to extrude line segments into triangles by touching the tool between the endpoints of the segment (first three panels, Fig. 4).

Geometry of Movement (ii): *Spinning*. Another mode of using hand movements to generate figures is *spinning*. Users can position an axis in space, select objects to rotate around the axis, and then spin a wheel to revolve the selected objects around the axis. Revolving objects in this way creates curves or surfaces of revolution.

110   J. Dimmel and C. Bock

Figure 5 shows a point in space (red dot, first panel) and an axis (white line) with a blue wheel affixed to it. The user spins the wheel (second panel) to revolve the point around the axis and create a circle (third panel). In Fig. 6, a segment that is parallel to the axis of rotation (first panel) generates a cylinder when revolved (second panel); the segment can be moved and the surface of revolution dynamically updates, yielding a truncated cone (third panel) and a twisted cone (fourth and fifth panels).

The *stretching* and *spinning* gestures that are the basis for constructing figures in *HandWaver* were designed to show how higher dimensional figures can be realized by spatial movements of lower-dimensional figures. By stretching points, line

Fig. 5 A point becomes a circle

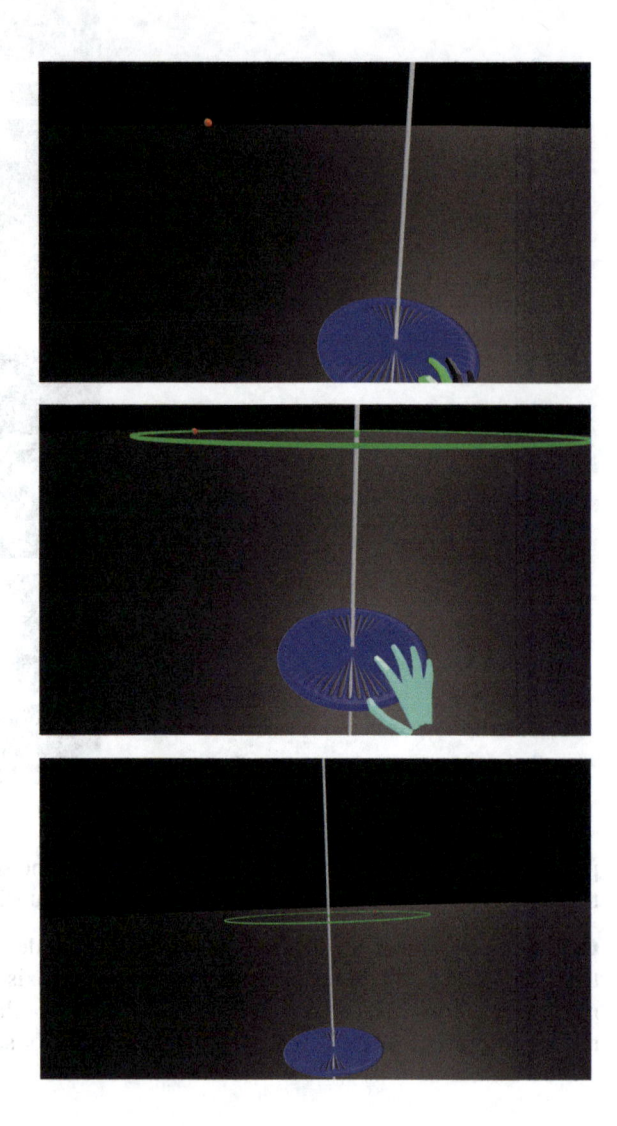

Fig. 6 Different cases of revolving a segment

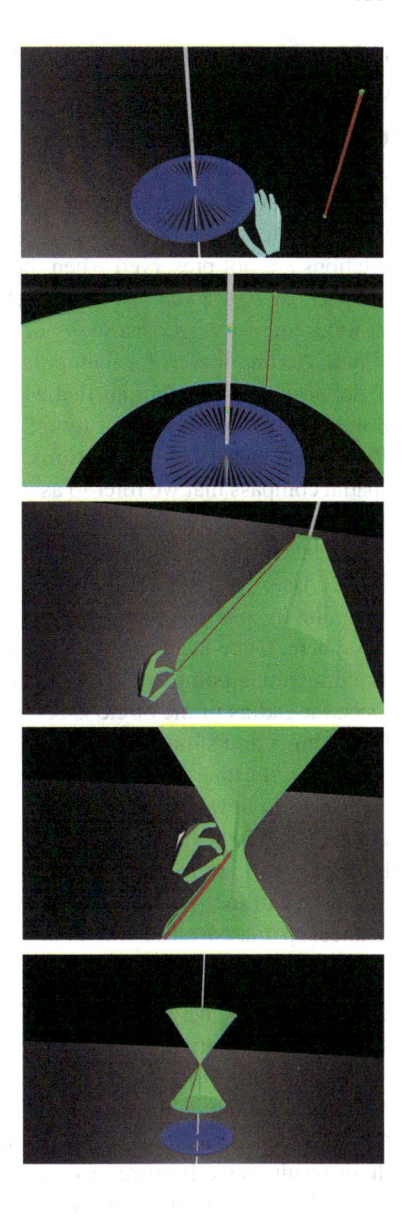

segments, and plane figures into line-segments, plane figures, or solids; or revolving points, curves, or surfaces around axes in space to create curves, surfaces, and solids, users can fluidly move from lower-dimensional shapes (e.g., circles) to their higher dimensional analogs (e.g., spheres). We are developing a related tool, *slice*, that will facilitate dimensional reasoning in the other direction. The *slice* tool is a plane that users can pinch to position in space to view cross sections of objects and explore the figures that can be cut from solids by varying the position and orientation of the cross-secting plane.

Geometry of *Stretched* and *Revolved* Figures. The current version of *HandWaver* employs a geometry solver that allows users to freely transform figures in space. Polygons are allowed to become skew-polygons (Coxeter, 1938)—i.e., polygons whose vertices are not all co-planar. Parallelism, perpendicularity, and congruence relations are not preserved when the vertices, edges, or faces of figures are moved in space. A figure such as a parallelogram may be manipulated by a user so that its sides are no longer parallel—or not even coplanar—by freely moving a single vertex. Adding modes for manipulating figures in *HandWaver* that will allow users to define geometric relations that will be preserved when the figures are transformed is a development priority for future releases.

Plane and Sphere Constructions. The spinning gesture is also used to operate a spatial compass that we refer to as *arctus*. This is a tool that allows users to generate a sphere with a given radius in space. The tool consists of a circle with a center point that can be anchored and radius that can be varied (see the first three panels in Fig. 7). Users position the tool by using a pinching gesture to grab it and move it to any point in space. The point where they set the center of the *arctus* is the center of the sphere. Once the center is set, the radius of the circle can be varied by pinching and dragging a point that is on the circle and moving it to any other point in space. Once the radius of the circle is set, a user can turn the circle through 360° in space by pinching and spinning its circumference. The spinning of the circle generates a sphere. *Arctus* thus allows users to inscribe spheres in space by using actions that are analogous to how circles can be inscribed with compasses on plane surfaces. Even though it would be possible to spawn spheres in space using simpler point-and-select logic (as in: point to select a center for the sphere, then point to select a point on the surface of the sphere), rendering spheres by spinning a spatial compass helps to connect spatial constructions to their planar analogues.

The spatial compass introduces the possibility of using *HandWaver* to investigate plane-and-sphere constructions, the three-dimensional analogs of compass-and-straightedge constructions. In the history of teaching geometry in the United States, the solid analogs of plane figures are "seldom developed" or "slighted...owing to their theoretic nature" (Franklin, 1919, p. 147). Mathematicians have characterized higher-dimensional generalizations of compass-and-straightedge constructions, but these results have been represented analytically, as opposed to diagrammatically. Three-dimensional dynamic geometry software (e.g., GeoGebra or Cabri 3D) has made it possible to engage in plane-and-sphere constructions, however the limitations of two-dimensional screens has constrained the practicability of doing so. On two-dimensional screens, the foreground and background layers of spatial constructions are compressed into one display pane, making it difficult to select or inspect or transform spatial constructions. But for users immersed in a three-dimensional space—where the user has natural control over the angle at which an object is viewed, is able to move and manipulate the object in space, and can readily select the com-

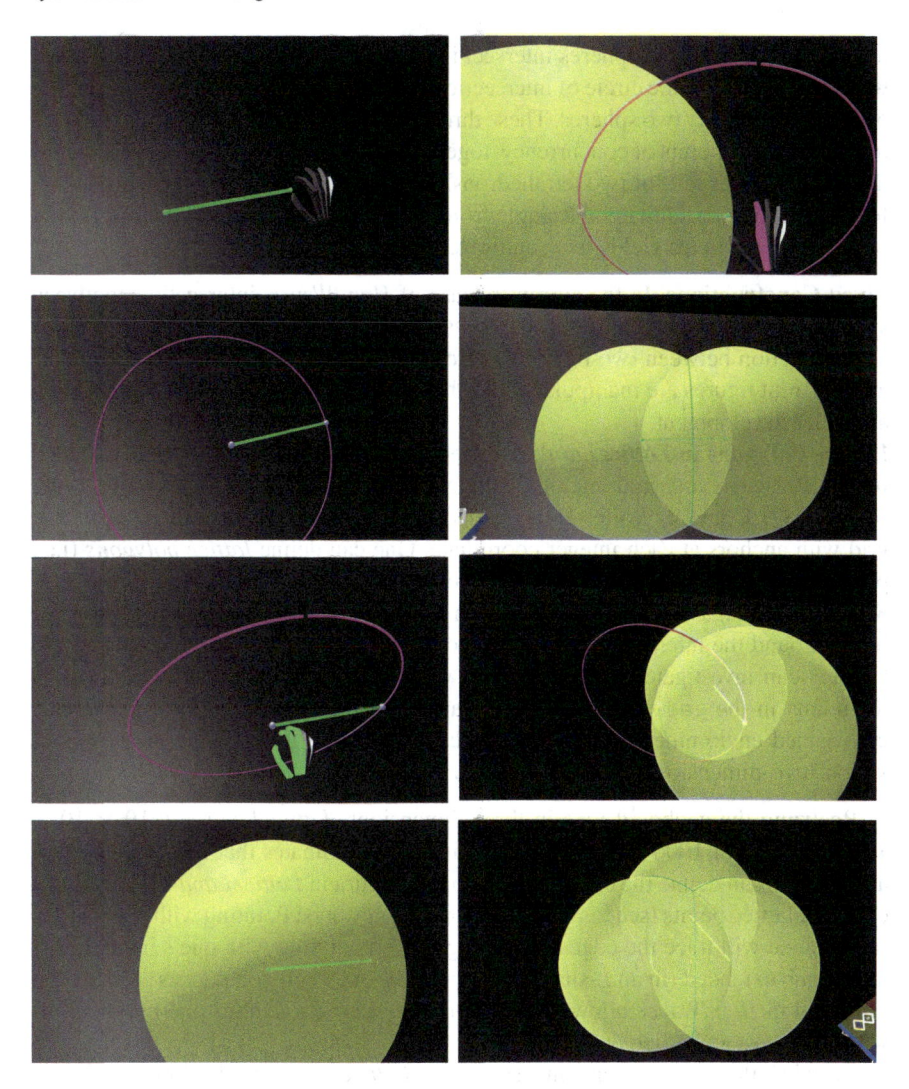

Dynamic Mathematical Figures with Immersive ... 113

Fig. 7 The *arctus* tool being used to inscribe a sphere

ponents of a figure to be incorporated into a new construction—three-dimensional constructive geometry becomes more feasible.

With *arctus* and a tool for constructing planes,[6] learners can complete solid geometry construction tasks that are inherently virtual, such as constructing a tetrahedron from three spheres, as shown in Fig. 7. The spheres shown in the panels all have the same radius. From an initial sphere, a second sphere is defined so that its center is a point on the surface of the initial sphere, and its surface contains the center point of

---

[6]This tool is still under development.

the initial sphere. These spheres intersect in a circle. Then, a third sphere is defined whose center lies on the circle of intersection such that its surface contains the center of either of the other two spheres. These three spheres will have two points of concurrency, and these points of concurrency, together with the centers of the three spheres, determine the vertices of two tetrahedrons. This is analogous to how the procedure for constructing an equilateral triangle from the centers and points of intersection of two congruent circles yields two solutions.

Fixed Constructions. In the current release of *HandWaver*, intersections between spheres are calculated on a 1-second interval, and these intersections are static. Once an intersection between two figures is calculated, that intersection does not update as the parent figures are manipulated. Methods for dynamic intersection calculation are under development.

Lattice Polygons in *LatticeLand*. Another mode for constructing figures included in *HandWaver* is a spatial analog of the geoboard (Kennedy & McDowell, 1998; Scott, 1987; Utley & Wolfe, 2004) called *LatticeLand*. A geoboard is an $n \times n$ grid with anchors at each integer coordinate. One can define *lattice polygons* (i.e., polygons whose vertices are at integer coordinates. See: Pólya, 1969; Poonen & Rodriguez-Villegas, 2000; Scott, 1987) on a geoboard by wrapping string or rubber bands around the integer-valued anchor points. Geoboards provide a setting where learners can investigate what figures can and can't be constructed by connecting the points in the grid. One of the affordances of a geoboard is that it provides a constrained environment where learners can investigate the defining properties of various two-dimensional figures (Kennedy & McDowell, 1998).

Realizing the geoboard idea in three dimensions, *LatticeLand* is a $10 \times 10 \times 10$ spatial grid of 1000 points, each with integer coordinates that are spaced 1 unit apart. Users can define the edges or faces of polyhedra in *LatticeLand* by selecting a circuit of lattice points (see Fig. 8) using pointing gestures: Pointing with one's index finger alone will trace the edges of a polygon, and pointing with one's index finger and thumb extended (in an *L*-shaped gesture) will trace polygonal faces. The spatial geoboard in *HandWaver* allows users to investigate the polygons and polyhedrons whose vertices are lattice points.

Bringing the geoboard up into space via *LatticeLand* could allow for deeper explorations into the mathematical structure of lattice polygons (Utley & Wolfe, 2004). Students might observe, for example, that, in the plane, the only regular polygon whose vertices are integers is a square. But in space, it is possible to define not only squares but also equilateral triangles and regular hexagons (Fig. 9). Students might observe that there are oblique cross sections of the spatial lattice that are equivalent to two-dimensional isometric geoboards—geoboards where the anchor points are not at integer coordinates but rather are spaced equidistant from each other. The differences between planar and spatial lattice polygons could be explored to help students appreciate how the dimensions of mathematical spaces constrain what figures are possible to create in those spaces. *LatticeLand* provides a means to begin such an investigation by grounding it in examples that students can readily access.

Fig. 8 Connecting the dots in *LatticeLand* to define the edges of a cube

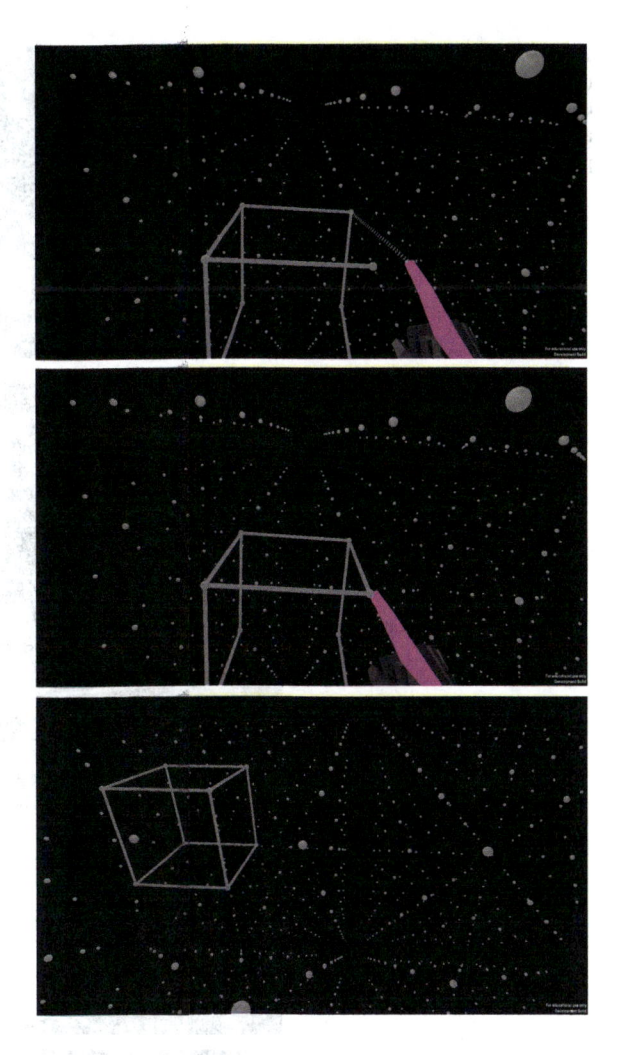

Gesture-Based Virtual Mathematics. Our vision in designing *HandWaver* is that learners at any stage of mathematical maturity should be able to quickly and easily construct and explore mathematical figures in space by making natural movements with their hands. The gesture-based tools for making figures we have created thus far are a starting point toward realizing this vision. Our work to refine these tools and add others is ongoing. In the next sections, we discuss our plans for research and consider the challenges and opportunities of bringing spatial display technologies into schools.

Fig. 9 A regular octahedron (first panel) with oblique equilateral triangle faces and a regular hexagon (last three panels) in *LatticeLand*

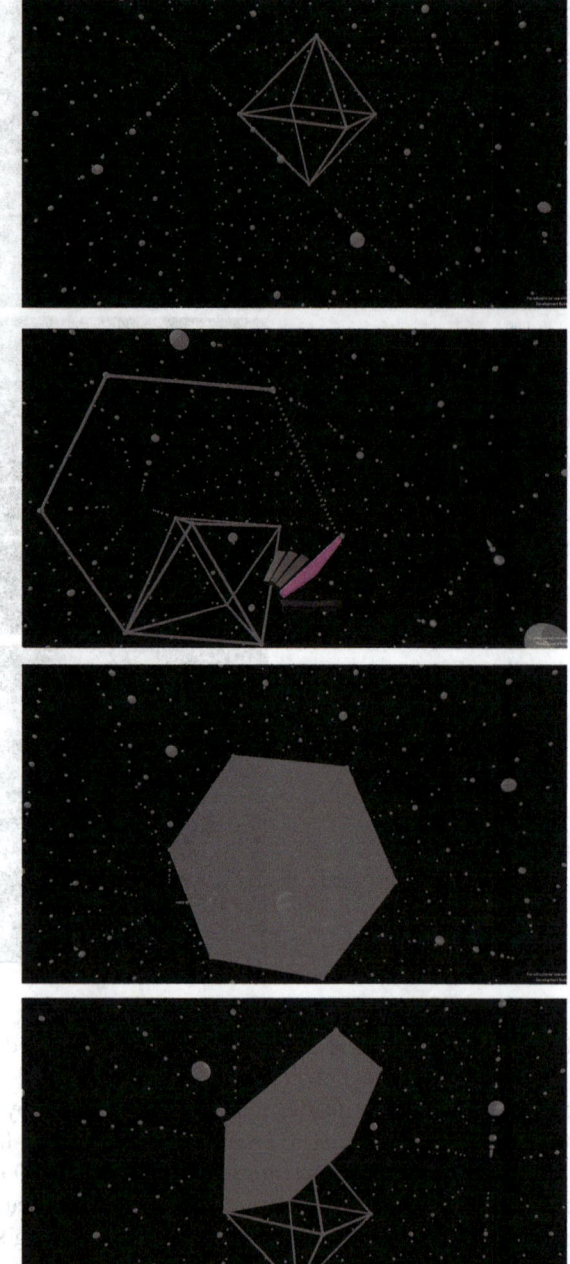

5 Research

Mathematical Explorations with Spatial Displays. We are building capacity to engage in parallel lines of research using *HandWaver*. One line of research concerns documenting student encounters with mathematical objects in the virtual space. The immersive nature of the environment, combined with the gestural interface, provides a level of control over perspective, orientation, and position relative to mathematical objects that is difficult to replicate with other display technologies. Even the relatively straightforward means for rotating the graphics view in the 3D version of GeoGebra is complicated when compared to moving one's head, walking around a figure, or examining it from several different angles in quick succession. How do students use the angle of their gaze, the position of their bodies relative to virtual mathematical figures, or the ability to quickly change the scale of figures—from something that one could hold in one's hands, to something that one could fit inside—to explore mathematical structures?

In one study (Bock & Dimmel, 2017), we used semi-structured interviews to investigate how people used non-measuring virtual tools to make and test mathematical claims about the volumes of pyramids. *Non-measuring virtual tools* refers to tools that were designed to facilitate comparing the spatial qualities of figures but that could not easily be used for taking measurements. Participants—three master's students pursuing certification as science teachers—were asked to think-aloud as they explored the volume of a pyramid. We selected science teachers because we were looking for non-mathematical experts who were interested in mathematical tasks but who would not be familiar with how to derive the expression for the volume of a pyramid.

Each participant completed the hour-long interview individually. They were provided with two red pyramids adjacent to each other on a virtual workbench (see Fig. 10). The red pyramids were unit pyramids—i.e., they had unit-area bases and were one unit high. Virtual tools available to the participants in the scene included a unit cube that could be displayed or hidden in line with the red pyramids on the virtual workbench; cubes around each of the red pyramids that could be displayed or hidden; rectangular grids that could be toggled to display on the faces of the cubes; and the ability to add up to four additional pyramids to the cube, each of whose apexes were the same as the initial pyramid (see Fig. 11). A *slice* tool allowed participants to view cross sections of the pyramids at any height, and an *explode* tool displayed the pyramids as stacks of rectangular or trapezoidal prisms—participants could use virtual sliders to vary the thickness of the slices in the stack and the gaps between them (the last two panels of Fig. 10).

The bottom-right corner of each image of Fig. 10 shows a third-person view of the resources available in the volume laboratory. The thin, green figure is a virtual placeholder for the participant that was in the scene. The second image shows the control panel where the user accessed the various virtual tools that were available. The apexes of the red pyramids could be moved by pinching (with index finger and thumb) and dragging them through space. The bases of the pyramids were constant

Fig. 10 Exploring the volume of a pyramid

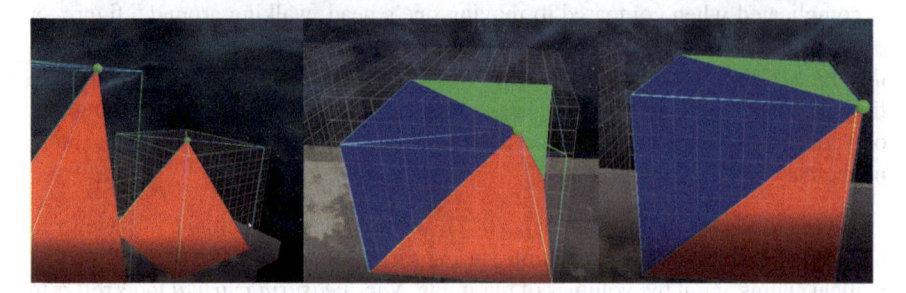

Fig. 11 Enclosing the pyramids in a unit cube, adding additional pyramids, and adjusting the pyramids by moving the apex

and fixed to the top of the workbench, as was the cube. Participants could lock the apex of either pyramid in the z-direction (shearing) or xy-directions (elongating) to control how the apex moved.

Participants were asked a series of open-ended questions to guide their explorations of the pyramids, such as: *How could you relate the volume of the red pyramid to the volume of the cube?* One anticipated strategy was that participants would fill the cube with the four additional pyramids and then investigate how the volumes of these pyramids were affected by shearing or elongating the apex of the pyramids. This anticipated strategy was not used by anyone. Instead, two of the participants attempted to reason about the volume of the red pyramid by analyzing how the surface areas of its faces were affected when its apex was moved in different ways. Even though we did not provide any measuring tools that would have facilitated calculating and comparing surface areas, this line of reasoning was compelling for these participants. We are planning studies that will investigate how different types of users (e.g., mathematics teachers, pre-service mathematics teachers, undergraduate mathematics students, and secondary mathematics students) build geometric figures with the *stretching* and *spinning* gestures and that compare virtual to non-virtual resources for investigating mathematics (e.g., a study of how people use *LatticeLand* or physical manipulatives like geoboards to make claims about lattice polygons).

Spatial Displays as Instructional Technologies. A parallel line of research pertains to issues of instructional implementation: How do practicing and preservice teachers imagine incorporating spatial display technologies into their teaching? What support do they need? What barriers do they anticipate? For this research, we are planning to develop multiplayer and partial immersion modes so that *HandWaver* could be used by a teacher with a whole classroom. The multiplayer mode will allow more than one user to be in the same virtual world at one time. The partial immersion mode will allow other users to view what is happening in the virtual world through a tablet. The partially immersed users will also be able to have some limited interactions with the virtual world, such as using gestures to control their angle of view, their position within the environment, or to construct figures. We are anticipating a time in the not-too-distant future when it will become feasible for a classroom to have multiple spatial displays that will allow students to work on problems in groups.

In such configurations, some students would be fully immersed in a virtual world and others would access the environment via a gesture-tracking tablet. We have a dedicated laboratory classroom space at the University of Maine where we will convene groups of teachers to study the instructional potential of teaching in a spatial display-enabled classroom. Groups of participating teachers will explore and critique the *HandWaver* environment. They will work with each other to devise plans for how such an environment could be used in their teaching and anticipate obstacles they would expect to encounter.

6 Spatial Displays and Schooling

Soon, children will routinely and increasingly incorporate spatial display technologies into their leisure activities. They will be playing games that require spatial reasoning and problem solving skills—imagine, for example, an immersive first-person version of *Monument Valley* (Ustwo, 2014)—but what will they be doing in schools? In US Schools, children's encounters with geometry in elementary schools are limited to shape recognition and naming tasks (Bruce & Hawes, 2015). Yet a growing body of research indicates that children have the interest and capacity to train their spatial reasoning skills (Hallowell, Okamoto, Romo, & La Joy, 2015; Taylor & Hutton, 2013; Whiteley, Sinclair, & Davis, 2015) and study meaningful mathematics (Newton & Alexander, 2013; Sinclair & Bruce, 2015) from the moment they enter the schoolroom door. New modes of interacting with virtual mathematical objects have the potential to expand children's access to deep geometric ideas. For all of their educational promise, however, immersive spatial display technologies are on track to follow the slow, complex process of technology acceptance and adoption that is standard in schools and that falls short of true integration (Ertmer, 1999; Inan & Lowther, 2010). Given how difficult it has been, historically, to incorporate promising technologies into classrooms at scale, there is every reason to believe that, without concerted effort, the educational potential of spatial displays will remain unfulfilled.

One of our practical motivations for developing *HandWaver* was to create a virtual environment that could introduce teachers to using spatial display technologies for mathematical investigations. When these technologies are as ubiquitous as smart phones, we want teachers to have access to mathematically and pedagogically sound instructional resources and also to be prepared to incorporate them into their teaching. There is a "scarcity of bold research on interactive mathematics learning" that "impedes the formulation of empirically based progressive policies concerning the integration of technological environments into educational institutions" (Abrahamson & Sánchez-García, 2016, p. 204). At the same time, consumer grade virtual reality (e.g., Oculus Rift, HTC Vive) is likely to usher a frenzy of development of commercial educational content. If such development follows the path of educational apps, a preponderance of the mathematics education content that is developed for spatial displays will amount to little more than immersive, visually engaging flashcards (Davis, 2015). By designing and developing the *HandWaver* environment, we are attempting to ensure that research-based ideas about the nature of productive mathematical activity are represented in this next generation of virtual learning environments.

References

Abrahamson, D., & Sánchez-García, R. (2016). Learning is moving in new ways: The ecological dynamics of mathematics education. *Journal of the Learning Sciences* (online first edition).

Alibali, M. W., & Nathan, M. J. (2012). Embodiment in mathematics teaching and learning: Evidence from learners' and teachers' gestures. *Journal of the learning sciences, 21*(2), 247–286.

Barrett, T. J., Stull, A. T., Hsu, T. M., & Hegarty, M. (2015). Constrained interactivity for relating multiple representations in science: When virtual is better than real. *Computers & Education, 81*, 69–81.

Bock, C., & Dimmel, J. K. (2017). Explorations of volume in a gesture-based virtual mathematics laboratory. In E. Galindo & J. Newton (Eds.), *Proceedings of the 39th annual meeting of the North American Chapter of the International Group for the Psychology of Mathematics Education* (pp. 371–374) Indianapolis, IN: Hoosier Association of Mathematics Teacher Educators.

Bruce, C. D., & Hawes, Z. (2015). The role of 2D and 3D mental rotation in mathematics for young children: What is it? Why does it matter? And what can we do about it? *ZDM Mathematics Education, 47*(3), 331–343.

Chen, C. L., & Herbst, P. (2013). The interplay among gestures, discourse, and diagrams in students' geometrical reasoning. *Educational Studies in Mathematics, 83*(2), 285–307.

Collins, H. (2010). *Tacit and explicit knowledge*. University of Chicago Press, Chicago, IL.

Coxeter, H. S. M. (1938). Regular skew polyhedra in three and four dimension, and their topological analogues. *Proceedings of the London Mathematical Society, 2*(1), 33–62.

Davis, B. (2015). Gumm(i)ing up the works? Lessons learned through designing a research-based "app-tutor". In *Proceedings of the 12th International Conference on Technology if Mathematics Teaching,* Faro, Portugal.

de Freitas, E., & Sinclair, N. (2012). Diagram, gesture, agency: Theorizing embodiment in the mathematics classroom. *Educational Studies in Mathematics, 80*(1–2), 133–152.

Dimmel, J. K., & Herbst, P. G. (2015). The semiotic structure of geometry diagrams: How textbook diagrams convey meaning. *Journal for Research in Mathematics Education, 46*(2), 147–195.

Duval, R. (2014). Commentary: Linking epistemology and semio-cognitive modeling in visualization. *ZDM Mathematics Education, 46*(1), 159–170.

Ebner, T., Feldmann, I., Renault, S., & Schreer, O. (2017). 46-2: Distinguished Paper: Dynamic real world objects in augmented and virtual reality applications. In *SID Symposium Digest of Technical Papers* (Vol. 48, No. 1, pp. 673–676).

Ertmer, P. A. (1999). Addressing first-and second-order barriers to change: Strategies for technology integration. *Educational Technology Research and Development, 47*(4), 47–61.

Franklin, P. (1919). Some geometrical relations of the plane, sphere, and tetrahedron. *The American Mathematical Monthly, 26*(4), 146–151.

Gnanadesikan, A. E. (2011). *The writing revolution: Cuneiform to the internet* (Vol. 25). Wiley.

Hallowell, D. A., Okamoto, Y., Romo, L. F., & La Joy, J. R. (2015). First-graders' spatial-mathematical reasoning about plane and solid shapes and their representations. *ZDM Mathematics Education, 47*(3), 363–375.

Harris, R. (1986). *The origin of writing*. Open Court Publishing.

Harris, R. (1989). How does writing restructure thought? *Language & Communication, 9*(2–3), 99–106.

Hart, V., Hawksley, A., Matsumoto, E. A., & Segerman, H. (2017a). Non-euclidean virtual reality I: Explorations of H^3. arXiv preprint arXiv:1702.04004.

Hart, V., Hawksley, A., Matsumoto, E. A., & Segerman, H. (2017b). Non-euclidean virtual reality II: Explorations of $H^2 \times E$. arXiv preprint arXiv:1702.04862.

Herbst, P., Fujita, T., Halverscheid, S., & Weiss, M. (2017). *The learning and teaching of geometry in secondary schools: A modeling perspective*. Taylor & Francis.

Hettinger, L. J., & Riccio, G. E. (1992). Visually induced motion sickness in virtual environments. *Presence: Teleoperators & Virtual Environments, 1*(3), 306–310.

Hollebrands, K. F. (2007). The role of a dynamic software program for geometry in the strategies high school mathematics students employ. *Journal for Research in Mathematics Education*, 164–192.

Inan, F. A., & Lowther, D. L. (2010). Factors affecting technology integration in K-12 classrooms: A path model. *Educational Technology Research and Development, 58*(2), 137–154.

Kennedy, J., & McDowell, E. (1998). Geoboard quadrilaterals. *The Mathematics Teacher, 91*(4), 288–290.

McNeil, N. M. (2008). Limitations to teaching children 2 + 2 = 4: Typical arithmetic problems can hinder learning of mathematical equivalence. *Child Development, 79*, 1524–1537.

Newton, K. J., & Alexander, P. A. (2013). Early mathematics learning in perspective: Eras and forces of change. In *Reconceptualizing early mathematics learning* (pp. 5–28). Springer Netherlands.

O'Halloran, K. L. (2005). *Mathematical discourse: Language, symbolism and visual images*. London and New York: Continuum.

Pólya, G. (1969). On the number of certain lattice polygons. *Journal of Combinatorial Theory, 6*(1), 102–105.

Poonen, B., & Rodriguez-Villegas, F. (2000). Lattice polygons and the number 12. *The American Mathematical Monthly, 107*(3), 238–250.

Sandoval, W. (2014). Science education's need for a theory of epistemological development. *Science Education, 98*(3), 383–387. https://doi.org/10.1002/sce.21107.

Scott, P. R. (1987). The fascination of the elementary. *The American Mathematical Monthly, 94*(8), 759–768.

Senner, W. M. (Ed.). (1991). *The origins of writing*. Lincoln, NE: University of Nebraska Press.

Sinclair, N. (2014). Generations of research on new technologies in mathematics education. *Teaching mathematics and its applications, 3*, 166–178.

Sinclair, N., & Bruce, C. D. (2015). New opportunities in geometry education at the primary school. *ZDM Mathematics Education, 47*(3), 319–329.

Taylor, H. A., & Hutton, A. (2013). Think3d!: training spatial thinking fundamental to STEM education. *Cognition and Instruction, 31*(4), 434–455.

Thurston, W. P. (1994). On proof and progress in mathematics. *Bulletin of the American mathematical society, 30*(2), 161–177.

Ustwo. (2014). Monument Valley [computer game]. Available at: https://www.monumentvalleygame.com/mv1.

Utley, J., & Wolfe, J. (2004). Geoboard areas: Students' remarkable ideas. *Mathematics Teacher, 97*(1), 18.

Whiteley, W., Sinclair, N., & Davis, B. (2015). What is spatial reasoning? In B. Davis (Ed.), *Spatial reasoning in the early years: Principle, assertions, and speculations*. New York, NY: Routledge.

Zuckerman, O., & Gal-Oz, A. (2013). To TUI or not to TUI: Evaluating performance and preference in tangible versus graphical user interfaces. *International Journal of Human-Computer Studies, 71*(7), 803–820.

WIMS: Innovative Pedagogy with 21 Year Old Interactive Exercise Software

Magdalena Kobylanski

1 Introduction

The use of digital technology is becoming more and more widespread in mathematics education at all levels. Large amounts of online resources are available, impacting mathematics teaching and learning. Among these resources, repositories of e-exercises are perhaps those that offer the most versatile use, in or out of the classroom, with a teacher's supervision or in autonomy. Cazes, Gueudet, Hersant, and Vandebrouck (2006) describe these repositories as consisting "mainly of classified exercises" and proposing "in addition to these exercises, an associated environment for each of them that can include suggestions, corrections, explanations, tools for the resolution of the exercise, and score" (p. 327). Besides WIMS on which this chapter focuses, Mathenpoche (http://mathenpoche.sesamath.net/), designed by a French association of secondary mathematics teachers and covering the secondary school mathematics program (Grades 6–12), Euler (https://euler.ac-versailles.fr), offering interactive pages for primary, secondary and tertiary mathematics, or emathematics.net, designed for K-12 grades, are examples of online exercise repositories. Research studies focusing on this kind of resource report an increase in learners' motivation leading to more intense activity (Ruthven & Hennessy, 2002; Cazes et al., 2006; Hersant & Vandebrouck, 2006). Immediate feedback, self-correcting facilities, and available hints are features that can explain these findings.

WIMS (Web Interactive Multipurpose Server) is one such web-based exercise repository designed for university mathematics students more than two decades ago. Nowadays, it is used also by secondary teachers and students and offers exercises not only in mathematics, but also in many other subjects such as physics (Berland, 2017), chemistry, biology, and French, among others. As the server becomes more

---

M. Kobylanski (✉)
Université Paris-Est Marne-la-Vallée (UPEM) LAMA-UMR8050, Université Paris-Est IDEA, Champs-sur-Marne, France
e-mail: magdalena.kobylanski@u-pem.fr

© Springer Nature Switzerland AG 2019
G. Aldon and J. Trgalová (eds.), *Technology in Mathematics Teaching*, Mathematics Education in the Digital Era 13,
https://doi.org/10.1007/978-3-030-19741-4_6

and more widespread, this chapter aims at filling to some extent the gap in literature about this technology.

This chapter is thus devoted to the presentation of the repository and questions the learner's possible activity when interacting with this technology. It first outlines a theoretical framework that underpins the design choices discussed subsequently. A large part of the chapter is dedicated to the presentation of WIMS affordances from both the learner's and teacher's points of view, so as to bring to the fore elements allowing a discussion of pedagogical interest and the added value of WIMS. Results of empirical studies involving WIMS are reported afterwards, before concluding and outlining perspectives for future developments.

2 Theoretical Framework

Research studies related to online exercise repositories, called e-exercise base (EEB) by Cazes et al. (2006), highlight an increase in students' motivation as one of the major impacts of EEB on students' behavior. We therefore start by outlining a state of the art on motivation.

2.1 Motivation

Research studies evidence an interrelation between motivation and the success of the learning process (Brophy, 2004; Chappaz, 1992; Viau, 2011). In the context of education, motivation is defined as a "set of dynamic factors that trigger, in a student or a group of students, a desire to learn" (Léon, 1972, p. 78). The concept being rather broad, Viau (2009) introduces the term "motivational dynamics" as

> a phenomenon that has its source in the student's perceptions of herself and her environment, and which has the consequence that she chooses to commit herself to accomplish the proposed pedagogical activity and to persevere in its accomplishment, and this in order to learn. (p. 12, our translation)

Viau (2011) identifies several factors that affect motivational dynamics of a student: social factors (e.g., culture, values); factors related to the student's life (e.g., family, friends, extra-school activities); factors related to school (e.g., school hours, rules); and factors related to the class (e.g., teacher, assessment, reward, punishment). Ryan and Deci (2000) distinguish "between different types of motivation based on the different reasons or goals that give rise to an action" (p. 55). They claim that

> the most basic distinction is between *intrinsic motivation*, which refers to doing something because it is inherently interesting or enjoyable, and *extrinsic motivation*, which refers to doing something because it leads to a separable outcome. (ibid.)

According to the authors, whereas "intrinsic motivation results in high-quality learning and creativity" and is therefore highly valued in education, extrinsic motivation is equally important "for educators who cannot always rely on intrinsic motivation to foster learning" (ibid.). Indeed, as the authors point out, there are different kinds of extrinsic motivation:

> Students can perform extrinsically motivated actions with resentment, resistance, and disinterest or, alternatively, with an attitude of willingness that reflects an inner acceptance of the value or utility of a task. In the former case—the classic case of extrinsic motivation—one feels externally propelled into action; in the latter case, the extrinsic goal is self-endorsed and thus adopted with a sense of volition. (ibid.)

These considerations are important for the design of mathematical tasks if we follow Ryan and Deci (ibid.) who claim that

> because many of the tasks that educators want their students to perform are not inherently interesting or enjoyable, knowing how to promote more active and volitional (versus passive and controlling) forms of extrinsic motivation becomes an essential strategy for successful teaching. (p. 55)

The authors also bring forward factors likely to enhance motivation. They claim that "*feelings of competence* during action can enhance intrinsic motivation for that action because they allow satisfaction of the basic psychological need for competence" (ibid., 58). They add, however, that "feelings of competence will *not* enhance intrinsic motivation unless they are accompanied by a *sense of autonomy*" (ibid.). Thus, for being intrinsically motivated, students need to "experience satisfaction of the needs both for competence and autonomy" (ibid.). Among factors that enhance extrinsic motivation, the authors mention the "feeling of relatedness", meaning in the classroom context that "students' feeling respected and cared for by the teacher is essential for their willingness to accept the proffered classroom values" (p. 64), as well as the "perceived competence", which implies that "supports for competence (e.g., offering optimal challenges and effectance-relevant feedback) facilitate internalization" (ibid.).

2.2 Learning, Memory, and Practice

According to Karpicke, traditionally, learning is considered as "the acquisition and encoding of new information" and "[t]ests [...] are used to assess what was learned in a prior experience but are not typically viewed as learning events" (2017, p. 487). Recent research in cognitive science has evidenced that tests are not neutral for the process of learning; on the contrary, they "can aid learning by providing feedback about what a person knows and does not know" (ibid.). The benefit of testing on learning can be explained, following the author, by the fact that the "'testing effect' is driven by the retrieval processes that learners engage in when they take tests, and thus the key phenomenon is referred to as *retrieval-based learning*" (ibid.). Practicing retrieval requires some effort from the learner, and "effortful retrieval of knowledge leaves that knowledge strengthened, increasing the likelihood that it can be accessed and used again in the future" (ibid., p. 492). Based on this idea, retrieval practice assumes that the learner should face difficulties to enhance learning. Indeed, "certain conditions that make initial learning slower and more difficult may result in very good long-term retention and transfer; hence, those conditions constitute desirable difficulties" (ibid.). This shows a positive impact that retrieval-based learning may have on memory and knowledge transfer. Retrieval-based learning can be practiced with short answer and multiple-choice tests. Indeed, Karpicke (ibid.) claims that both test formats may lead to strengthen learning provided that they engage the learner in a retrieval effort and include feedback that is crucial to prevent the learner from creating misconceptions. In the case of multiple-choice questions, it has been shown that when these are constructed in a way that they contain "plausible alternatives", learners need not only retrieve the correct answer but also retrieve reasons why other alternatives are incorrect, which makes them engage in more retrieval effort than questions with an obvious correct answer.

Let us note that the anthropological theory of didactics (Chevallard, 1998) considers "practicing and testing" as two among the six didactic moments that constitute the study of any mathematics task, thus acknowledging the importance of training and (self-)assessing a newly acquired knowledge in a variety of contexts and situations. These considerations lead us to address a competency model that informs assessment.

2.3 Competency Models

Our purpose is not to elaborate a state of the art on the issue of competence and competency models. Rather, we present a model that enables us to reflect on how a student develops competency in order to be able to sustain this development.

According to Getha-Taylor, Hummert, Nalbandian, and Silvia (2013), competency models are related to the concepts of mastery and transfer. For Ambrose et al. (2010), in order to develop mastery, "students must acquire component skills, practice integrating them, and know when to apply what they have learned" (p. 95). Competency models can thus help students identify "what they have learned" and reflect on when and how to apply it. Referring to other research, Getha-Taylor et al. (2013) claim that mastery includes two key dimensions: competence and consciousness. The authors suggest the following stages of competence development (Table 1).

As the authors explain,

> the mastery developmental process begins in a state where students not only lack competence but also are generally unaware of what they do not know. This situation may result in inflated initial self-assessments. As students progress in their education, it is expected that both their consciousness and competence will develop to help them identify what they are learning and what they still have to learn. True mastery occurs only when the initial stage of unconscious incompetence progresses to the final stage of unconscious competence. We want our students to have competence that can be used automatically and instinctively. (ibid., p. 144)

Table 1 Competency model (Getha-Taylor et al., 2013, p. 144)

Level	Stage	Description
1	Unconscious incompetence	Students do not know what they do not know
2	Conscious incompetence	Students are aware of what they need to learn
3	Conscious competence	Students have competence but must act deliberately
4	Unconscious competence	Students exercise skills automatically or instinctively

We close this section by recalling Bloom's taxonomy of educational objectives that depicts levels of knowledge and skills that still inform teaching. Bloom (1956) explains the main categories of the taxonomy as follows (pp. 201–207):

- *Knowledge* "involves the recall of specifics and universals, the recall of methods and processes, or the recall of a pattern, structure, or setting."
- *Comprehension* "refers to a type of understanding or apprehension such that the individual knows what is being communicated and can make use of the material or idea being communicated without necessarily relating it to other material or seeing its fullest implications."
- *Application* means the "use of abstractions in particular and concrete situations."
- *Analysis* consists in the "breakdown of a communication into its constituent elements or parts such that the relative hierarchy of ideas is made clear and/or the relations between ideas expressed are made explicit."
- *Synthesis* involves the "putting together of elements and parts so as to form a whole."
- *Evaluation* involves "judgments about the value of Materials and methods for given purposes."

In the next section, we briefly present WIMS before describing its main affordances with a rationale drawing on the presented theoretical framework.

3 WIMS

WIMS is a collaborative, open source e-learning platform. Its main specificity is to host online, interactive, random, self-correcting exercises in many different fields such as mathematics, chemistry, physics, biology, French, and English, among others.

It is used mostly in France,[1] in mathematics classes at high school level or in the first years of higher education.[2]

WIMS was created in 1997 by Xiao Gang (1951–2014), professor of mathematics at the University of Nice (France). Ten years after WIMS' first release, the association WIMS EDU[3] was founded that comprised a small community of developers. The association, whose main goal is to support distribution of WIMS, organizes, among other events, a biyearly colloquium attended by more than a hundred participants.

3.1 WIMS Design Choices and Affordances from the Students' Perspective

WIMS has been created to support the development of students' competency by providing them with the opportunity to practice and test their knowledge in a wide range of exercises. This section is devoted to the presentation of affordances WIMS offer to learners.

3.1.1 A Large Scale of Exercises

WIMS is a Learning Management System that offers different kinds of interactive exercises. It offers an environment in which the teacher can create interactive exercises such as multiple-choice questions, matching, drag-and-drop exercises, numerical or algebraic exercises, as well as graphic exercises. Formal calculus or drawing figures can be done quite simply. Some original exercises that are rarely offered in traditional textbooks are proposed in WIMS, particularly graphic exercises, such as the one shown in Fig. 1. An aim, for instance, can be to become skilled in quickly and accurately visualizing a mathematical transformation, a skill often taken for granted and therefore rarely taught.

Exercises published on the platform correspond to different educational needs and to different information processing in the sense of Bloom's taxonomy, in line with the retrieval-based learning principles. Some correspond for example to the first level, i.e., to the basic memorization of the course, the restitution (knowledge). These exercises, although basic, are useful because they allow elementary manipulations, and in doing so the student engages in a first level of familiarization with the concepts. They thus make it possible to remove ambiguities. For example, the exercise

---

[1]Interactive map showing where WIMS is used. http://downloadcenter.wimsedu.info/download/map/map2.html.

[2]Enquête auprès des utilisateurs WIMS [WIMS user survey]. http://moin.irem.univ-mrs.fr/groupe-wims/Enquete [consulted 2017/11/14].

[3]http://wimsedu.info.

WIMS: Innovative Pedagogy with 21 Year Old Interactive Exercise ... 129

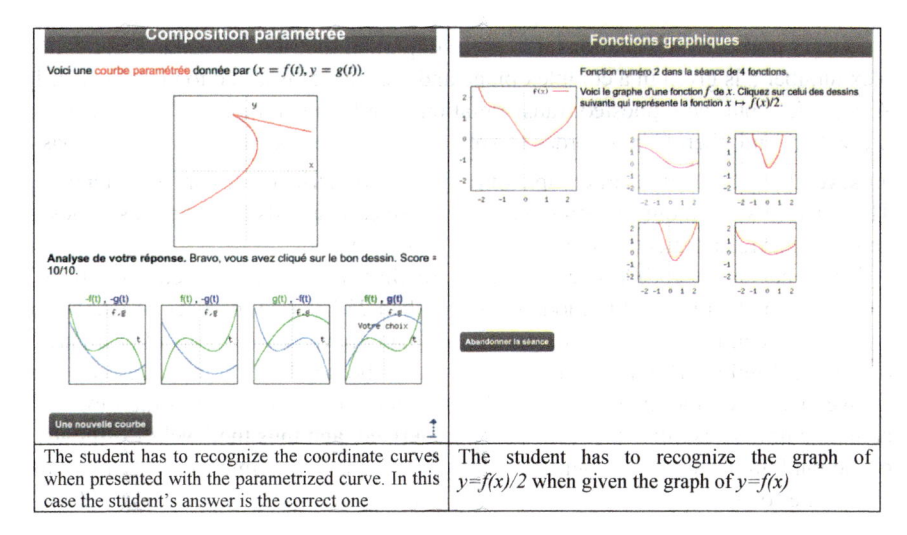

Fig. 1 WIMS graphical exercises

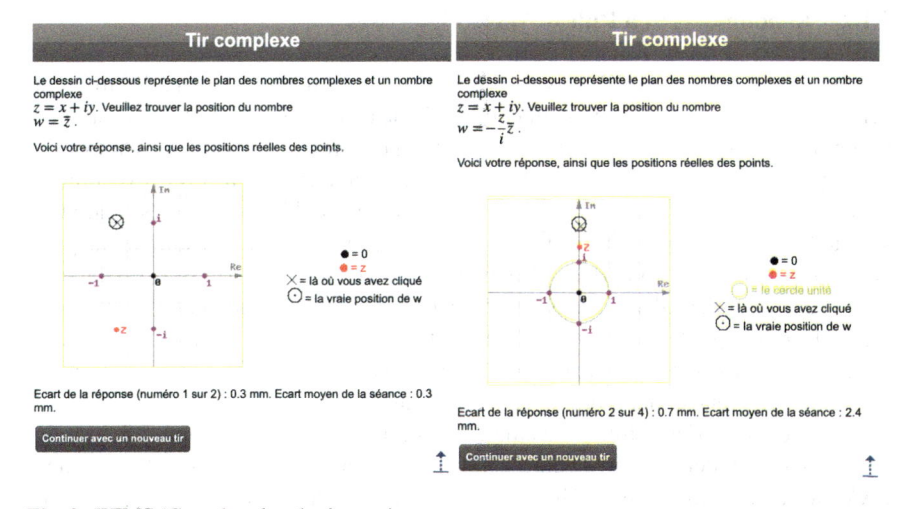

Fig. 2 WIMS 'Complex shooting' exercise

$$\text{Is the sub-set } [1, a) \setminus cup\, (b, +\setminus infty) \text{ an interval? (with parameters } a, b$$
$$> 1 \text{ drawn randomly)}$$

allows manipulating of the notion of interval.

Some exercises use a large number of registers (Duval, 1993) such as the 'complex shooting (Tir complexe)' exercise (Fig. 2).

The proposed transformations start with simple symmetries or translations: a complex number z is given in a complex plane and the student is asked to place $-z$, iz, $1 + z$, $z - i$..., but the requested transformations can be much more difficult, such as z^2, $iz^2/\bar{z}$... and, therefore, in order to solve the exercise, the student needs to master several representations of complex numbers (geometric, Cartesian, exponential). Thus, in terms of Bloom's taxonomy, the exercise corresponds to a level of synthesis and is of particular interest, appreciated by many teachers.

There are about 15,000 exercises that are shared and can be used directly by students, which are stored in modules. More than 250 shared modules of exercises exist in French, the most used language, but many have been translated or created directly in English, Dutch, Chinese and Italian. About half of the modules contain an average of twenty random exercises around a theme. About half of the modules contain a unique exercise that can be parameterized, and thus the level of difficulty of the random draw can be controlled. The randomness is an important feature that allows the focus on "types of exercises" instead of "exercises" (Chevallard, 1998).

3.1.2 Randomness as a Rule of WIMS Exercises

The main feature of the WIMS is its randomness, which allows a slightly different version of the exercise to be immediately available and allows the student to continue his or her learning. The surface strategy of typing the previous answer does not work, and copying the answer of a nearby student is also doomed to fail. In a WIMS exercise, we do not program question $2 + 2$ but a + b, where a and b are variables randomly selected by WIMS from values determined by the programmer, for example integers between 1 and 5, or 1 and 10, or 11 and 49 etc. When a student starts a session of a WIMS exercise, the software presents the exercise with the values a and b randomly drawn from the predetermined set. Thus, with a few lines of programming, we can obtain an exercise that will have a large number of versions (Fig. 3), which encourages the student to engage in a more in-depth learning strategy.

The necessity of multiplying examples with different types of representation in order to allow students to form a correct representation of a problem is well explained in Cordier and Cordier (1991). The authors show that students who were only exposed to the use of Thales' theorem in cases where the parallel lines were at the same side of the intersection of the two secant lines had a lesser chance of understanding the entire generality of the theorem than those exposed to a greater amount of different cases. WIMS clearly meets this purpose. Its random features push the creator of an exercise to imagine a variety of examples and to encode them from the beginning. Moreover, sharing exercises among teachers (see below) further reinforces the variety of exercises.

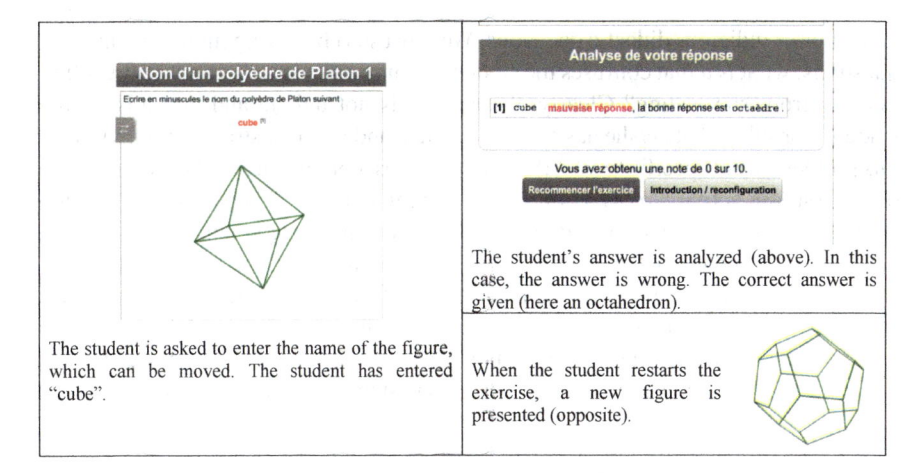

Fig. 3 WIMS 'recognize and name a Plato's polyhedron' exercise

3.1.3 Feedback and Assessment

Proposing WIMS exercises very early makes it possible for each student to self-assess themselves. Poor understanding of a notion can thus be detected at an early stage by the learner and corrected quickly, which is necessary in order to avoid misconceptions manifesting themselves.

WIMS also allows direct access to errors: once the answer is given, the learner can immediately know if his/her answer is correct or not, with a score also provided. As mentioned previously, on time and relevant feedback contributes to kindle students' feelings of autonomy and competence, and thus sustain their motivation and engagement when working on WIMS exercises. In most cases, the correct answer, when unique, is already programmed in the WIMS exercise program and can be displayed on request. In addition, the work students do and the time they spent on WIMS should be somehow rewarded. An easy way to do so is to use the WIMS grade as part of the overall grade of the course. Repeated experience shows that it is necessary to sustain students' engagement; when the only benefit of the training is the training itself, studies show that only 5% of students continue to work up to the end of the course. This reward does not need to be important; it can take the form of a bonus on the final grade, or be some part of the formative grade. We will come back to this point later.

However, there is sometimes more than one correct answer, and displaying a correct answer in this case is more difficult than checking whether the solution given by a student is correct. It is, however, possible to schedule additional feedback as part of a WIMS exercise. This feedback can, for example, explain a way to find the solution. This kind of feedback is meant to engage the learner in a retrieval effort. It is natural to wonder how WIMS should or can help a student who is unable to solve an exercise. The answer is not simple; this is a difficult task, which takes time and requires judicious didactic decisions. We must start by asking more fundamental questions: what is it that confuses the student in this particular case, and is preventing him/her from succeeding? Clearly, the answer is not unique and depends on the student's profile, what he/she has already learned, and learning strategies. Obviously, this problem is not specific to WIMS. Since the most common use of WIMS is as part of a classroom activity, or as a homework assignment, it is also possible to provide feedback by discussing with students the difficulties they encounter.

It should be noted that an exercise alone has a rather poor meaning as a learning activity. An exercise is part of a subject which is formed through a set of exercises. Proposing exercises to cover a topic can follow different design models. In WIMS, exercises are chosen by the teacher and proposed in the form of an exercise sheet in a specific way, which can be considered as consistent with retrieval-based learning.

3.1.4 Self-regulation and WIMS

Self-regulation is considered as a predictive indicator of how students would succeed in their learning and in life later on. By self-regulation we mean the capacity of displaying the necessary endeavor to change one's inner state and patterns of thinking and acting, which is closely related with students' motivation, either intrinsic or extrinsic. Engaging in learning has different types of costs. First the learner has to give up other activities and decide to dedicate his/her attention to learning rather than letting it be diverted by other activities. One has to be able to arbitrate between different tasks so as to preserve and support the original intention of learning. A second obstacle occurs when the learner is facing a difficulty; at this moment, s/he needs enhancing of his/her energy and attention to identify his/her mistakes, so as to analyze them and search for new strategies. According to Cosnefroy (2011), there are four conditions required for taking control of one's own learning: an initial motivation that is strong enough and sustainable, a clear aim, the ability to use different strategies, and the capacity of self-observation. Clearly, WIMS activities do not respond to all these requirements for self-regulation: the regulation is partly external. However, when well designed, training with WIMS fosters motivation and persistence. As the difficulty is chunked, feedback is provided right away, and the correct answer is at hand, the energy needed to persist in learning is lower. Today, the platform does not propose clues for self-observation, but it could be done. Moreover, self-regulation skills can be addressed in class.

3.2 WIMS Design Choices and Affordances from the Teacher's Perspective

Besides being a learning tool for students, WIMS turns out to also be an efficient teaching tool thanks to the affordances for teachers that are presented in this section.

3.2.1 Creating Own Exercises

One of the outstanding features WIMS offers to teachers is the opportunity to create their own exercises, in respect of pedagogical freedom. From an organizational point of view, the only person in charge of a class is the teacher. S/he decides the exercises to propose to students, how to set them, the grading scale. S/he can retrieve and modify the exercises published by others and modify them to adapt them to her/his students and educational goals. S/he can also develop his/her own resources. This makes it possible to adapt to a very wide variety of users.

The teacher can also program feedback in a given exercise. Let us consider the exercise shown in Fig. 1. A type of feedback can be, for example, to systematically refer to the definition of the notion (i.e., interval). A more ambitious and perhaps sometimes more useful type of feedback would depend on the question asked and the student's answer, and would consist in proposing a second question as synthesized in the following table (Table 2).

Programming WIMS exercises may require a good mastery of randomness and didactic variables. Indeed, some exercises may not be random enough, e.g., solve the equation $ax^2 + bx + c = 0$ only in the case where $a = 1$ and b and c are integers such that $b > 2c$ is not a sufficient frame in terms of exploration of the possibilities. The even more particular case where $b^2 - 4c$ is the square of an integer is also not sufficient. It is nevertheless interesting because it allows the setting up of fast procedures, finding for example a particular solution and deducing the other one. At the other end of the spectrum, proposing only to solve $ax^2 + bx + c = 0$ for a, b, c decimal numbers presents a technical difficulty that is not necessary.

Table 2 Feedback depending on the student's answer

Student's answer/expected answer	E is an interval	E is not an interval
E is an interval	**Additional question**: Represent the interval E in the graphic interface	Recalling the definition of an interval and explaining a graphic counterexample
E is not an interval	Recalling the definition of an interval and displaying the expected answer in natural language and graphically	**Additional question**: Give two points, a and b from E, and a point c that does not belong to E and such that a< c<b in the graphic interface

Moreover, if only this variability is proposed, it will almost never allow the student to confront the case $b = 0$ or $c = 0$. But not knowing how to treat these falsely simple cases is detrimental to the control of the resolution of a second degree equation. A student will be exposed to these cases only if the teacher pays attention to them in the setting of the worksheet.

These different exercises are very simple to program; they can be seen as a parameterisation of the same exercise. This parameter can be considered as a didactic variable and a student should be exposed to a sufficient number of examples containing various values of these didactic variables.

For example, the exercise

$$\text{Is the solution set of the inequality } ax^2 + bx + c > 0 \text{ an interval?}$$
$$\text{If so, which one?}$$

asks a more complex question of the student. It can be considered, in terms of Bloom's taxonomy, as a question of synthesis between the resolution of an inequality and the notion of interval. From the programming point of view, this exercise is not much more complex than that of solving the equation of the second degree. The same applies to the exercise

$$\text{What is the definition domain of the function } f(x) = \ln(ax^2 + bx + c)?$$

The answer is indeed the same, but the number of registers to be manipulated by the student is still higher from the didactic point of view. However, from the point of view of programming, this is not the case.

3.2.2 Organizing Exercises into Sheets

A sheet of exercises is composed of a number of exercises that are selected and organized together. This possibility given to teachers contributes to the scenario development of online lessons.

However, the length of a sheet should not be excessive. Our experience shows that a sheet of exercises should not exceed 7–15 exercises. Above this number, the task tends to appear discouraging. Before beginning the sheet of exercises, the student has at his/his disposal a global representation of the sheet with its name and a series of small squares (Fig. 4a). Each exercise is presented by a small square. Thus, visually, the student sees a certain amount of work to be done, which is not overwhelming. The square changes the color to green or red as the student succeeds or fails to solve the corresponding exercise. However, a student may wish to succeed and turn the square to green, but what is hidden behind a square is not a single exercise but an exercise string. This string can be chosen by the teacher to correspond to a unique item of one exercise or to a bunch of several items of exercises. For instance, a string may be parameterized as three random versions of the same exercise, with a

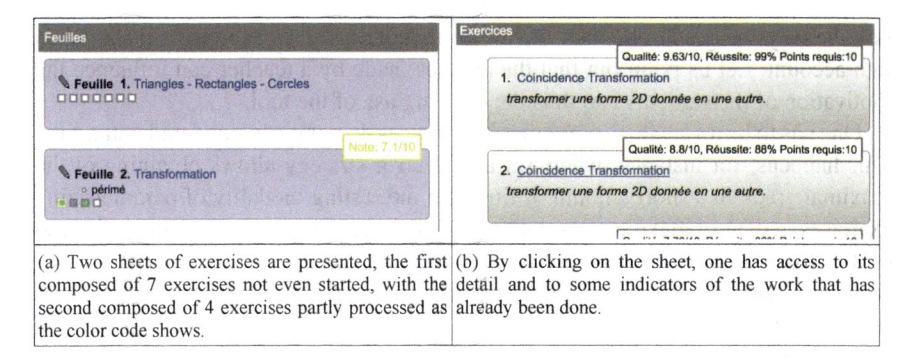

WIMS: Innovative Pedagogy with 21 Year Old Interactive Exercise ... 135

(a) Two sheets of exercises are presented, the first composed of 7 exercises not even started, with the second composed of 4 exercises partly processed as the color code shows.	(b) By clicking on the sheet, one has access to its detail and to some indicators of the work that has already been done.

Fig. 4 WIMS sheet of exercises

grade given at the end of the string. Thus, completing a square cannot be achieved without some persistence. The string of exercises can also be chosen to be gradual, first presenting a lower difficulty, then increasing it. The time needed to complete the whole sheet should not be excessive, neither for good students nor for average students. Yet, giving one or two more challenging exercises may be a good choice. This stimulates students; it creates an opportunity to discuss the exercises between them and in class. Obviously, the sheets should be pedagogically aligned with the scope of the course and with the curriculum. Alignment means that the aim of the WIMS exercises meets the aims of the course and allows the acquirement of skills that will be indeed necessary to pass the final exam. In particular, they should cover at least part of the exam requirements.

Moreover, the work on WIMS is more beneficial if it is clearly articulated in the class and even in the institution. Student engagement is much easier to obtain when the work on WIMS is presented as part of the curriculum. This engagement is sustained if teachers regularly check student's work on WIMS, answer questions addressed by students and take time, especially when an exercise is difficult, to discuss it in class, thus giving the students a feeling of relatedness. Student engagement and efficiency could increase even more when, at some appropriate moments of the course, the teacher gives some meta-cognitive comments on how to use WIMS.

3.2.3 Building the Scenario with Exercise Sheets in WIMS

Of course, the class can be customized, exercises can be imported locally and changed, added, and taken away. It is also possible to restore a class built previously and backed-up. A search engine can also propose full exercise sheets corresponding to a keyword and to a level, and thus whole parts of the class structure can be selected and imported in a fast way. The search engine can also help select exercises one by one. An important task, which has many pedagogical implications, is the choice of evaluation and grading. As we described above, several parameters have to be set up, such as the severity (one mistake with a strong impact) or the grading scale, i.e., the way the grade obtained in several repetitions of an exercise will be taken into account. Let us point out that the choices made by a teacher can influence the motivation of the students and thus the ongoing use of the tool.

One has to be aware that sometimes high grades do not mean good understanding. This happens, for instance, when a trial and error strategy allows obtaining of the maximal grade in a short amount of time. An interesting modality of parametrizing exercises is then to make strings, as explained above. The grade is given at the end of the string. Relying on trial and error to fulfil a string of exercises is no longer a winning strategy: it takes too much time. Taking the time for deep understanding becomes a time gaining strategy.

The teacher can also define the weight of an exercise in his/her exercise sheet, as well as the weight of exercise sheets in the global average. The task of choosing the exercises and the parametrization of a class is an occasion of didactical reflection. Sometimes it is time-consuming. After all, composing a classical exercise sheet can also be long. Of course, if there are no resources corresponding to one's curriculum in WIMS, one has always the choice of developing them. This, however, requires even more time.

The second task is to enroll students. Several modalities are possible. One of them consists simply of providing students with the address of the server and the name and the code of the class (the teacher chooses this code while creating the class). The students can then enroll by creating their private username and password, or the teacher may register the students by creating usernames and passwords. It is also possible to use directly a Central Authentification Service (CAS), for instance the one used in the standard e-learning environment of the school or university.

3.3 Pedagogical Alignment

Following Grubb and Cox (2005), the challenge of a learning environment is to align student needs, teacher approach, course content, and institutional settings. Teachers have at least two tasks: first, to build his/her class resources, and then to enroll students. Creating resources consists mostly in choosing or creating and organizing exercises in sheets of exercises (see above). There are ways for doing this: for instance, if a so called "classe ouverte" (open class) corresponds to the teacher's aim, it can be easily copied and privatized for her own use. Each teacher can profit from the large WIMS community of teachers, as WIMS allows and encourages sharing of exercises. After some review of the code and the content, the exercise enters the common base and is published under a free license. Anyone will then be able to not only use it but also register the code of the exercise in his/her own class and change or modify it.

A sheet of exercises proposes pedagogical progression in a particular institution. When this progression is well designed, the learner enters a flow. S/he is fully absorbed by the activity in a feeling of energized focus, full involvement, and enjoyment. We underline that creating such a sheet design is not easy. As outlined by some studies (Giner & Kobylanski, 2017; Jacquemin, 2017, 2018), during our experimentation at UPEM,[4] we succeeded mostly in the "classe de prérentrée" (revision course before the beginning of a school year). This class corresponded mainly with the revision of basic algebra proposed to students. It is a subject that students had learnt; we now wanted them to train in order to integrate this type of calculation, so as to use it quickly and correctly. The design of this particular class, drawing on Pilet's (2012) Ph.D. research, had the chance to benefit from the expertise of the didactic group directed by Brigitte Grugeon. According to Jacquemin (2017, 2018), students took great pleasure and most of them entered a flow while practicing in this specific WIMS class. The alignment of student needs and teachers' intentions in this particular course was effective.

Moreover, students need to have a feeling of controllability, an important condition for sustaining motivation. It has to be present in the design. In order to achieve this, one has to be consistent in the quantity of work given to students, and consistent with the opening and closing of the sheets, with enough time to get through the exercise. Within such a framework, a student knows how long she needs to complete her work on WIMS and can organize him/herself. Furthermore, depending on the score, one cannot pass to the next exercise without having completed a level defined by the teacher in the current exercise.

3.4 WIMS Analytics

The student, when s/he enters his/her class, sees the sheets (one can have a very precise overview by entering a 'classe ouverte' (open class) on a WIMS server). The sheets are often organized by chapters. At the bottom of the sheet, one sees a toolbar composed of little squares, each corresponding to an exercise (or a string of exercises) that, as we described above, will be green once s/he has completed the corresponding exercise (cf. Figure 4). Thus, in a glimpse of an eye, the learner can see where s/he stands, what s/he has achieved, and what has still to be done. This helps him/her become a "conscious incompetent" (see Table 1), a necessary stage to evolve into "conscious competent". By clicking on one sheet, s/he accesses the list of exercises, while one final click will present them with the exercise, where the work begins. Of course, at any moment, s/he can also consult "my grades" ("mes notes").

The teacher can also see in her class the results of each student (Fig. 5), first by a global average, by an average on each sheet, or detailed in one sheet exercise by exercise. Other statistics of the class are available. One of the very meaningful ones is the difficulty index of an exercise ("indice de difficulté d'un exercice"), which indicates the average number of times necessary to complete the exercise. Clearly, if this indicator is between 1 and 2, the exercise is not difficult. Experience shows that when this indicator is above 3, the teacher should consider explaining the solution to the exercise to the class.

---

[4]Université Paris-Est Marne-la-Vallée, one of the universities in Paris (France).

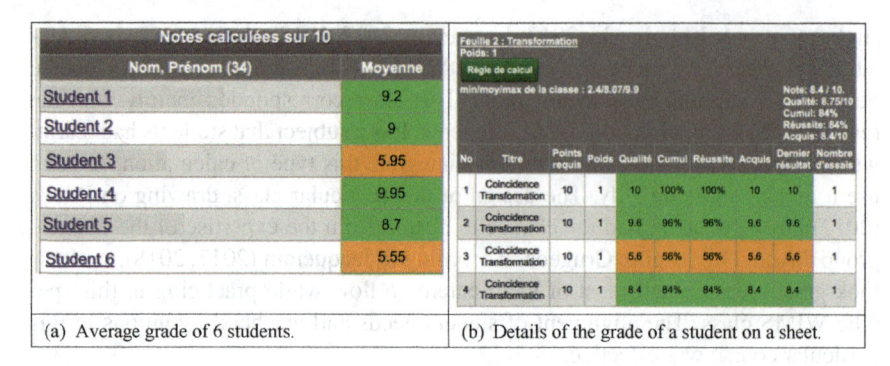

(a) Average grade of 6 students. (b) Details of the grade of a student on a sheet.

Fig. 5 Display of the class grades and of a student grading

The teacher can also set up groups by defining technical variables ("variables techniques"). Groups, such as A, B, and C, can be given specific exercises, with the deadlines set to open or close a sheet specified depending on the groups. Of course, analytics can also be sorted by these variables.

In the next section, we briefly present several empirical studies involving WIMS.

4 Experimenting WIMS

4.1 Survey with Students

In November 2014, the community of Universities Paris Est (UPE) started a project[5] aiming at experimenting WIMS. Working groups were set up to experiment WIMS in pedagogical settings. This project was an opportunity to develop and test the use of WIMS during the first year of university mathematics teaching. There are four major mathematics courses (modules) in the first year of study, each corresponding to 6 ECTS (European Credits Transfer System) as well as to two hours of lectures and three hours of tutorials per week over a period of 12 weeks. The main aim of the project was to build series of exercises corresponding to each learning module.

There exists a huge number of exercises in the common base of WIMS corresponding to the first year of university mathematics and we could begin by relying on these resources. Elsewhere, hundreds of exercises have been developed, especially basic ones, which were not needed. For each chapter of the courses, two sheets of exercises were created, a basic and a standard one. Each sheet is composed of 8–15 strings of exercises. Within the basic sheet, exercises allow the opportunity to directly manipulate elementary notions of a chapter, while the standard sheet aims to propose

---

[5] The project called IDEA was funded by the French Research National Agency (ANR) in the context of "Initiative of excellence in innovative training" (Initiative d'Excellence en Formations Innovantes, IDEFI) and of the Future Investments Programme (Programme Investissements d'Avenir, PIA).

exercises that correspond to the course level. It is important to note that, since WIMS does not train in writing a proof, more time can be spent in the classroom practicing this competency with part of the training done on WIMS. Evaluation of exercise strings was set to high severity, and the best of the string success counted for a grade. The sheets were opened during two weeks; after closure students could continue to practice, but the grade was frozen. We used WIMS as a formative assessment, with its grade counting for a part of the continuous assessment. The final grade was formed by the maximum between the exam mark on the one hand and the average of the mark of continuous assessment and the exam mark on the other hand. The work on WIMS had to be done outside the classroom. When asked, teachers answered questions and used a video projector to address some examples. A tutorship system was organized, at first compulsory during the two weeks before the university year started and then available daily from noon throughout the academic year. Mentorship was provided by students from previous years.

It appears from statistics that students worked on WIMS for two to three hours a week on average. The amount of time spent on WIMS did not depend on the student's level; good students finished all with the maximum grade whereas average students may have had trouble completing all the exercises.

An anonymous survey was proposed to participating students. In the first semester of 2016–2017, we obtained 82 answers from 250 students. The survey dealt not only with WIMS but all aspects of the course. It turns out that users were convinced by the merits of WIMS. More specifically, 84.2% respondents answered that WIMS goals had been reached, while a large majority thought that evaluation was clear and fair. There were some complaints about the time during which the sheets were opened; indeed, at the end of the semester, this information was not provided clearly enough due to some overflow of the teaching staff. However, a large majority of respondents (70.7%) declared themselves satisfied with WIMS. Concerning learning methods, we are faced with students lacking in working methods. This can be inferred from the way they engage in learning: only 7% answered that they opened the course notes or the lecture notes shortly after the course, about 54% opened them while preparing an evaluation and 49% read their notes while working on WIMS. This leads us to think that WIMS may be a tool that fosters working on the course itself. 78% of respondents used scrap paper while completing WIMS exercises; hence, WIMS invites students to mobilize appropriate tools to build their thinking paths and answers. This observation has to be set against the fact that only 36.6% of respondents declared paying sustained attention to the reading of the text of the exercises, while half of the students did not appreciate feedback given by WIMS. Further investigation has to be made to understand why.

According to 83.3% of students, WIMS helped them develop competencies in mathematics. Some gave testimony from which it seems that it is through WIMS that they understood the main concepts in the course and begun to expand their mathematical thinking. According to 43% of students, WIMS also helped them develop meta-competencies. The area that is mostly mentioned is time management. The results of this survey seem to show first that WIMS was a truly effective tool for structuring time during which students had the occasion to mobilize resources of the course. Second, students often indicated that WIMS allowed them to be rigorous in calculus or in reasoning. Third, exercise paths could still be optimized as students stipulated that sometimes the exercises were repetitive and the time required to complete the exercise sheet was sometimes too long.

4.2 Survey with Teachers Using WIMS

A survey[6] was released in the form of an online questionnaire in winter of 2015 and made public to teachers through the WIMS EDU association and published on sites hosting a WIMS server. The questionnaire obtained more than 600 responses including nearly 200 complete answers; these were analyzed and a brief analysis was carried out by the WIMS group during spring of 2016. The questionnaire combined single or multiple-choice questions (6 items) as well as open-ended questions (6 items).

The following results came from the single or multiple-choice questions. The majority of respondents are mathematics teachers (nearly 80%), with 10% physics teachers, and others including French language and natural sciences teachers. 60% of teachers teach at upper secondary school level (Grades 10–12), while 27% are at tertiary level. WIMS is also used at the lower secondary school level (15%) and more rarely in primary school. Unsurprisingly, the users who responded to the survey are more often 'veterans' in the use of WIMS: more than 20% have been using it for 2–5 years and more than 20% for more than 5 years. But a significant part of the answers came from more recent users: 18% using it for 1 year and 12% for 2 years. More than 32% of responding teachers use WIMS weekly, while 24% use it monthly. Almost 21% use it only a few times a year and 8 users (3.8%) say that they use it on a daily basis. More than 12% consider WIMS as "essential" for their teaching and 40% find it "useful", while WIMS is only an additional resource for 25% and "unnecessary" for 3% of respondents. WIMS is mostly used out of school (69%), but also in a school computer lab (one or two students per computer: 53%). Rarer is its use by video projector in front of an entire class (16%). Teachers most often organize exercise sessions (72%) or propose homework (43%) with WIMS. 16% of respondents reported using WIMS for assessment and 11% for viewing interactive courses. Some teachers specified that they use WIMS to familiarize students with definitions and simple exercises. It is also for them a tool for revision or remediation. Some declared that they individualize (or customize) the work thanks to WIMS, and one teacher said they use it in support of a flipped classroom. 35% of respondents specified that they modify WIMS resources before using them with their students and 30% create new WIMS resources. Elsewhere in the survey, open questions addressed strengths and weaknesses of WIMS, among other issues.

Diversity, variety, abundance, and wealth of the bank of exercises are the most appreciated features of WIMS. The variety of levels and types of exercises (mul-

---

[6]https://moin.irem.univ-mrs.fr/groupe-pion/Enquete.

tiple choice, association, graphic exercises) are also mentioned. Among the most frequently cited keywords to highlight strengths of WIMS are autonomy, individualization of learning, and adaptation to the rhythm of each student. Thus, the possibility of implementing a differentiated pedagogy is highly valued, as well as the ease for the student to work at home. Another advantage that appears very often in the answers is the attractiveness of WIMS for students due its playful or innovative character. Its interactivity and immediate feedback (grade and correct answer) are also valued. From the technical point of view, the most cited strength is the existence of random parameters, which allows the renewal, and therefore the repetition, of the same exercise, but also a differentiation of the same exercise according to students' level. An important feature for the teachers is the follow-up (work and achievement) of the students; indeed, the tool includes statistical analysis of the results. Automatic correction, the notion of score, and the possibility of using WIMS for the assessment of students are other features that teachers appreciate.

The two most cited weaknesses of WIMS are its ergonomics (interface) and the difficulty to create or modify exercises (or the time this takes). Some respondents complain about a lack of online documentation, technical support, and even follow-up in teacher training, difficulty in sharing or removing resources, and lack of flexibility in classroom management. Others criticize the lack of resources for certain topics, or for certain levels, especially at vocational or technical high school levels, as well as the difficulty to search for and find exercises for a given theme, and sometimes the redundancy of resources. These opinions point out a lack of a good search engine and of a peer exercise assessment protocol. From a more technical point of view, some respondents sometimes regret a calculation of scores that is not very transparent, or is sometimes perceived as unfair, a lack of feedback, and difficulties in monitoring students.

These results tend to confirm that, despite some critics, which the WIMS developers are attempting to address, WIMS is a pedagogical tool appreciated by both teachers and students.

5 Concluding Remarks and Perspectives

Let us recall that WIMS is more than 20 years old. In 1997, it was a visionary tool, and it is still highly performing, although some parts would gain from being updated. Among the greatest strengths of WIMS, that the teacher survey confirms, are the community built around it and the free and open model. This allows teachers to use and share exercises and possibly whole classes of exercises. We underline that there is something in the values shared that fosters a great engagement with the volunteers developing exercises and the software itself.

The design of WIMS seems to be an indisputable success considering the number of teachers who integrate it into their class and the demands from students asking for the tool to be set up when they have the opportunity to use it with a teacher who manages it well. From the surveys conducted with teachers and students, it also appears that WIMS is a pedagogical resource of high quality created and used by teachers to help and enhance students' learning. The development of resources, however, requires important skills in programming and a significant time investment. Resource sharing was planned fairly quickly, as well as simpler programming languages making resource creation easier. However, the design of sharing facilities is not yet completed, nor the web 2.0 aspect, which is at most embryonic. Thus, the software has an important technical debt, which is mostly due to the fact that its development relies on a small group of software developers including a few computer scientists.

A new national project, "WIMS-evolution", has been proposed at the national level in France and will be funded for two years. It aims to form a reasoned and coordinated strategy for the evolution of WIMS software and communities. First, this project allows us to audit uses of WIMS using sociological methods through numerous teacher interviews and class observation. In parallel, thanks to a synergy created in our university with the designer of the PL platform,[7] we are working on a redesign of the software, preserving and even further affirming the advantage of the free software model. Our motto is 'by teachers for teachers', with the starting point of the design of this platform the central place left to the teacher: it is him/her who is responsible for the design of his/her teaching in a class. A novelty is to encourage the teacher to define the objectives of the course, activity, or exercise in terms of intended learning outcomes. This makes it possible to objectively define the conditions for student success, facilitate pedagogical alignment, and allow better referencing of resources. The new design facilitates the creation of original exercises, taking advantage of contemporary computer developments. What we are looking for is to make it possible and easy to create exercises that allow for varied manipulations and changes of register and allow creativity on the part of students. Finally, we aim at facilitating the teacher's work on selection in the design of his/her teaching. To do this, we need to build criteria to classify exercises and activities.

Another work is under way in order to welcome different communities: teacher communities, by subject matter, level and geographical area, researcher communities (the new software will allow experimentations of POCs[8] (Proof of Concepts), and implementation of POCs in the courant distribution), and last but not least a free community (all will be done to welcome it so that it takes care of part of the soft-

---

[7] Platform first aimed at testing code for the learning of a first programming language. The first language project, launched by Dominique Revuz, aims to provide an easy-to-use self-correction exercise platform, https://github.com/plgitlogin/premierlangage/. The structure of this platform, written in Python and Django and interoperable, fully meets the current standards of software development and free software. It is based on a vision similar to that of WIMS: the teaching continues to be done in class, with pupils and teacher remaining the main actors. The platform allows for individual training with immediate feedback to the student and the teacher. This platform must therefore be able to respond to all tasks already answered by WIMS, in a better way.

[8] A proof of concept (POC) is, in software engineering, the first step in the software implementation process. In the case of software development for teaching, it gives the teacher access to the entire system environment, documentation, and architecture. POC makes it possible to test the software in the real conditions of use, to clarify the needs in terms of development and the expectations in terms of configuration.

ware development). The articulation of these different communities will be achieved through the implementation of an editorial process, the purpose of which is the creation, validation, and proof of the pedagogical effectiveness of teaching resources, in the form of a body of exercises and coherent and well-argued use scenarios available to teachers and students.

An important work is therefore under way concerning the design and architecture of the software, facilitating writing of original exercises, designing educational activities involving sets of exercises, and monitoring the activity of learners.

Acknowledgements This chapter would never have been written without the kind and skilled support of Jana Trgalová et Gilles Aldon. The author expresses to them her warmest thanks.

References

Ambrose, S. A., Bridges, M. W., DiPietro, M., Lovett, M. C., & Norman, M. K. (2010). *How learning works: 7 Research-based principles for smart teaching.* Jossey-Bass.

Berland, C. (2017) *E-learning platform WIMS for bachelor first year courses on electric circuit.* Presented at the European Association for Education in Electrical and Information Engineering (EAEEIE) Annual Conference, June 2017, Grenoble (France).

Bloom, B. S. (1956). *Taxonomy of educational objectives, handbook 1: The cognitive domain.* New York: David McKay.

Brophy, J. (2004). *Motivating students to learn* (2nd ed.). New Jersey: Lawrence Erlbaum Associates Publishers.

Cazes, C., Gueudet, G., Hersant, M., & Vandebrouck, F. (2006). Using e-exercise bases in mathematics: Case studies at university. *International Journal of Computers for Mathematical Learning, 11,* 327–350.

Chappaz, G. (1992). Peut-on éduquer la motivation? *Cahiers pédagogiques, 300.*

Chevallard, Y. (1998). Analyse des pratiques enseignantes et didactique des mathématiques: l'approche anthropologique. In *Actes de l'université d'été Analyse des pratiques enseignantes et didactique des mathématiques* (pp. 91–120). IREM de Clermont-Ferrand.

Cordier, F., & Cordier, J. (1991). L'application du théorème de Thalès. Un exemple du rôle des représentations typiques comme biais cognitifs. *Recherche en Didactique des Mathématiques, 11*(1), 45–64.

Cosnefroy, L. (2011). *L'apprentissage autorégulé: Entre cognition et motivation.* Grenoble: Presses universitaires de Grenoble.

Duval, R. (1993). Registres de représentations sémiotiques et fonctionnement cognitif de la pensée. *Annales de Didactique et de Sciences Cognitives, 5,* 37–65.

Getha-Taylor, H., Hummert, R., Nalbandian, J., & Silvia, C. (2013). Competency model design and assessment: Findings and future directions. *Journal of Public Affairs Education, 19*(1), 141–171.

Giner, E., & Kobylanski, M. (2017). *Retour sur l'expérimentation WIMS.* Research Report. UPEM.

Grubb, W. N., & Cox, R. D. (2005). Pedagogical alignment and curricular consistency: The challenges for developmental education. *New Directions for Community Colleges, 129,* 93–103.

Hersant, M., & Vandebrouck, F. (2006). Bases d'exercices de mathématiques en ligne et phénomènes d'enseignement-apprentissage. *Repères-IREM, 62,* 71–84.

Jacquemin, L. (2017). *WIMS: une ressource comme les autres mais en mieux. Enquête sociologique auprès d'utilisateurs enseignants de WIMS sur leurs usages et pratiques effectives.* Research Report. UPEM.

Jacquemin, L. (2018). *Enquête sociologique auprès des étudiants utilisateurs de WIMS à Marne-la-Vallée.* Research Report. UPEM.

Karpicke, J. D. (2017). Retrieval-based learning: A decade of progress. In J. H. Byrne (ed.), *Learning and Memory: A comprehensive reference* (2nd Ed., pp. 487–514). Elsevier Ltd.

Léon, A. (1972). La motivation chez les élèves de l'enseignement technique. *Psychologie scolaire, 9*, 78.

Pilet, J. (2012). *Parcours d'enseignement différencié appuyés sur un diagnostic en algèbre élémentaire à la fin de la scolarité obligatoire: Modélisation, implémentation dans une plateforme en ligne et évaluation*. Thèse de doctorat, Université Paris-Diderot, Paris 7.

Ruthven, K., & Hennessy, S. (2002). A practitioner model of the use of computer-based tools and resources to support mathematics teaching and learning. *Educational Studies in Mathematics, 49*(2–3), 47–86.

Ryan, R. W., & Deci, E. L. (2000). Intrinsic and extrinsic motivations: Classic definitions and new directions. *Contemporary Educational Psychology, 25,* 54–67.

Viau, R. (2009). *La motivation en contexte scolaire*. Paris: De Boeck.

Viau, R. (2011). *La motivation condition essentielle de réussite* (Edition révisée). In J. C. Ruano-Borbalan (Ed.), *Éduquer et Former* (pp. 113–121). Paris: Éditions Sciences Humaines.

Design and Evaluation of Digital Resources for the Development of Creative Mathematical Thinking: A Case of Teaching the Concept of Locus

Mohamed El-Demerdash, Jana Trgalová, Oliver Labs and Christian Mercat

1 Introduction

Promoting innovation skills and creativity is a central aim of the P21's Framework for 21st Century Learning[1] (2015):

> Learning and innovation skills increasingly are being recognized as those that separate students who are prepared for a more and more complex life and work environments in the 21st century, and those who are not. A focus on creativity, critical thinking, communication and collaboration is essential to prepare students for the future (p. 3).

Likewise, the European Union (EU) considers "creativity, innovation and risk-taking" as part of the key competencies for lifelong learning aiming at personal and social empowerment for EU citizens (EC, 2006). Creative mathematical thinking (CMT) is considered as a highly valued asset in industry (Noss & Hoyles, 2010) and as a prerequisite for meeting current and future economic challenges.

Creative mathematical thinking is seen as an individual and collective construction of mathematical meanings, norms and uses in novel and useful ways (Sternberg,

---

[1] http://www.p21.org/our-work/p21-framework.

M. El-Demerdash (✉)
Menoufia University, Al Minufiyah, Egypt
e-mail: m_eldemerdash70@edu.menofia.edu.eg

J. Trgalová
S2HEP, Université Claude Bernard Lyon 1, Villeurbanne, France
e-mail: jana.trgalova@univ-lyon1.fr

O. Labs
MO-Labs, Ingelheim, Germany
e-mail: oliver@mo-labs.com

C. Mercat
Université Claude Bernard Lyon 1, Villeurbanne, France
e-mail: christian.mercat@math.univ-lyon1.fr

© Springer Nature Switzerland AG 2019
G. Aldon and J. Trgalová (eds.), *Technology in Mathematics Teaching*,
Mathematics Education in the Digital Era 13,
https://doi.org/10.1007/978-3-030-19741-4_7

2003). Exploratory and expressive digital media provide users with access to and potential for engagement with creative mathematical thinking in unprecedented ways (Hoyles & Noss, 2003). Yet, new designs are needed to provide new ways of thinking about and learning mathematics and to support learners' engagement with creative mathematical thinking using dynamic digital media.

The MC Squared European project[2] aimed at contributing in this direction by developing an innovative technology, the so-called C-book technology, to support stakeholders from creative industries producing educational content to engage in collective forms of creative design of appropriate digital media. C-book technology offers an authoring environment for collaborative design of digital resources, called c-books, aiming at fostering the development of creative mathematical thinking. In this chapter, we share our experience with the design of a c-book devoted to the teaching of the concept of geometric locus through which we discuss design choices leading to strengthening CMT potential of c-book resources.

The chapter is organized as follows. We start by discussing what creativity, and especially creativity in mathematics, is and present the conceptualisation of creative mathematical thinking adopted within the MC2 project (Sect. 2). The C-book authoring environment is briefly described in Sect. 3. Section 4 focuses on the design of a specific c-book, called "Experimental geometry", highlighting the design choices and the resource affordances to foster creative mathematical thinking in its users. In Sect. 5, we report the results of two a priori analyses of the c-book CMT potential realised by researchers and by secondary teachers respectively. Concluding remarks summarizing the C-book technology affordances and bringing forward factors stimulating creativity in digital resources collaborative design are proposed in the final Sect. 6.

2 Creativity in Mathematics

In this section, we elaborate on the concept of creativity and especially creativity in mathematics and present the operational definition of creative mathematical thinking adopted within the MC2 project, which constitutes the main theoretical frame of our study.

2.1 Creativity

Hennessey and Amabile (1988) stress the difficulty to capture the essence of creativity in a definition. They highlight a diversity of approaches to this problem of definition and point out two main trends that view creativity either as a process or as a product:

---

[2] http://mc2-project.eu.

> Although many contemporary theorists think of creativity as a process and look for evidence of it in persons […], their definitions most frequently use characteristics of the *product*[3] as the distinguishing signs of creativity (p. 13).

Torrance (1969, 1988) is perhaps the best-known defender of the process approach to creativity. He considers it broadly as a process of approaching a problem, searching for possible solutions, drawing hypotheses, testing and validating, and communicating the results to others. Drawing on previous studies, the author brings to the fore the *novelty* or *newness* as a criterion that is common to most of the definitions of creativity, the novelty being considered at the level of the thinker her/himself: "creative thinking may take place in the mind of the humblest woman or in the mind of the most distinguished statesman, artist, or scientist" (pp. 43–44). According to Reuter (2007), the process approach assumes "that creativity is a trait normally distributed in the population. This assumption implies that every person is creative. The question remains how creative a person is" (p. 80). The process approach therefore leads to the development of instruments to measure creativity, such as the well-known Torrance Tests of Creative Thinking (TTCT), which, drawing on Gilford's work (1950), score the following four cognitive abilities: (1) fluency, i.e., the total number of meaningful and relevant ideas generated in response to a given problem, (2) flexibility, i.e., the number of different categories of relevant responses, (3) originality, i.e., the statistical rarity of the responses, and (4) elaboration, i.e., the amount of detail in the provided responses. The product approach, in contrast to the process approach, sees creativity in "exceptional real-world creative production, which very few individuals are able to achieve" (Reuter, 2007, p. 80).

The process and product approaches echo more recent conceptions of creativity that can be grouped in two categories: the so called "high" or "Big-C" creativity and "ordinary" or "little-c" creativity. "Big-C" creativity refers to productions and/or persons manifesting a non-conventional way of thinking and having a substantial contribution to the advancement of our knowledge of the world. On the other hand, "little-c" creativity considers creativity as a character or a potential all people can display, and which can guide choices and route-finding in everyday life (Craft, 2008). These paradigms are in line with the distinction between absolute and relative creativity (Lev-Zamir & Leikin, 2011), the former relating to great historical (mathematical) works and achievements, while the latter refers to discoveries by a certain person in a specific reference group. Applicable to both paradigms are the definitions by Sternberg and Lubart (2000) who see creativity as the ability to predict non-predictable conclusions that are useful and applicable, or by Tammadge (cited in Haylock, 1997) who defines creativity as the ability to see new relationships between previously unrelated ideas.

---

[3] Stressed by the authors.

2.2 Mathematical Creativity

The issue of defining mathematical creativity is an old and still unresolved one in a sense that "[t]here is not a specific conventional definition of mathematical creativity" (Nadjafikhah et al., 2012, p. 286). As it is the case of creativity in general, some conceptualisations of mathematical creativity focus rather on the process while others emphasise the product.

Along the process line of thought, Hadamard (1945) takes Wallas' (1929) model of creative process based on the accounts of famous inventors and mathematicians, in which four stages can be distinguished: (1) preparation consisting in investigation the problem at hand from all perspectives, (2) incubation that is a period of unconscious processing during which no direct effort is exerted on the problem, (3) illumination that is a sort of flash of insight leading to a new idea, and (4) verification that is a conscious and deliberate effort of testing the validity of the idea. Liljedahl (2013) extends Wallas' and Hadamard's model by adding the AHA! experience phenomenon. Ervynck (1991) sees mathematical creativity as the ability to solve problems and/or to develop thinking in structures, considering the peculiar logic-deductive nature of the discipline. Liljedahl and Sriraman (2006) refer to it as (1) the process resulting in an unusual (novel) and/or insightful solution to a given problem, and/or (2) the formulation of new questions and/or possibilities that allow an old problem to be regarded from a new perspective.

The product approach to creativity focuses on the outcomes that result from creative processes. It assumes that, in order to deem a process or activity as creative, one has to discern the existence of some creative outcome. An example is the suggestion by Chamberlin and Moon (2005) to see creativity as the generation of novel, desired and useful solutions to (simulated or real) problems using mathematical modelling.

Considering creative process that leads to creative products, it is worth raising the question whether there is a considerable input of mathematical knowledge to the development of mathematical creativity. Mann (2006) argues that there is a strong relation between mathematical experience (knowledge and abilities) in a school setting and mathematical creativity. On the contrary, Sriraman (2005), among others, emphasizes that there is not necessarily a relationship between mathematical abilities and creativity, implying thus that mathematical creativity can be developed in students if properly supported. Likewise, Silver (1997) sees creativity as a disposition toward mathematical activity that can be fostered in the school population. This view suggests that teaching toward creativity might be conducive for a broad range of students, and not merely for gifted individuals.

2.3 Creative Mathematical Thinking in the MC2 Project

In the MC2 project, we have adopted a 'little-c" creativity paradigm leading us to assume, in line with Silver (1997), that mathematical creativity can be developed in students through appropriate learning situations. Based on this assumption, we first agreed upon a definition reflecting our vision of creative mathematical thinking (CMT) that defines it as an intellectual activity generating new mathematical ideas or responses in a non-routine mathematical situation. Drawing on Guilford's (1950) model of divergent thinking and Torrance TTCT, we consider that the process of generation of new ideas shows the abilities of fluency (ability to generate many responses to a problem at stake), flexibility (ability to generate different categories of responses), originality (ability to generate new and unique ideas that are different from those others have produced), and elaboration (ability to provide details in a response or to redefine a problem to create others by changing one or more aspects). We then searched for conditions and characteristics of situations likely to foster the development of CMT in students. Drawing on research results, we agreed upon the following characteristics of situations or problems that we deem appropriate to engage students in creative mathematical activity:

- Situations based on the interplay between problem-posing and problem-solving (Silver, 1997);
- "Problematic situations" serving as the context for learning (Torp & Sage, 2002) or open-ended situations that are not solved easily and require a combination of various approaches and knowledge (El-Demerdash & Kortenkamp, 2009; El-Demerdash, 2010; Trgalová, El-Demerdash, Labs, & Nicaud, 2018);
- Students seen as active problem-solvers and learners; teachers acting as cognitive and metacognitive coaches (Torp & Sage, 2002);
- Social interactions in problem-solving processes (Sriraman, 2004);
- Intrinsic motivation implying enjoyment and own interest in engaging in the mathematical activities (Hennessey & Amabile, 1988).

In this chapter, we address the following research question: what affordances of the C-book technology can be exploited in the design of resources intended to enhance CMT in students? To bring to the fore such affordances, we present, in the next section, one of the resources designed with the technology, discuss the design choices and highlight the affordances that made them possible.

3 C-Book Technology and C-Book Resources

3.1 C-Book Authoring Environment

MC Squared project allowed to design and develop an innovative software system, the so-called C-book[4] environment (Fig. 1). This environment provides an authorable tool including diverse dynamic widgets, an authorable data analytics engine and a tool supporting collaborative design of resources called "c-books". The project aimed at

---

[4]C-book (capital C) designates the authoring environment, whereas c-books (lowercase c) designate resources designed in the C-book environment.

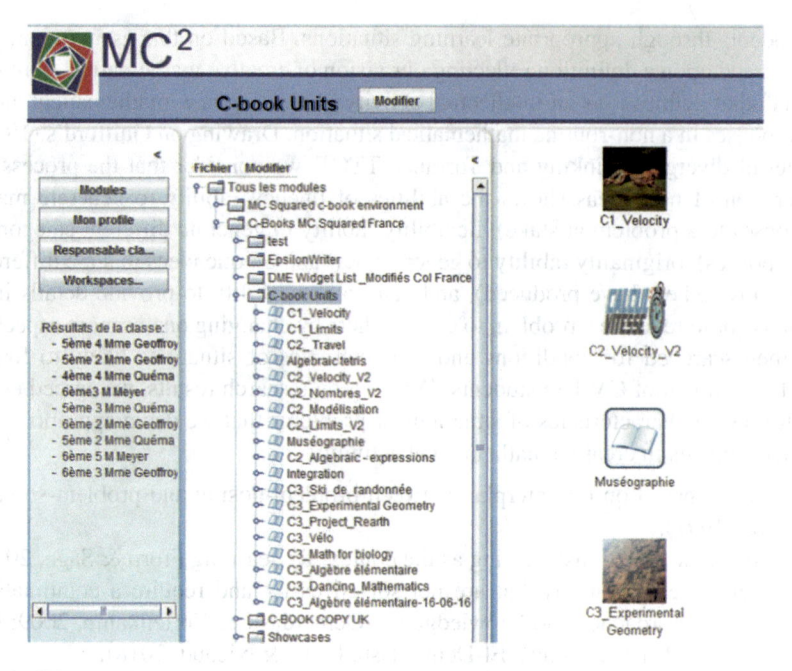

Fig. 1 C-book technology environment

studying the processes of collaborative design of c-books intended to enhance CMT in students.

The authoring environment of the C-book technology has three main elements integrated together to build its infrastructure:

- **C-book widgets** that are small pieces of dynamic software (Fig. 2), which can be included into the c-book resources to allow interactive content.
- **C-book widget instances** that are widgets inserted into a c-book page. Many of the widget instances can still be configured by the c-book author to fulfil the specific needs of the page. For example, to visualize a graph of a function, the c-book author can specify the ranges to be used, etc.
- **C-book widget factories** are software systems, often developed independently of the C-book environment over many years, allowing to produce C-book widgets easily. Examples of C-book widget factories are GeoGebra, Cinderella or EpsilonWriter (Fig. 2).

An advantage of the fact that the C-book environment comes with a set of widget factories is that it is easy for a c-book author to create new widgets for the specific needs of the c-book she is currently developing. The quickest way to do this is by adapting existing widgets to specific needs of a currently written c-book. For example, a widget allowing certain geometric constructions by changing the available tools, add some geometric objects, etc. But in addition to this, a c-book author can develop a completely new widget from scratch using one of the widget factories; it will

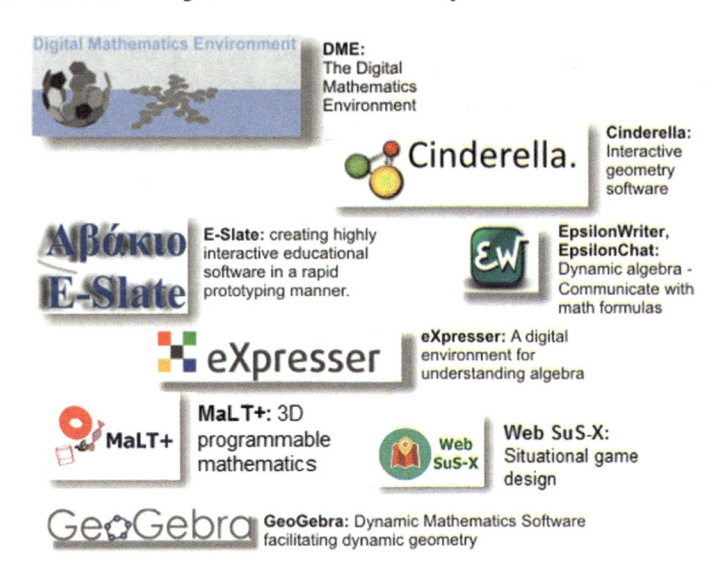

Fig. 2 Some of the c-book widget factories

automatically work within the C-book environment via the interfaces implemented on both sides.

3.2 C-Book Resources

Resources that the C-book technology allows to create are called c-books (c for creative). They consist of pages that can embed text, several interoperable interactive widgets, links towards external online resources, videos etc. Their design within the MC2 project aimed at the development of CMT in students as their end-users. More than 60 c-books[5] have been created over the three years of the project (Fig. 3).

4 Design of the "Experimental Geometry" C-Book

The "Experimental geometry" c-book is devoted to the teaching and learning of geometric locus of points. In the following section, we explain why this concept has been chosen.

---

[5]The c-books can be accessed at the MC2 website, http://mc2-project.eu/index.php/technology-and-production/c-books.

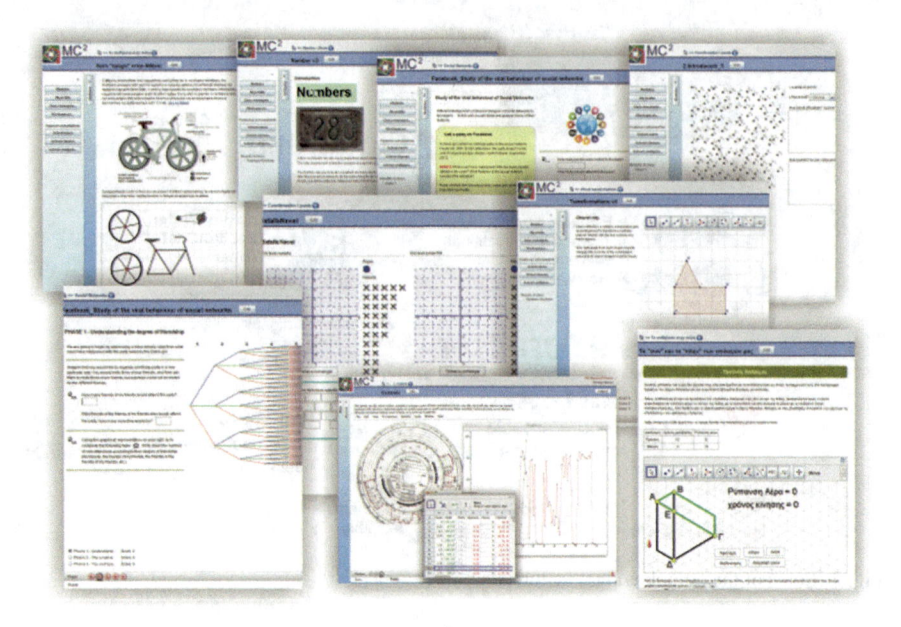

Fig. 3 Examples of c-book resources produced in the framework of the MC2 project

4.1 Geometric Locus of Points

Historically, mathematical curves essentially occurred as loci, e.g. a parabola as the locus of all points having the same distance from a given fixed line and a given fixed point outside the line. The concept of a curve within a coordinate system was only quite recently developed by Descartes, Fermat, and other mathematicians in the 17th century—about 2000 years after the Greeks had performed quite detailed studies of various curves and had created interesting and important new curves for specific purposes, purely via their description as loci (Boyer, 2012).

The curves in Descartes' "La Geometrie" (1637, see Boyer, op. cit., pp. 74–102) then arise naturally as implicit curves, as a result of solving systems of implicit equations where each equation represents a condition on the geometric objects involved. In the subsequent centuries, curves and the special cases of graphs of functions in one variable were being studied deeply. But at the beginning of the 20th century, when Felix Klein was working on the question of how to teach mathematics, implicit equations still played a very important role for him. For example, in the first section on algebra in his "Mathematics from a Higher Standpoint" (Klein, 1924, part I, pp. 93–109), where he discusses simple examples such as graphs of quadratic functions, he immediately uses implicit equations as well, namely some discriminants describing the reality of the roots of the functions. It was only later that some others misinterpreted his ideas to focus on the importance of functions in a too narrow way, namely only to functions from R to R.

Besides the historical importance of loci and their often implicit equations, there are many reasons for using them at school level. The following is an important one: The description of a curve as a locus gives a more intrinsic description and a more operational description than an equation. Thus, even for curves appearing as graphs of functions, looking at them as geometric loci often deepens their understanding. Moreover, we have experienced that using loci and implicit equations early in teaching is a good preparation for studying implicit equations later in linear algebra, e.g. planes and the classification of quadratics. It is also a good companion to the implicit equation of a circle which is otherwise a quite isolated example of such an equation in many cases (sometimes, ellipses are also mentioned, at least in their standard form $x^2/a^2 + y^2/b^2 = 1$, but the fact that implicitly described curves are a most natural thing to consider is rarely mentioned).

The importance of implicit loci arises even more at the undergraduate level: for example, a curve might be associated not only with an explicit equation, a function graph, a parametrized curve, or an implicit algebraic equation but also with solutions of differential equations. It is important to understand function graphs, implicit equations and parametrized curves as loci in order to be able to fully grasp differential geometry tools such as tangent line or plane, curvature, osculating circles or ellipsoids. In real life mathematics or engineering, most objects are loci of some sort. Control theory for example deals with trying to keep a mobile position not far from a target trajectory, with the help of integral and differential calculus. The investigation of soft loci[6] with a dynamic geometry system (Healy, 2000; Laborde, 2005) is very helpful in building this picture in the mind of students.

The flexible production of loci is of paramount interest in industry and design to define curves and surfaces used in computer aided design, such as Bézier curves and their variants (see Peigl & Wayne, 2013). The main feature of those is the fact that they can be described in many different ways: as parametrized curves, as implicit curves, and also as loci. All those descriptions have their advantages for the application at hand such as: Through parameterization, many properties of the curves can be studied easily; with the implicit equation, it is straightforward to decide if a given point is on the curve, or on one side or the other side; and the description as a locus provides a numerically robust and quick way to compute points on the curve, draw it or 3D-print it.

---

[6]Dynamic geometry allows obtaining a locus of points with a corresponding functionality that makes the locus appear at once. This way of constructing a locus may hinder the point-wise perception of the curve. A soft locus, that is the trace of a (free) point dragged by the user attempting to preserve the condition that defines the locus while dragging, allows generating the locus point by point under the control by the student, which helps make sense of it. An example of a soft locus is shown in Fig. 5a: a circle obtained as a trace of a (free) point M that is dragged in a way to keep it at the same distance from a given point O.

4.2 Presentation of the "Elementary Geometry" C-Book

The notion of geometric loci of points is the topic of the "Experimental Geometry" c-book presented in this section. According to Jareš and Pech (2013), this notion is difficult to grasp, and technology can be an appropriate media to facilitate its learning. The authors suggest using dynamic geometry software to "find the searched locus and state a conjecture" and a computer algebra system to "identify the locus equation".

The challenge in designing this c-book was to exploit C-book technology affordances to propose a comprehensive study of geometric and algebraic characterization of some loci within the c-book. We decided to create activities aiming at studying loci of important points in a triangle. These loci (for example locus of the orthocentre) are generated by the movement of one vertex of a triangle along a line parallel to the opposite side (see Fig. 4). These are classical problems from the field of geometry of movement that were proposed for teaching purposes even before the advent of dynamic geometry (Botsch, 1956). Elschenbroich (2001) revisits the problem of locus of the orthocentre in a triangle with a new media, dynamic geometry software. El-Demerdash uses this example to promote CMT among mathematically gifted students at high schools (El-Demerdash, 2010; El-Demerdash & Kortenkamp, 2009).

The c-book invites students to experiment geometric loci generated by intersection points of special lines of a triangle while one of its vertices moves along a line parallel to the opposite side (see Fig. 4). The activity can give rise to several various configurations, which makes it a rich situation for exploring, conjecturing, experimenting, and proving.

The c-book is organized in three sections, each offering an independent mathematical activity.

The first section proposes the main activity called 'Loci of special points of a triangle'. It starts by inviting the students to explore, with a widget created from the Cinderella factory,[7] dynamic geometry software, the geometric locus of the ortho-

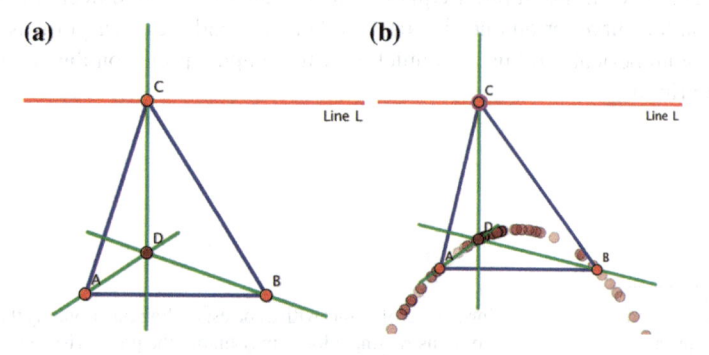

Fig. 4 a Geometrical situation proposed with Cinderella (Act. 1, page 1), b visualizing the trace of D while C moves on the red line (Act. 1, page 2)

---

[7] http://www.cinderella.de.

center of a triangle while one of its vertices moves along a line parallel to the opposite side (Fig. 4a). The students are asked to explore the situation, formulate a conjecture about the geometrical locus of the point D (page 1), and test the conjecture (page 2) by visualizing the trace of the point D (Fig. 4b). On page 3, the students are asked to find an algebraic formula of the locus, which is a parabola. The formula is to be written with a widget created from the EpsilonWriter factory,[8] dynamic algebra software, and the interoperability between the two widgets allows the students to check whether the provided formula fits the locus or not. The next pages invite the students to think of, explore, and experiment the geometrical loci in other similar situations, such as the locus of the circumcenter (intersection of the perpendicular bisectors), the in-center (intersection of the angle bisectors) or the centroid (intersection of the medians). Other situations can be generated by considering the intersection of two different lines, for example a height and a perpendicular bisector. Twelve such situations can be generated. For each case, one page is devoted offering to the students:

- Cinderella widget with a triangle ABC such that the vertex C moves along a line parallel to [AB] and a collection of tools for constructing intersection point, midpoint, line, perpendicular line, angle bisector, locus, as well as the tool for visualizing the trace of a point;
- EpsilonWriter widget enabling a communication with the Cinderella widget;
- EpsilonChat widget enabling remote communication among students.

The second section called 'The concept of geometric locus' introduces the concept of locus of points. It starts by the activity leading the students to "discover" the fact that a circle can be characterized as a locus of points that are at the same distance from a given point (page 1). The students first experiment a 'soft' locus (Healy, 2000; Laborde, 2005) of a point A placed at the distance 6 cm from a given point M (Fig. 5a): they are expected to drag the point A while attempting to maintain the distance of 6 cm from the point O. Then they verify their conjectures by observing a 'robust' construction of the circle centered at A with a radius 6 cm (Fig. 5b), i.e., they create a point A at 6 cm from the point O and drag it; unlike the soft locus, in this case, the position of A is controlled by the geometric constraint.

The next page is constructed in a similar way to allow the students to explore perpendicular bisector as a geometric locus of points that are at a same distance from two given points. Finally, the page 3 proposes a synthesis of these two activities and provides a definition of the concept of geometrical locus of points.

The third activity, 'Algebraic representation of loci', proposes a guided discovery of algebraic characterization of the main curves that can be generated as loci of points as those in section 2 of the c-book.

---

[8]http://epsilonwriter.com/en/.

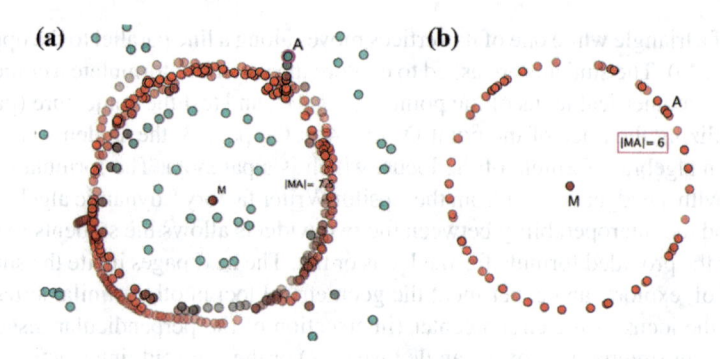

Fig. 5 Circle as a locus of points that are at a given distance from a given point, **a** 'soft' locus, and **b** 'robust' locus

4.3 Design Choices and Rationale

4.3.1 Personalized Non-linear Path

The c-book is designed to allow students going through it according to their knowledge and interest. They are invited to enter by the main activity in section 1 of the c-book. However, the concept of geometric locus is a prerequisite. In case this knowledge is not acquired yet, or the students need revising it, they can reach the section 2 via an internal hyperlink from various places of the main activity. Similarly, section 3, which allows the students learning about the algebraic characterization of some common curves, is reachable from the main activity. Thus, the students can "read" the c-book autonomously, in a non-linear personalized way, depending on their knowledge about geometric or algebraic aspects of loci of points according to their needs.

4.3.2 Promoting Creative Mathematical Thinking

The c-book is designed in a way to support the development of creative mathematical thinking through promoting its four components (fluency, flexibility, originality, and elaboration[9]) among upper secondary school students. First, the main activity, as it is designed, calls for students' *elaboration*: they are invited to modify the initial situation by considering various combinations of special lines in a triangle, whose intersection point generates a locus to explore. *Fluency* and *flexibility* are fostered by providing the students with a rich environment in which they can explore geometric situations and related algebraic formulas while benefitting from feedback

---

[9]In MC Squared project, we adopted the four components of the CMT in this order because of the CMT assessment. Indeed, fluency, i.e., the total number of solutions or approaches to solve a problem, is the first component to be evaluated. These solutions or approaches are then classified into categories, which gives flexibility. Among these categories, rare ones allow evaluating originality.

allowing them to control their actions and verify their conjectures (see learning analytics below). Specific feedback is implemented toward directing students to produce different and varied situations and help them break down their mind fixation by considering yet different configurations, such as two different kinds of special lines in a triangle passing through the movable vertex (e.g., a height intersecting with an angle bisector), and then the intersection of two different lines that do not pass through the movable vertex. The c-book provides the students not only with digital tools enabling them to explore geometric and algebraic aspects of the studied loci separately, but also with a so-called 'cross-widget communication' affordances of Cinderella, dynamic geometry environment, and EpsilonWriter, dynamic algebra environment, which makes it possible to experimentally discover the algebraic formula that matches the generated locus in a unique way; this feature may contribute to the development of original approaches by the students (*originality*).

4.3.3 Constructivist Approach

The c-book activities in sections 2 and 3 are developed based on the constructivist learning theory principles enabling students to create new experiences and link them to their prior cognitive structure supported with learning opportunities for conjecturing, exploring, explaining and communicating mathematics. The feedback drawing on learning analytics (see below) is designed to allow students solve the proposed activities autonomously and thus construct the target knowledge.

4.3.4 Meta-cognition—Learning by Reflecting on One's Own Action

All c-book sections end up with a meta-cognitive activity that has been designed to encourage students to reflect about their own activity and learning and enable them further to understand, analyze and control their own cognitive processes. These activities have also been designed to develop students' written mathematical communication skills using EpsilonChat, a widget for communicating mathematics.

4.3.5 Multiple Representations of the Loci

An outstanding feature of the C-book environment is the fact that it does not only come with many existing widgets in the mathematical context from several different European developer teams, but it also comes with so-called widget factories, one from each of the developer teams allowing authors to generate their own specialized widgets, if they want. The interesting point of this is that all these diverse widgets work perfectly together with the back-end of the environment and they can even collaborate with each other within pages. For example, the dynamic algebra system EpsilonWriter is an interesting tool for manipulating formulas and equations via a unique drag and drop interface (right part of Fig. 6). But it neither has a built-

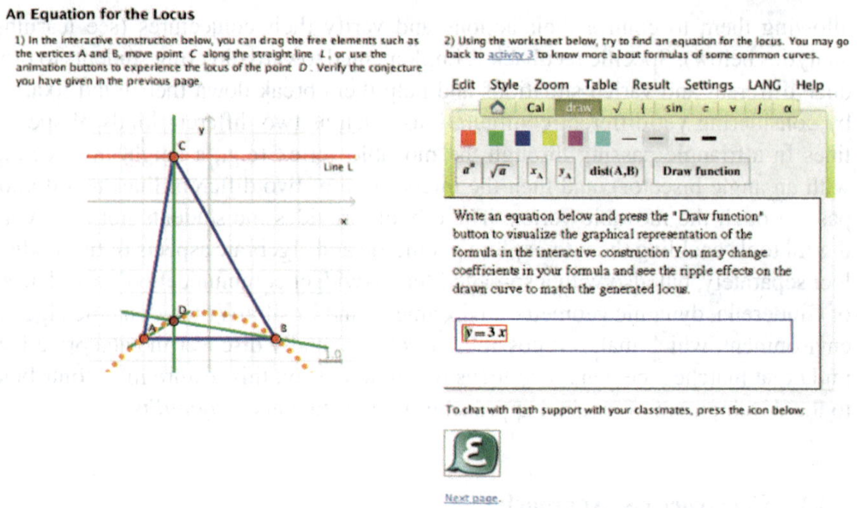

Fig. 6 A screenshot of a c-book page showing three widgets: Cinderella, EpsilonWriter and EpsilonChat

in function graphing tool nor geometric construction capability. These aspects are some of the specialties of the programmable dynamic geometry system Cinderella (left part of Fig. 6).

Later, when working with the c-book, a student may have produced a reasonable equation for a function within EpsilonWriter, and she can visualize it by using the 'draw' tab. The graph of the function will be shown in the Cinderella construction at the right. For the student, this is visually clear and intuitive; but technically a lot is happening in the background. First, the equation will be sent from the Epsilon-Writer software via a standardized protocol to the c-book environment and from there to the Cinderella software, which finally visualizes it as a part of the interactive construction. All this is possible within the c-book player running in a web-browser.

As the example above illustrates, cross-widget communication is a quite powerful feature. In this case, it opens the opportunity for the c-book author to make explicit connections between different representations of a mathematical object: a curve represented as a geometric locus, its formula or equation with the ability to modify it dynamically, and a geometric figure combining both the construction as a locus and the visualization of the curve given by the equation. Within the C-book environment, such opportunities exist in other branches of mathematics as well, e.g., via this mechanism statistics and probability widgets may be connected to geometry, algebra, a number theory widget or even to a logo programming widget, to name just a few more use cases.

4.3.6 Learning Analytics and Feedback

One of the important aspects in the design of this c-book is to decide which of the student's activities should be logged to a database while she is studying the c-book (Labs, 2015). There have been many different types of logs implemented in this c-book. These logs enable the teacher to capture the student's path in studying the c-book, e.g., whether the student starts from the c-book main activity, what pages she goes through while studying the c-book, how far she goes through the additional two activities, whether she goes back and forth through the c-book pages and activities and when, whether she uses the provided internal and external hyperlinks to look for further information, how she uses the available hints and how many levels of hints etc.

Moreover, logs were implemented to trace the student's trails or attempts while she is using the provided Cinderella tools to construct a configuration to elaborate the given problem situation: the time the student spends on each page and each activity as an indicator of motivation; the number of student's trials for each page and each activity of the c-book; the student's use of EpsilonChat as a social aspect of creativity and collaborative work with others whether in pairs or groups.

Two types of feedback are provided to students, while they are studying the c-book to guarantee their smooth move from page to page and switch between the c-book activities: mathematical or educational feedback and technical feedback. Mathematical or educational feedback includes hints and comments oriented toward solving the given problem or developing creative mathematical thinking. This type of feedback is in the form of a message sent in a pop-up window, of a hyperlink or of an internal link. Technical feedback aims at helping students master the available widgets so that technical issues do not become obstacles to the problem-solving processes. This type of feedback is in the form of hints or instructions about how to use Cinderella or EpsilonWriter provided tools, or hints regarding the use of cross-communication between the two widgets.

5 A Priori Evaluation of the C-Book

The "Experimental geometry" c-book was evaluated twice. The first evaluation focused on the c-book CMT potential and was done by experts, persons involved in the MC2 project but not engaged in the c-book design. The second evaluation aimed at gathering users' opinions about the c-book. This evaluation was done by secondary mathematics teachers. For both evaluations, we present the methodology and the main results.

5.1 Evaluation of the CMT Potential

5.1.1 Methodology

The c-book CMT affordances were evaluated by three researchers involved in the MC2 project. This a priori evaluation was guided by the following two research questions:

RQ1—Which of the four cognitive components of CMT: fluency, flexibility, originality and elaboration, and social and affective aspects have been better integrated and promoted through the design of the c-book? That is, what affordances are perceived by the evaluators as enhancers of these components?

RQ2—Is there any correlation among the cognitive components of CMT, as perceived by the evaluators?

In order to answer the above research questions, we used an evaluation tool called the "CMT affordance grid" (see Appendix 1). This tool was developed and refined within the MC Squared project (for more details, see Trgalová, 2016). The grid contains three sections. The 13 first items aim at evaluating the c-book affordances towards the development of mathematical creativity in users/students. These items address the c-book affordances such as nature of the activities or variety of representations of mathematical concepts at stake and ask the evaluators to what extent these affordances are likely to enhance the user's cognitive processes (fluency, flexibility, originality, elaboration). The second and third sections deal with social and affective aspects of the c-book that are likely to impact the users' intrinsic motivation and thus enhance their mathematical creativity.

As for the first aspect, the responders were asked to evaluate the items in relation to each one of the four cognitive components of mathematical creativity in a scale from 1 (weak affordance) up to 4 (strong affordance). There was an extra option called N/A in case the affordance was not applicable for the specific item.

The evaluation of the mathematical creativity affordances of this c-book was done by three experts in the field of mathematics education, a senior researcher, a postdoctoral researcher, and a Ph.D. student who were not involved in its design. It was organized in three steps. First, the evaluators had to play with the c-book to get acquainted with its affordances. Second, a teleconference was organized by the main designer of the c-book to address possible evaluators' needs for understanding and further clarification. Third, the evaluators rated the c-book affordances based on the grid using an online form prepared for this purpose.

6 Results

The chart, shown in Fig. 7, represents the evaluation of the cognitive components of CMT from the experts' point of view. The height of the bars represents the mean value of each component (fluency, flexibility, originality and elaboration), while the

Design and Evaluation of Digital Resources for the Development ...

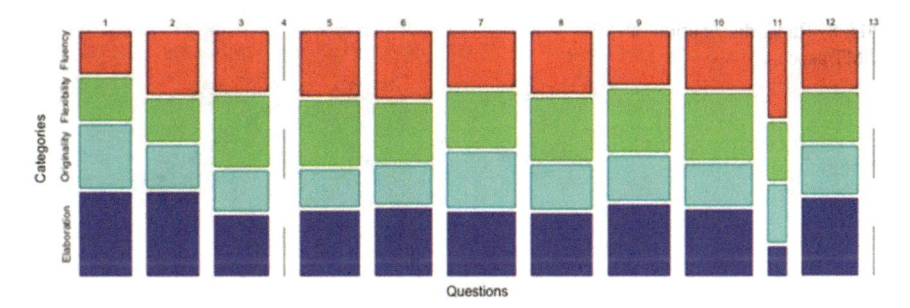

Fig. 7 Evaluation of CMT cognitive components from the experts' point of view

Table 1 CMT evaluation summary

Fluency	Flexibility	Originality	Elaboration	Social	Affective
2.53	2.46	1.96	2.92	2.3	1.6

thickness represents the mean between the four aspects for each question. From the evaluators' point of view, there are no affordances on the items 4 and 13, which means that the c-book does not establish connections between different knowledge areas and mathematics (item 4) and it does not include half-baked constructs that call for intervention (item 13). On the other hand, the evaluators consider that the c-book encourages exploratory activity and user experimentations (item 7) and encourages also generalizing mathematical phenomena, going from concrete cases to general ones or generalizing real world phenomena using mathematics (item 10).

In Table 1, we present the quantitative data for each component computing the mean from No Affordance (scored 1) to Strong Affordance (scored 4). From the scale defined to evaluate the c-book we got the following values for each component, as shown in the Table 1 above: Fluency = 2.53, Flexibility = 2.46, Originality = 1.96 and Elaboration = 2.92. Except the originality component, all other components are in the range of "weak to possible" affordances. The originality got a value of 1.96 which means "no affordance". However, the value is quite close to "weak" affordance.

The highest value for this c-book in terms of cognitive aspects was elaboration for which the value achieved the rank of "good affordance". It means that, in general, the c-book is judged to have a potential to boost the students' development of their ability to redefine a problem to create others by changing one or more aspects. Fluency and flexibility are the components with lower values, meaning that the evaluators perceive a c-book slight potential to foster students' ability to provide many responses (fluency) or to come up with diverse strategies to solve a mathematical problem or challenge (flexibility).

The radar chart (Fig. 8) shows the distribution of the evaluation among the evaluated categories. This chart shows which component of CMT is most likely to be

Fig. 8 Radar distribution of CMT aspects

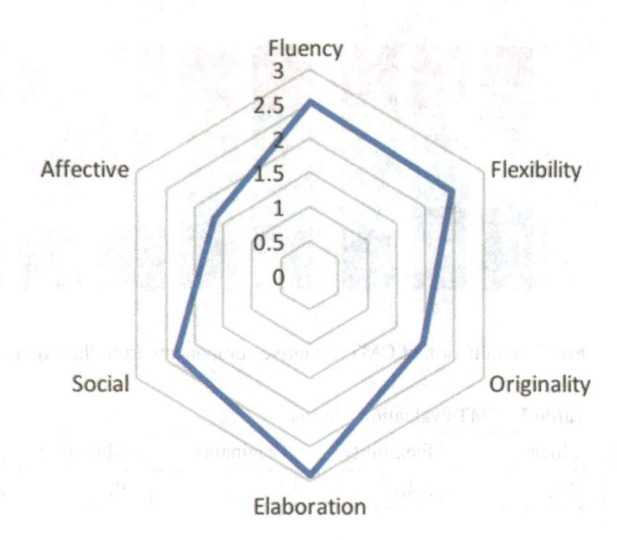

Table 2 Correlation values of CMT components

	Fluency	Flexibility	Originality	Elaboration
Fluency	1.00	0.94	0.83	0.89
Flexibility	0.94	1.00	0.82	0.84
Originality	0.83	0.82	1.00	0.92
Elaboration	0.89	0.84	0.92	1.00

enhanced using the c-book. In the case of this c-book it is the elaboration aspect, followed by fluency and flexibility.

Table 2, collating the 13 questionnaire items, shows correlations among the four cognitive components of CMT. We can notice that the correlations are strong between some cognitive aspects. It means that, considering a significant value of $r > 0.80$ ($p = 0.05$), we may conclude that fluency, flexibility and elaboration can be fostered at the same time. In the case of originality, there is no statistical evidence that supports the hypothesis that this component can be fostered by the other ones.

We can conclude that the experts seem to perceive the CMT affordances the designers were aiming at. Thus even though the c-book main activity is designed to call for students' elaboration (they are invited to modify the initial situation by considering various combinations of special lines in a triangle, whose intersection point generates a locus to explore), fluency and flexibility are fostered by providing the students a rich environment in which they can explore geometric situations and try out algebraic formulas whereas benefitting from a feedback system allowing them to control their actions and verify their conjectures. Specific feedback is implemented toward directing students to produce different and varied situations and help them to break down their mind fixation by considering yet different configurations.

6.1 Evaluation by Teachers

6.1.1 Methodology

An empirical study in a form of case studies was conducted with four secondary mathematics teachers to gather their opinions about the c-book, its potential to develop creativity in students and its relevance for a classroom use. A questionnaire (see Appendix 2) was designed for this purpose aiming at gathering (a) teachers' representations of creativity in mathematics, and (b) their opinions about the c-book.

6.1.2 Results

Regarding the c-book affordances, the teachers appreciate the opportunity the c-book creates for experimenting, conjecturing, testing conjectures, which, according to them, promotes creativity:

T1   In my opinion, the fact that the c-book promotes the mode "Conjecture/Validation" fosters creativity without generating fear of error for the student (especially as she is facing a computer rather than a teacher). [...] DGS promotes experimentation if the student is not able to conjecture ... she will play with elements she chooses in order to observe the impact which will feed into new ideas and thus foster her creativity.

They also value the freedom the students are given in exploring the c-book activities, the articulation of various representations, as well as the possibility of collaboration among students:

T2   Freedom of strategies, communication and collaboration [...] the combination of different mathematical fields.
T3   If the students had worked all together, they could have chatted "in live", that can be interesting.

These responses tend to show that the teachers vouch the design choices. However, the teachers also show concerns related to potential students' difficulties:

T1   [Soft locus] seems to me hardly exploitable by the students because it is very difficult to keep with the mouse the point A at 6 units from the fixed point M in the case of the circle and the point P at the same distance from the endpoints of the line segment.
T2   Regarding the equation of the circle, it bothers me to have to write the equation in the form of $y = f(x)$ as the circle does not represent a function [...] Grade 10 students will never succeed to guess what this equation may look like since they do not know at all the equation of geometric object.

Feedback features, such as colour scaffolding in the case of soft locus (see Fig. 5a), were implemented to handle such difficulties; however, the teachers' opinions make us think more in-depth what appropriate scaffolding could be in such cases. Nevertheless, the teachers' feedback seems to be rather positive on the interest of using the c-book in a classroom.

7 Conclusion

In this chapter we wanted to share our experience with the design of digital resources aiming at fostering the development of creative mathematical thinking in students, as well as a few reflections on mathematical creativity and the role of technology in promoting creative thinking in mathematics.

Creativity is one of the key competencies in lifelong learning and is highly valued in professional and social spheres. Research has shown that creativity is not just a gift a few talented individuals have, but everyone can manifest creativity at her own level. School has therefore an important role to play in helping pupils and students become creative.

MC2 project explored ways of designing technology and related digital resources that would empower teachers as creative designers of appropriate resources engaging students in activities that invite creative mathematical thinking. Innovative technology has been designed allowing for producing resources offering to students a rich exploratory environment with carefully devised scaffolding supporting students' learning mathematics as well as their creative approach to problems at hand. The use of such resources in classroom requires rethinking didactic contract favorable to collaborative students' work on open-ended problems (Emin et al., 2015). Moreover, to assess the development of CMT in students while using the c-book, a specific methodology is required for measuring the four CMT components (fluency, flexibility, originality, and elaboration). As the MC2 project focused on social creativity among the c-book designers and the evaluation of the CMT potential of the produced c-books by researchers, this remains an open avenue for a continuation of the project.

The c-book presented in this chapter is the result of a collaborative work of a group of designers coming from various professional backgrounds, as the group comprises researchers in mathematics, mathematics education and computer science, as well as educational software developers (Trgalová et al., 2016; El-Demerdash et al., 2017). Without the synergy among those group members, several design choices would have remained in a hypothetical state, namely the technological advances in terms of cross-widget communication and learning analytics features. The design of the c-book has thus become a driving force in the c-book technology development, and in return, the unique c-book technology features enabled the creation of a resource with affordances promoting creative mathematical thinking.

Acknowledgements The research leading to these results has received funding from the EU 7th Framework Programme (FP7/2007-2013) under grant agreement n° 610467—project "M C Squared", http://mc2-project.eu. The C-book technology is based on the widely used Freudenthal Institute's DME portal and is being developed by a consortium of nine partner organizations, led by CTI and Press Diophantus. This publication reflects only the authors' views and the European Union is not liable for any use that may be made of the information contained therein.

Appendix 1: CMT Affordances Grid

The four main criteria proposed in the literature to characterise CMT are taken into consideration:

- **Fluency**: The students' ability to provide many responses or to come up with many strategies to solve a mathematical problem or challenge.
- **Flexibility**: The students' ability to provide different/varied responses or to come up with different/varied strategies to solve a mathematical problem or challenge.
- **Originality**: The students' ability to come up with unique (original) responses (solutions, strategies, representations, etc.) to a mathematical problem or challenge.
- **Elaboration**: The students' ability to describe, substitute, combine, adapt, modify, magnify, extend the usability, eliminate or rearrange mathematical situations.

A c-book unit environment (that includes activities, orchestration, context of use, etc.) is considered fostering CMT if it stimulates and provides feedback supporting the above-mentioned students' abilities.

This tool aims to evaluate the c-book unit affordances towards the development of CMT in users/ students, relating the nature of activities to the users' cognitive processes (fluency, flexibility, originality, elaboration). The properties or characteristics of the c-book unit can be perceived as affordances that might foster students' abilities in terms of fluency, flexibility, originality or/and elaboration. In the table certain attributes of the c-book unit design possibly allow developing students' skills and abilities.

Affordances	Fluency					Flexibility					Originality					Elaboration					Comments
	N/A	1	2	3	4	N/A	1	2	3	4	N/A	1	2	3	4	N/A	1	2	3	4	
(1) The c-book unit includes mathematical open problems or questions.																					
(2) The c-book unit includes devising and formulating problems by the students / users themselves or formulating new questions to extend the investigation of the initial problem (problem posing tasks).																					
(3) The c-book unit includes calling for students' / users' constructions that call for mathematical thinking.																					
(4) The c-book unit provides users with opportunities to establish connections between different knowledge areas and mathematics (interdisciplinary/cross-disciplinary/external connections).																					
(5) The c-book unit provides users with opportunities to establish connections between different mathematical fields / concepts (internal connections).																					
(6) The c-book unit provides users with opportunities to establish connections between various representations of the mathematical concepts at stake (e.g. through a combination of widgets offering various representations).																					

Design and Evaluation of Digital Resources for the Development … 167

(7) The c-book unit stimulates/encourages users' exploratory activity and users' experimentation.							
(8) The c-book unit stimulates/encourages users to formulate and check their mathematical conjectures.							
(9) The c-book unit stimulates/encourages users to search and find multiple solutions / multiple strategies to solve a mathematical problem.							
(10) The c-book unit stimulates/encourages users to think about, reflect, summarize and evaluate the mathematical work already developed.							
(11) The c-book unit stimulates/encourages to generalize mathematical phenomena, going from concrete cases to general ones or to generalize real world phenomena through the use of mathematics							
(12) The c-book unit includes non-standard problems calling for mathematical solutions.							
(13) The c-book unit includes half-baked constructs (which are partially constructed or intentionally incorrect constructs) that call for intervention							

Social aspects	1	2	3	4	N/A	Comments
(S1) The c-book unit stimulates / encourages user's collaboration / cooperation / interaction with other users.						
(S2) The c-book unit stimulates / encourages the students to develop their mathematical communicative skills.						
(S3) The c-book unit provides users with opportunities to stimulate competition.						

Affective aspects	1	2	3	4	N/A	Comments
(A1) The c-book unit actively promotes engagement by generating a perception of usefulness of mathematics, either in everyday life, or inside the mathematical context.						
(A2) The c-book unit actively promotes engagement by generating a feeling of pleasure / fun / challenge (narratives, game features, feeling of flow/immersion in the activities, etc.).						
(A3) The c-book unit actively promotes engagement by generating a feeling of aesthetic pleasure from their contact with the mathematical concepts.						

Appendix 2: Questionnaire for Teacher Evaluation of the C-Book

1. **Representations of creativity and creativity in mathematics**
   - What does the term creativity mean for you?
   - How would you characterize creativity in mathematics?
   - Rate the following statements from 1 (strongly disagree) to 5 (strongly agree):

	1	2	3	4	5
Creativity is an innate quality, i.e., an individual is or is not creative	○	○	○	○	○
Only very talented individuals are creative	○	○	○	○	○
An individual can be creative while doing mathematics.	○	○	○	○	○
Mathematical creativity can be developed in each individual.	○	○	○	○	○
Mathematics teachers can help students develop their mathematical creativity.	○	○	○	○	○
Creativity in mathematics is the ability to approach a problem from various perspectives.	○	○	○	○	○
Creativity in mathematics is the ability to find an unusual solution to a given problem.	○	○	○	○	○
Creativity in mathematics is the ability to find several strategies to solve a given problem.	○	○	○	○	○
Mathematical creativity always produces new mathematical knowledge (theorems, proofs, definitions…).	○	○	○	○	○
The community of researchers in mathematics is the only instance that can decide whether a mathematical idea is creative or not.	○	○	○	○	○

2. **Opinions about the c-book unit affordances for developing creativity**
   - How would you position the c-book potential to foster creativity in mathematics on a scale from 0 to 10 (0 no potential—10 extremely high potential)? Explain your answer.
   - In your opinion, which aspects of the c-book are likely to contribute to the development of creativity in mathematics? (For example: freedom of strategies, possibility to elaborate, collaboration, challenge, feedback, freedom in studying the c-book, the way the work will be organized by the c-book, the combination of different mathematical fields, etc.)
   - Do you think the c-book is engaging for the students? Please explain.
   - Do you think the c-book offers enough opportunities for collaboration among students? Please explain.
   - Please write any comment or suggestion about the c-book and its potential to foster creativity in the students.

3. **Opinions about the c-book unit feedback features**
   - Do you think the feedback features provided by the c-book (help messages, prompts…) are likely to help the students solve the mathematical problems proposed by the c-book? Please comment.

- Do you think the feedback features provided by the c-book (help messages, prompts...) are likely to foster the students' creativity in approaching the mathematical problems proposed by the c-book? Please comment.
- Would you like to suggest other feedback, intended for students, teachers or both?

4. **General opinions about the c-book unit**
   - How would you estimate the school level for which the c-book is relevant?
   - Do you think the c-book offers learning opportunities to students to acquire the concept of geometric locus? Please explain.
   - Do you think the c-book fosters students' creativity in mathematics? Please explain.
   - Do you think the c-book is user-friendly (easy to manipulate...)? Please comment.
   - Do you have suggestions how to improve the interface to foster the user-friendliness of the c-book?
   - Do you think the c-book is in adequacy with the French mathematics curriculum?
   - Would you use the c-book with your students? If so, please describe briefly how you would use it (in a whole class or in groups, in class or outside (e.g., homework) ...). If no, explain why.

5. **Suggestions**
   - From your point of view, what is missing in the c-book in order to learn about geometric loci?
   - From your point of view, what is missing in the c-book in order to foster mathematical creativity?
   - What would you like to change in the c-book (e.g., add, withdraw, modify, and develop further...)?
   - Please write any other comments.

References

Botsch, O. (1956). *Bewegungsgeometrie*. Reinhardt-Zeisberg, Band 4b. Moritz Diesterweg.
Boyer, C. B. (2012). *History of analytic geometry*. Courier Corporation Publications.
Chamberlin, S. A., & Moon, S. M. (2005). Model-eliciting activities as a tool to develop and identify creatively gifted mathematicians. *The Journal of Secondary Gifted Education (JSGE), 17*(1), 37–47.
Craft, A. (2008). Studying collaborative creativity: Implications for education. *Thinking Skills and Creativity, 3*(3), 241–245.
Descartes, R. (1637). La Géométrie. Appendix to Discours de la méthode.
EC. (2006). Recommendation 2006/962/EC of the European Parliament and of the Council of 18 December 2006 on key competences for lifelong learning. *Official Journal of the European Union, L 394*, 10–18.

El-Demerdash, M., Trgalová, J., Labs, O., Mercat, C. (2017). Teaching locus at undergraduate level: A creativity approach. In J. Trglova & G. Aldon (Eds.), *Proceedings of the 13th international conference on technology in mathematics teaching (ICTMT 13, 3 to 6 July 2017)* (pp. 347–356), Lyon—France, École Normale Supérieure de Lyon/Université Claude Bernard Lyon 1.

El-Demerdash, M. (2010). *The Effectiveness of an enrichment program using dynamic geometry software in developing mathematically gifted students' geometric creativity in high schools* (Doctoral dissertation, University of Education Schwaebisch Gmünd, Germany).

El-Demerdash, M. & Kortenkamp, U. (2009). The effectiveness of an enrichment program using dynamic geometry software in developing mathematically gifted students' geometric creativity. In *Paper Presented at the 9th International Conference on Technology in Mathematics Teaching—ICTMT 9*, Metz, France, July 6–9, 2009.

Elschenbroich, H. J. (2001). Dem Hoehenschnittpunk auf der Spur. In Herget, W. et al. (Eds.), *Medien verbreiten Mathematik (Proceedings)* (pp. 86–91), Berlin: Verlag Franzbecker, Hildesheim.

Emin, V., Essonnier, N., Lealdino, P., Mercat, C., & Trgalová, J. (2015). Assigned to creativity: didactical contract negotiation and technology. In *Proceedings of the 9th congress of european research in mathematics education (CERME 9)*, Czech Republic. 2015, Thematic Working Group 7: Mathematical Potential, creativity and talent.

Ervynck, G. (1991). Mathematical creativity. In A. M. Thinking & D. Tall (Eds.), *Advanced mathematical thinking* (pp. 42–53). Dordrecht, the Netherlands: Springer.

Guilford, J. P. (1950). Creativity. *American Psychologist, 5,* 444–454.

Hadamard, J. (1945). *The psychology of invention in the mathematical field*. New York: Dover Publications.

Haylock, D. (1997). Recognizing mathematical creativity in schoolchildren. *ZDM Mathematics Education, 27*(2), 68–74.

Healy, L. (2000). Identifying and explaining geometrical relationship: interactions with robust and soft Cabri constructions. In T. Nakahara & M. Koyama (Eds.), *Proceedings of the 24th Conference of the International Group for the Psychology of Mathematics Education* (Vol. 1, pp. 103–117). Hiroshima: Hiroshima University.

Hennessey, B. A., & Amabile, T. M. (1988). The conditions of creativity. In R. J. Sternberg (Ed.), *The nature of creativity: Contemporary psychological perspectives* (pp. 11–38). Cambridge University Press.

Hoyles, C., & Noss, R. (2003). What can digital technologies take from and bring to research in mathematics education? In A. J. Bishop, et al. (Eds.), *Second international handbook of mathematics education*. Dordrecht: Kluwer Academic Publishers.

Jareš, J., & Pech, P. (2013). Exploring loci of points by DGS and CAS in teaching geometry. *Electronic Journal of Mathematics and Technology, 7*(2), 143–154.

Klein, F. (1924). *Elementarmathematik vom höheren Standpunkte aus, Band I. Dritte Auflage.* Springer.

Laborde, C. (2005). Robust and soft constructions: two sides of the use of dynamic geometry environments. In *Proceedings of the 10th Asian Technology Conference in Mathematics* (pp. 22–35), Korea National University of Education.

Labs, O. (2015). D5.3: Tool and technical report on data analysis and configuration. In Deliverable of the MC2 project, FP7—Information and Communication Technologies, Grant Agreement no: 610467 (http://mc2-project.eu/index.php/dissemination).

Lev-Zamir, H., & Leikin, R. (2011). Creative mathematics teaching in the eye of the beholder: focusing on teachers' conceptions. *Research in Mathematics Education, 13*(1), 17–32.

Liljedahl, P. (2013). Illumination: An affective experience? *ZDM—Mathematics Education, 45*(2), 253–265.

Liljedahl, P., & Sriraman, B. (2006). Musing on mathematical creativity. *For the learning of mathematics, 26*(1), 20–23.

Mann, E. L. (2006). Mathematical creativity and school mathematics: Indicators of mathematical creativity in middle school students (Doctoral dissertation, University of Connecticut).

Nadjafikhah, M., Yaftian, N., & Bakhshalizadeh, S. (2012). Mathematical creativity: Some definitions and characteristics. *Procedia—Social and Behavioral Sciences, 31,* 285–291.

Noss, R., & Hoyles, C. (2010). Modelling to address techno-mathematical literacies in work. In R. Lesh, P. L. Galbraith, C. R. Haines, & A. Hurford (Eds.), *Modelling students' mathematical modelling competencies.* New York: Springer.

Peigl, L. A., & Wayne, T. (2013). *The NURBS book: Monographs in visual communication.* New York: Springer.

Reuter, M. (2007). The biological basis of creativity. In A.-G. Tan (Ed.), *Creativity: A handbook for teachers* (pp. 79–100). Singapore: World Scientific Publishing Co., Pte. Ltd.

Silver, E. A. (1997). Fostering creativity through instruction rich in mathematical problem solving and problem posing. *ZDM—Mathematics Education, 3,* 75–80.

Sriraman, B. (2005). Are giftedness and creativity synonyms in mathematics? *The Journal of Secondary Gifted Education (JSGE), 17*(1), 20–36.

Sriraman, B. (2004). The characteristics of mathematical creativity. *The Mathematics Educator, 14*(1), 19–34.

Sternberg, R. J. (2003). The development of creativity as a decision-making process. In R. K. Sawyer, V. John-Steiner, S. Moran, R. J. Sternberg, D. H. Feldman, J. Nakamura, & M. Csikszentmihalyi (Eds.), *Creativity and development* (pp. 91–138). New York: Oxford University Press.

Sternberg, R. J., & Lubart, T. I. (2000). The concept of creativity: Prospects and paradigms. In R. J. Sternberg (Ed.), *Handbook of creativity* (pp. 93–115). Cambridge, UK: Cambridge University Press.

Torp, L., & Sage, S. (2002). *Problems as possibilities: Problem-based learning for K-16 education* (2nd ed.). Alexandria, VA: Association for Supervision and Curriculum Development.

Torrance, E. P. (1988). The nature of creativity as manifest in its testing. In R. J. Sternberg (Ed.), *The nature of creativity: contemporary psychological perspectives* (pp. 43–75). Cambridge: Cambridge University Press.

Torrance, E. P. (1969). Prediction of adult creative achievement among high school seniors. *Gifted Child Quarterly, 13,* 223–229.

Trgalová, J., El-Demerdash, M., Labs, O., & Nicaud, JF. (2018). Design of digital resources for promoting creative mathematical thinking. In L. Ball, P. Drijvers, S. Ladel, R.H.S., Sille, M. Tabach, C. Vale, (Eds.), *Uses of technology in primary and secondary mathematics education.* ICME-13 Monographs. Springer, Cham.

Trgalová, J. (2016). D6.3: Report on the creative design process of CoI—Cycle 3. Deliverable of the MC2 project, FP7—Information and Communication Technologies, Grant Agreement no: 610467. (http://mc2-project.eu/index.php/dissemination).

Trgalová, J., El-Demerdash, M., Labs, O., & Nicaud, J. (2016). Collaborative design of educational digital resources for promoting creative mathematical thinking. In *Paper presented at the 13th International Congress on Mathematical Education—ICME 13,* Hamburg, Germany, July 24–31, 2016.

Wallas, G. (1929). *The arts of thought.* London: Jonathan Cape.

Part III
Mathematics Teachers' Education for Technological Integration: Necessary Knowledge and Possible Online Means for its Development. Introduction to the Section

Ana Isabel Sacristán
Center for Research and Advanced Studies (Cinvestav-IPN), Mexico;
asacrist@cinvestav.mx

Integrating digital technologies for learning, including in mathematical education, is a foremost issue in the twenty-first century. In relation to that, there tends to be a consensus that a requirement for successful integration of technology in schools is the mathematics teacher's knowledge, skills, and competencies in that area. This implies the importance of teacher education for technology integration. However, the rapid development of technology in the past decades has made it difficult for mathematics teachers' professional development to keep up. This constant technological flux makes it difficult to develop proper teacher training programs in a timely manner so that contents remain useful and valid. Nonetheless, it is essential, on the one hand, to identify the basic knowledge and skills that are needed for teachers to harness the affordances and potential of digital technologies in their practice—perhaps those that are more fundamental, general, and detached from particular technologies, and thus are more enduring. This identification of teachers' basic knowledge is necessary both for enhancing their teaching as well as for providing their students with different and hopefully, more successful learning experiences. On the other hand, technological developments also make it possible to change the mechanisms in which teacher training takes place.

The following three chapters address these issues. First, Tabach and Trgalová (Chap. 8) aim to better understand what specific knowledge and skills mathematics teachers need to efficiently use digital technology in their classes. Then, the two other chapters—by van den Bogaart, Drijvers, and Tolboom (Chap. 9) and Aldon et al. (Chap. 10)—present investigations dealing with mathematics teachers' education, analysing ways for developing the knowledge for adequate technological integration, in both cases from online professional development perspectives: van den Bogaart et al.'s chapter focuses on the design and use of online materials for blended learning; and Aldon et al.'s chapter analyses international experiences with

MOOCs, in Italy and France, focusing on design principles related to participants engagement and collaboration.

Towards Understanding and Defining Needed Skills and Competencies for Technological Integration in Teachers' Practice

In the first paper, Tabach and Trgalová begin by pointing to the lack of standards for defining the knowledge and skills that mathematics teachers need to efficiently use digital technology in their classes. They follow from an ICME topical survey report in which they participated (Hegedus et al., 2017), where an identified gap between teachers' needs and teacher educators' contents underlined the importance of better understanding teachers' necessities for successful integration of technology and resulted in a call to the mathematics education international community for elaborating ICT standards for mathematics teacher education. In order to advance towards defining standards for mathematics teacher training, Tabach and Trgalová, in their particular chapter in this book, take the following approach: First, they review current trends in research on teacher education in order to attempt to identify what professional competencies are considered by researchers as important for mathematics teachers to integrate technology in their practice. In doing that, they distinguish between two types of research approaches: content-driven (focussing on the knowledge/skills that are needed to teach with technology a particular mathematical concept or area) and tool-driven (investigating what knowledge/skills are required to use a particular technological tool). They then analyse a selection of international and national documents that may represent teacher ICT competencies standards. At the same time, Tabach and Trgalová provide an important discussion of theoretical frameworks for researching (and guiding) the teaching of mathematics with technology: They contrast the widely used TPACK model defined by Mishra and Koehler (2006), with two frameworks they use in their own study: Thomas and Hong's (2005) *Pedagogical Technological Knowledge (PTK)* and Haspekian's (2011) *double instrumental genesis* (personal and professional), both of which are based on the instrumental approach and thus take into account the instrumental genesis of the teachers with respect to technology. They also modify Ball's mathematics knowledge for teaching (MKT) framework into what they call the Mathematics Digital Knowledge for Teaching (MDKT) model, in order to include: the specialized digital content knowledge (SDCK)—which they consider is closely linked with the personal instrumental genesis; knowledge of digital content and students (KDCS); knowledge of content and teaching (KDCT); and knowledge of digital-content and curriculum (KDCC).

In their discussion of TPACK, Tabach and Trgalová provide a brief overview of how different researchers use and interpret that framework—such as citing Bowers and Stephens (2011) who consider that TPACK may constitute more of an orientation than sets of skills or knowledge. An important point that Tabach and Trgalová make, is that the TPACK framework is non-specific to mathematics—it was because of this that, soon after TPACK was proposed, members of the US Association of Mathematics Teacher Educators (AMTE) developed the Mathematics TPACK or M-TPACK (Niess et al., 2009), which Tabach and Trgalová analyse. In contrast to

TPACK, as the authors of the chapter explain, "PTK was developed based on research related specifically to mathematics teachers" and takes into account the mathematical knowledge for teaching (MTK); furthermore, "PTK explicitly sees teachers' orientation to technology as an important component" and also takes into account teachers' personal orientations which include "beliefs about the value of technology and the nature of learning mathematical knowledge, and other affective aspects, such as confidence in using technology." This latter emphasis on affect is one that Tabach and Trgalová consider is important in their analysis: they call it the personal orientation perspective—one that has to do with teachers' affective aspects and beliefs regarding mathematics, teaching mathematics and technology. They also explain that the PTK model acknowledges the necessity for the teacher to appropriate technology before its use in the classroom through processes of instrumental geneses. It is in that regard that Tabach and Trgalová integrate into their work the double instrumental genesis because, as they explain, this theory considers that teachers need first to acquire basic skills to master the technology they use and develop utilization schemes related to it, as well as develop an understanding of how to support students' mathematics learning in a technological environment.

I consider this overview and discussion of the different theoretical frameworks, as a very valuable and significant contribution of Tabach and Trgalová's chapter.

It is by using PTK, the double instrumental genesis, and the MDKT, that Tabach and Trgalová analyse their selection of documents: three documents by international organizations or groups—one by UNESCO; another derived from a collaboration between the P21 organization, the NCTM and the Mathematical Association of America; and finally the Niess and colleagues' (2009) M-TPACK framework; and two national level documents, from Australia and France, describing the institutional situation in each of those countries. Although a varying degree of knowledge and competencies were identified in the different documents in terms of the MDTK, it seems there is still a need to operationalize the standards for teacher education programs and for building support systems for teachers; and to identify specific goals to help teacher educators devise professional development programs for technological integration.

Tabach and Trgalová observe that some of the frameworks in the analysed documents suggest several stages for teacher professional development, where teachers first need to perceive the potential of technology and undergo a personal instrumental genesis before considering its classroom integration. Although these stages may be necessary (i.e., that personal instrumental genesis needs precede professional genesis), I would caution against how this could be interpreted, either by teachers or by designers of professional development programs. Taking it as an unavoidable premise could be a dangerous pitfall that could constrain both professional development and technological integration in the classroom: I've seen far too many teachers refrain from integrating technological tools in their teaching practice because they feel insufficiently prepared in terms of the technology (Trigueros & Sacristán, 2008). I argue that the personal instrumental genesis can develop simultaneously with the professional one, and I have long advocated that teachers—akin to how one needs to jump in the water in order to learn to swim—

should dare to use the technological tools even if they have insufficient knowledge and skills both in regards to technology and in terms of what Tabach and Trgalová would call MDKT. Likewise, even if professional development programs are designed in stages, at some point professional in-classroom experience will be necessary to develop both the personal and the professional instrumental genesis. Independently of my thoughts on this, Tabach and Trgalová consider that the specific knowledge and skills to be developed at each stage still need to be identified and formalized, and consider that it could be a goal of the mathematics education research community.

Another, and in my view, most important contribution of Tabach and Trgalová's chapter is that, in addition to teachers' personal instrumental geneses, they consider teachers' personal orientations (values, attitudes and confidence in technology use) discussed above, as crucial for teachers' adoption and integration of technology. This is something of which I have evidence in my own research (Trigueros & Sacristán, 2008). Thomas and Palmer (2014) also consider those personal orientations in their PTK model, although they deem that they may be under-estimated. In that respect, they conclude by describing how the MDKT framework adapts the PTK framework, by emphasizing the role of mathematical-knowledge-for-teaching-with-technology components in relation to teacher orientations, as well as personal and professional instrumental genesis.

Whereas Tabach and Trgalová's chapter addresses the skills considered essential for teachers to develop and that teacher educators should consider, the other chapters in this section explore ways in which to design and implement teacher education programs, particularly integrating online means.

Considering the Design and Implementation Processes of Online and Blended Teacher Education

The chapter by van den Bogaart and colleagues describes a project in which teacher educators in the Netherlands engaged in a co-design process of developing and testing the implementation of open online learning units for mathematics and science didactics. More specifically, they research the potential of online and blended learning (face-to-face education combined with the online one) and their affordances for teacher education, and for domain specific didactical courses. They consider that, in spite of the immense increase in online educational resources and courses for the learning of specific subject knowledge—often produced by co-design teams of teachers, designers or researchers, and which then have to be selected and re-designed for their implementation–, the affordances of online learning and blended learning in teacher education for didactics are underused and largely unexplored. They attribute this underuse to the complexities and challenge of designing online learning units that facilitate the transfer from theory to practice (they refer, in particular, to the difficulties in addressing, in an online setting, the mixture of skills, knowledge and attitudes that are part of the learning goals of didactical courses.). They also refer to other challenges such as issues of time and of financial resources that are necessary to produce them.

One way that they proposed to address the effort and financial challenges in order to produce quality didactical materials, was to form a collaboration between different teacher education institutes for the design (i.e. a co-design) of online learning units for STEM for pre-service teachers. What is interesting is that both the design process and the design product involved the blended approach that combined face-to-face sessions with online aspects.

More specifically, in their chapter, van den Bogaart and colleagues describe the co-design process, implementation, and evaluation of two online learning units for pre-service secondary teachers' mathematics education: a generic one on mathematical thinking; and a more specific one on statistics education, which had to follow a new curricular approach. For their evaluation and research, they focus on three aspects: the features of the designed online learning units; the organization of the co-design process; and the experiences derived from the implementation of those learning units in teacher education practice. So, in fact, they look at several dimensions: from the perspective of the organization of the design process, to the features and experiences of the design product. From this study process, they are able to identify salient and transferable design features and heuristics.

The co-design process was carried out by small-size teams of experienced educators from different institutes, who designed blended online learning units, mixing face-to-face meetings (short intensive collaborative "boot camps" for co-design) and ongoing online collaboration. The face-to-face sessions allowed the design teams to collaborate intensively, and had an interesting positive "side-effect": that the design participants became a community of practice (Wenger, 1998) in which knowledge and experiences were co-created and shared, through in-depth discussion of educational content and didactics that lead to an increase in knowledge, and to more coherent views on teacher education across participating institutions.

The design of the learning units used a mix of resources, including video clips (e.g., with team members discussing an exercise before and after using it with pupils; or showing students working on it and interviewed about the strategies they used). The video resources were one of the contents found valuable by the educators. Also, van den Bogaart and colleagues consider that giving students a voice through the video recordings is very important for pre-service teachers to gain insights into the way students think (in contrast to their own thinking) and leading them to reflect on possible didactical interventions.

A second heuristic that van den Bogaart and colleagues consider, and that relates to Tabach and Trgalová's chapter, is how to address the prerequisite content subject knowledge that is necessary for a didactical approach. This is something they faced in the design of the statistics didactics unit, because many mathematics teachers only have limited knowledge about statistics. They dealt with this issue by intertwining, in the learning units, both content knowledge and didactical knowledge, thus providing opportunities for teacher students to extend their knowledge.

A third and very important point that they make in terms of the design heuristics is that learning units should not be stand-alone or provide fixed materials or learning trajectories for students, but rather should keep in mind different target groups (e.g., student teachers and the teacher educators) and thus be flexible: i.e., instead of suggesting building blocks for activities that provide teacher educators with autonomy so they can use their didactical expertise, allow the type of interactions they prefer, and are adaptable to what is used in each institution; as well as also serve not only the educators but also both pre-and in-service mathematics teachers. This flexibility of the units is, in fact, an aspect that was appreciated by the educators.

In terms of the products of the co-design, a challenge that was faced by the participants was how to articulate online and face-to-face activities into the learning units for an optimal educational design, with the idea of having curriculum innovation and content design impact classroom practice.

Finally, another highlight of the results of how educators and students experienced the use of the online units is that both educators and students appreciated these learning units, particularly the availability of new types of resources and activities for students to work on independently online. However, some educators did struggle with finding ways to embed the online units in their courses. In this regard, an important point that van den Bogaart et al. make is that the actual use by educators, and incorporation in their existing educational practices of the materials, required considerable time and effort; particularly, educators needed time to prepare that integration. Another time issue identified was the limited face-to-face teaching time.

From Small-Scale Online Teacher Education to MOOCs

Whereas the work by van den Bogaart et al. looks at a small scale design for mathematics teacher education, the third chapter in this section by Aldon and colleagues looks at mathematics teachers' professional development at a large scale, through MOOCs, specifically focusing on the design and assessment aspects (trying to determine design principles useful for mediating online teachers' professional courses and assessing the impact of such courses on teachers' engagement). Therefore, like the previous paper, it also focuses on presenting and analysing the design of teacher education courses involving online means.

Despite the exponential growth of MOOCs in the past few years, Aldon and colleagues claim that the use of MOOCs for teacher professional development is still rare; I argue, however, that this area is growing fast, as demonstrated by the increasing number of papers and research in this topic—for example, at the PME-NA 39 conference in 2017 (Galindo & Newton, 2017), there were many works that focused on MOOCs for mathematics educators and/or online professional development, including a plenary paper (Hollebrands, 2017) and a working group (Choppin, Amador, Callard, & Carson, 2017). Nevertheless, this growth is hindered by challenges, such as those related to: the transfer from theory to practice (mentioned in the previous chapter by van den Bogaart et al.); the skills, knowledge and attitudes needed to be developed by teachers (discussed in the chapter by Tabach and Trgalová); the new roles, in these massive courses, of educators and participants, which imply carefully considering the structure of the pedagogical design (as Aldon and colleagues mention); as well as issues of assessment of the learning that takes place.

In their chapter, Aldon and colleagues focus on three MOOCs for their analysis: one from France (with, on average, over 2500 participants) and two from Italy (with close to 300 participants), which aim at supporting mathematics teachers in new curricular or innovative practices, particularly in the integration of digital resources.

An important aspect of the work by Aldon and colleagues is that the MOOCs were designed, as well as described and analysed, in a highly structured way: first, by using aspects (of the theme of contexts and features) of the frame proposed by Robutti et al. (2016) on mathematics teachers working and learning through collaboration; specifically the four dimensions of collaborative work that consider the initiation, aims, composition and scales of collaborations and their ways of working. They also use as main theoretical frameworks, Arzarello et al.'s (2014) theory of Meta-Didactical Transposition (MDT), and that of Wenger's (1998) Communities of Practice (CoPs).

In terms of MDT, they use the distinction between didactical and meta-didactical praxeologies, where the first type aims to model the mathematical activity when solving a didactical task (such as when teaching a particular mathematical topic); whereas, meta-didactical praxeologies concern meta-didactical tasks (such as reflecting on possible praxeologies for teaching a particular concept).

Through the analysis of praxeologies associated to tasks, they identify essential topics regarding MOOCs, such as the relationship between design principles and professional development; relationships between trainees and trainers; assessment strategies; and are able to assess the mathematics teachers' engagement.

In terms of design issues, they point to aspects such as the time variable where designers had to decide either how much time to devote to a MOOC module or how much material participants could cover in a fixed period of time (such as in one week).

By adapting the meta-didactical lens, the analysis by Aldon and colleagues of their chosen MOOCs, also centred on the collaboration processes. Those MOOCs not only focus on instruction, but they seek to create collaborative contexts for teachers to work and share their practice experiences. Thus, both types of MOOCs were conceived as collaborative experiences (rather than having trainees watch videos and carry out related activities), using a platform (the French case used a professional social network) or forums and/or other online tools for participants to collaborate. Aldon et al. point out that, as in the French case, the availability of social network-like tools allowed trainees themselves to regulate their tasks and collaboration. In both the French and Italian MOOCs, trainers included videos and pdfs for trainees to be familiarized with the technological and collaborative tools. But the authors emphasise that a real involvement and engagement of trainees in collaborative work cannot be considered spontaneous, so it needs to be triggered; thus, designers need to make it possible through multi-techniques, and have it supported by suitable tools (e.g., the technological collaborative tools). But they point out that, after their analysis, the question is still open as to which devices are best for improving active collaboration among trainees.

Aldon and colleagues identify two communities (trainers and trainees) in the analysed MOOCs, which interact and evolve over time. The MOOCs experiences offered teachers a possibility to collaborate in small groups within a wider online community, and there was a goal of making these groups evolve into lasting communities of practice. One of the approaches taken towards that goal was in accompanying the teachers in producing resources and in reflection. Whereas in the Italian MOOCs, trainers were the ones involved in the design, in the French MOOC trainees (the enrolled teachers) were invited to work on the proposed activities in a collaborative way (either asynchronously or synchronously), by forming small groups around common interests; also, in the latter case, during MOOC activities a trainer joined the trainees as a group member, in order to encourage the collaborative work and help the group become a community of practice.

In terms of the issue of how to assess the impact of the MOOC courses on teachers' learning and engagement, Aldon and his colleagues remind us that the massiveness of such courses makes assessment of the progress of each trainee very difficult. One approach used by both teams, was through weekly tests; this was later abandoned in the Italian MOOCs when it was perceived as a work overload and those MOOCs relied instead on "end module badges" obtained when trainees finished work with specific resources. Another more profound approach that helps assess the trainees' engagement, and is used by both the Italian and the French teams, is a project-based methodology (where trainees design a classroom activity through which they demonstrate acquired teaching competencies and expertise), either individual (in the Italian case) or collective one (the French case), using grids with guiding questions (French case) or for revision (Italian case). In both cases, the participants' projects were peer reviewed (with, in the French case, a second evaluation by the trainers). As Aldon et al. point out, the peer review is a formative assessment that helps stimulate collaboration among trainees.

It is thus that the following section provides important elements for the development of teacher training programmes, in this fast evolving era of technological developments: From the attempt by Tabach and Trgalová to identify the basics skills and competencies that teachers require to useful design heuristics and principles, as well as challenges (such as those related to time issues), exemplified and analysed by both van den Bogaart et al. and Aldon et al. that provide guidelines and aspects to take into account in the design and implementation, at both small and large scale, of future mathematics teacher education programs for technological integration by the mathematics education community. All three chapters also provide important theoretical frameworks for designing such programmes as well as for analyzing them. They also exemplify the usefulness of multifaceted approaches for optimizing the implementation and knowledge development.

References

Arzarello, F., Robutti, O., Sabena, C., Cusi, A., Garuti, R., Malara, N., & Martignone, F. (2014). Meta-didactical transposition: A theoretical model for teacher education programs. In A. Clark-Wilson, O. Robutti, & N. Sinclair (Eds.), *The mathematics teacher in the digital era: An international perspective on technology focused professional development* (Vol. 2, pp. 347–372). Dordrecht, The Netherlands: Springer.

Choppin, J., Amador, J., Callard, C., & Carson, C. (2017). Designing and researching online professional development (Working group). In E. Galindo & J. Newton (Ed.), *Proceedings of the 39th Annual Meeting of the North American Chapter of the International Group for the Psychology of Mathematics Education* (pp. 1481–1488). Indianapolis, IN: Hoosier Association of Mathematics Teacher Educators. Retrieved from http://www.pmena.org/pmenaproceedings/PMENA%2039%202017%20Proceedings.pdf.

Galindo, E., & Newton, J. (Eds.). (2017). *Proceedings PME-NA 39 2017*. Indianapolis, IN: Hoosier Association of Mathematics Teacher Educators. Retrieved from http://www.pmena.org/pmenaproceedings/PMENA%2039%202017%20Proceedings.pdf.

Hegedus, S., Laborde, C., Brady, C., Dalton, S., Siller, H.-S., Tabach, M.,Trgalová, J., Moreno-Armella, L. (2017). *Uses of technology in upper secondary mathematics education* (S. Hegedus, C. Laborde, C. Brady, S. Dalton, H.-S. Siller, M. Tabach, … L. Moreno-Armella, (Eds.)). Cham: Springer International Publishing. doi:10.1007/978-3-319-42611-2_1.

Hollebrands, K. F. (2017). A framework to guide the development of a teaching mathematics with technology massive open online course for educators (MOOC-Ed). In E. Galindo & J. Newton (Ed.), *Proceedings of the 39th Annual Meeting of the North American Chapter of the International Group for the Psychology of Mathematics Education* (pp. 80–89). Indianapolis, IN: Hoosier Association of Mathematics Teacher Educators. Retrieved from http://www.pmena.org/pmenaproceedings/PMENA%2039%202017%20Proceedings.pdf.

Robutti, O., Cusi, A., Clark-Wilson, A., Jaworski, B., Chapman, O., Esteley, C., … Joubert, M. (2016). ICME international survey on teachers working and learning through collaboration. *ZDM Mathematics Education, 48*(5), 651–690.

Trigueros, M., & Sacristán, A. I. (2008). Teachers' practice and students' learning in the Mexican programme for teaching mathematics with technology. *International Journal of Continuing Engineering Education and Life-Long Learning, 18*(5/6), 678. doi:10.1504/IJCEELL.2008.022174.

Wenger, E. (1998). *Communities of practice: Learning, meaning, and identity*. Cambridge: University Press.

The Knowledge and Skills that Mathematics Teachers Need for ICT Integration: The Issue of Standards

Michal Tabach and Jana Trgalová

1 Introduction

Digital technology has permeated society as a whole, including education. Whether information and communication technology (ICT) should be used in the classroom is no longer an issue. Rather, the question now focuses on how technology can be used more efficiently in the classroom and how the most benefit can be derived from its use. Converging research studies indicate that teacher training is one of the key elements in responding to this question. Indeed, in a literature review of "barriers to the uptake of ICT by teachers", Jones (2004) highlights that "there is a great deal of literature evidence to suggest that effective training is crucial if teachers are to implement ICT effectively in their teaching" (p. 8). This statement clearly addresses both pre-service teacher education (TE) and in-service teacher professional development (TPD) regarding the use of digital technology.

According to the research literature, ICT integration has a profound impact on teacher practices. Technology facilitates new approaches to teaching and learning, among them direct manipulation or visualization of mathematical objects and collaborative learning. Teachers need to develop new knowledge and skills to enable them to design relevant technology-mediated tasks, monitor student work and assess student learning using technology. The literature discusses several initiatives in teacher education or teacher professional development. Hegedus et al. (2017) pointed out that in a number of cases, researchers report being disappointed with the outcomes of these initiatives. Emprin (2010) identified the discrepancy between teachers' needs and TE-TPD contents as one of the main reasons for this disappointment. Hence,

M. Tabach (✉)
Tel-Aviv University, Tel Aviv, Israel
e-mail: tabachm@post.tau.ac.il

J. Trgalová
S2HEP, Université Claude Bernard Lyon 1, Villeurbanne, France
e-mail: jana.trgalova@univ-lyon1.fr

© Springer Nature Switzerland AG 2019
G. Aldon and J. Trgalová (eds.), *Technology in Mathematics Teaching*,
Mathematics Education in the Digital Era 13,
https://doi.org/10.1007/978-3-030-19741-4_8

teacher educators require a better understanding of what teachers need to know in order to use ICT effectively. This raises the issue of ICT competency standards.

In our previous research (Hegedus et al., 2017; Tabach & Trgalová, 2017), we began searching for institutional frameworks that specify what knowledge teachers require to teach mathematics with technology. We were surprised to find very few such standards for mathematics teachers, or even for teachers in general, at both the national and international levels. Most of those we did find were not geared specifically either to subject matter or to school level. Therefore, we recommended that "[e]laboration of ICT standards for mathematics teacher education might become one of the goals of the mathematics education international community" (Hegedus et al., 2017, p. 30).

In this chapter, we expand on our previous studies with the aim of gaining a better understanding of the specific knowledge and skills needed by mathematics teachers to use ICT efficiently in mathematics classrooms. We begin by reviewing recent research on teacher education that focuses on ICT-related knowledge and skills. In the subsequent sections, we provide a rationale for our choice of theoretical frameworks and describe our investigation method. In the main part of the chapter, we provide detailed analyses of a selection of both international and national documents that stipulate teacher ICT competency standards. In the concluding section we summarize the results and suggest a further research agenda to examine the theme of ICT competency standards for teachers.

2 Current Trends in Research on Teachers' ICT Competencies

2.1 General Versus Specific Knowledge

In theorizing about the unique knowledge needed for teaching with digital technology in general, Mishra and Koehler (2006) introduced the Technology, Pedagogy and Content Knowledge framework (TPCK or TPACK). TPACK incorporates the knowledge and skills teachers need to integrate technology meaningfully into instruction in specific content areas. Among the frameworks that address teachers' ICT knowledge and skills, TPACK is used most frequently as it offers "a helpful way to conceptualize what knowledge prospective teachers need in order to integrate technology into teaching practices" (Bowers & Stephens, 2011, p. 286).

TPACK introduces an additional body of knowledge to the Pedagogical Content Knowledge (PCK) model proposed by Shulman (1986): *technological knowledge* (TK), which partially overlaps *content knowledge* (CK) and *pedagogical knowledge* (PK). The TPACK framework depicted in Fig. 1 offers a theoretical lens through which researchers can analyse teachers' professional knowledge at a general level. TPACK is used by many researchers, and today several interpretations have been accepted (Voogt et al., 2012): T(PCK) as extended PCK; TPCK as a unique and

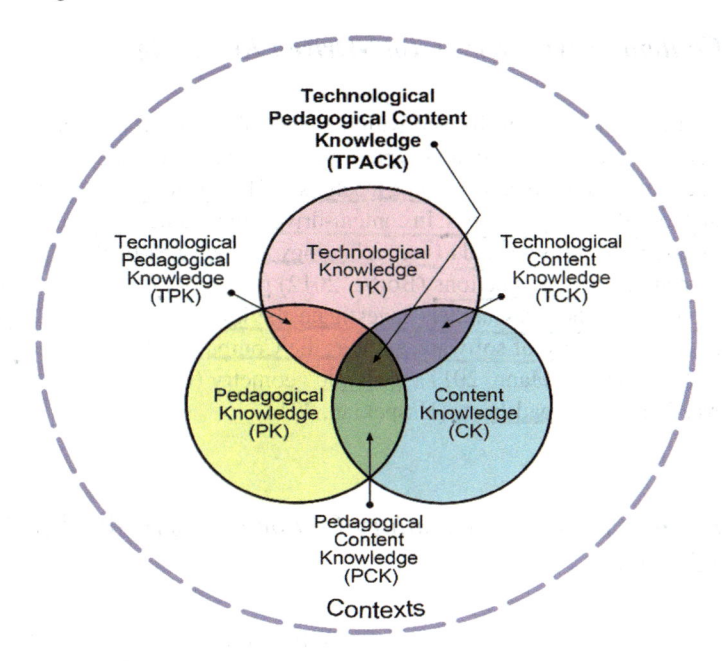

Fig. 1 The TPACK framework (with permission from TPACK.org)

distinct body of knowledge; and TP(A)CK as the interplay between seven bodies of knowledge—three knowledge domains and their intersections.

Nevertheless, the TPACK framework allows for a variety of interpretations. While some authors attempt to define specific TPACK knowledge, others consider TPACK to be an orientation that enables teacher educators "to develop a greater sense of how to plan and focus instruction for prospective math teachers" (Bowers & Stephens, 2011, p. 301). The knowledge definition approach, for example, has been adopted by Robová (2013), who draws on TPACK to define what she calls "Specific Skills for work in GeoGebra". She proposes a set of such skills instantiated for functions (e.g., "making functions visible (on the screen)" or "using dynamic features of GeoGebra"). The orientation approach is advocated by Bowers and Stephens (2011), who claim, based on a literature review, that

> teachers need not acquire one particular expertise or pick one particular role; instead, teachers (and prospective teachers) need to become aware of how to design rich tasks that integrate technology into the classroom discourse so that technology-based conjectures and arguments become normative (p. 290).

Other rather general forms of knowledge and skills have also been considered, such as being able to support student problem-solving in a technological environment (Lee, 2005), analyse digital resources in order to evaluate their pedagogical affordances and relevance (Trgalová & Jahn, 2013), encourage students to use a tool of their choice to observe the mathematical relations at stake (Bowers & Stephens, 2011), or use ICT to develop reasoning capacities in students (Zuccheri, 2003).

2.2 Content-Driven Versus Tool-Driven Knowledge

Most research reports are positioned within a particular context that links a specific mathematics domain to a particular type of technology. Nevertheless, two different approaches can be identified in defining teachers' ICT knowledge and skills: content-driven and tool-driven approaches. In content-driven approaches, researchers analyse the knowledge and skills needed to use technology to teach a particular mathematical concept or area, such as functions (Borba, 2012) or algebra (Clay et al., 2012). In tool-driven approaches, researchers investigate the knowledge and skills required to use a particular piece of software, such as the Computer Algebra System (CAS) (Ball, 2004; Zehavi & Mann, 2011), dynamic geometry (Robová, 2013; Robová & Vondrová, 2015) or spreadsheets (Haspekian, 2014).

2.3 From Students' 21st Century Competencies to Teachers' ICT Skills

The ICT skills teachers require can be inferred from the competencies students need to develop throughout their schooling to enable them to "adapt flexibly to a rapidly changing and highly interconnected world" (EU, 2006). The growing presence of technology in the lives of adults—for their personal use, as citizens and as part of the requirements of their jobs—has made it clear that students need to be educated towards these future needs. The P21 organization[1] developed a structure for 21st century learning skills that includes both the knowledge and the skills expected of students as they complete their K-12 education. In general, these skills transcend any specific subject matter, yet should be nurtured in each core subject, including mathematics. These skills are sometimes referred to as the "four Cs":

- *Critical Thinking and Problem Solving*, e.g., the ability to effectively analyse and evaluate evidence, arguments, claims and beliefs and to solve different kinds of non-familiar problems in both conventional and innovative ways.
- *Communication*, e.g., the ability to articulate thoughts and ideas effectively using oral and written communication skills in a variety of forms and contexts.
- *Collaboration*, e.g., the demonstrated ability to work effectively and respectfully with diverse teams.
- *Creativity and Innovation*, e.g., the ability to use a wide range of idea creation techniques to generate new and worthwhile ideas.

A naïve view is that these Cs are not related to ICT. Yet it is clear that in the presence of technology, each of these skills involves a new level of complexity. Students are expected to be able to navigate among numerous e-sources, critically evaluate their reliability and make sense of and integrate available information. They need to learn how to apply technology effectively, that is, use technology as a tool

---

[1] http://P21.org.

to search for, organize, evaluate and communicate information. They must learn to apply communication and networking tools and social networks appropriately to be able to access, manage, integrate, evaluate and create information and to function successfully in a knowledge economy. Moreover, they must demonstrate a fundamental understanding of the ethical and legal issues surrounding the access and use of information technologies.

Because mathematics is considered a core subject, mathematics educators must become familiar with these ideas and appropriate them into their teaching. This brings us back to where we began: the need to define the knowledge and skills expected of mathematics teachers in the digital era that will enable them to support the development of their students' skills.

3 Theoretical Perspective

Several researchers have proposed theoretical perspectives to study the work of teachers who integrate technology into their teaching practice. Some of these, like the aforementioned TRACK framework, were developed without considering any particular subject matter. Others are rooted within mathematics education, among them the double instrumental genesis and the Pedagogical Technological Knowledge (PTK) frameworks. These two perspectives, which serve as the guiding theoretical perspectives in this study, are discussed in the next section.

3.1 Double Instrumental Genesis

The theoretical construct of the *double instrumental genesis* (Haspekian, 2011) was developed in accordance with the instrumental approach (Rabardel, 2002) and encompasses both the personal and professional instrumental geneses of teachers who use ICT. Whereas the personal instrumental genesis is related to the development of a teacher's *personal instrument* for a mathematical activity from a given artefact, the professional instrumental genesis yields a *professional instrument* for a teacher's didactical activity. In the context of the TPACK model, these two processes mobilize knowledge of the artefact (TK) and the abilities to solve mathematical problems using this knowledge (TCK) within the personal genesis, while the teacher's abilities to orchestrate ICT-supported learning situations (TPK) and to teach mathematics with ICT (TPACK) are mobilized in the professional genesis.

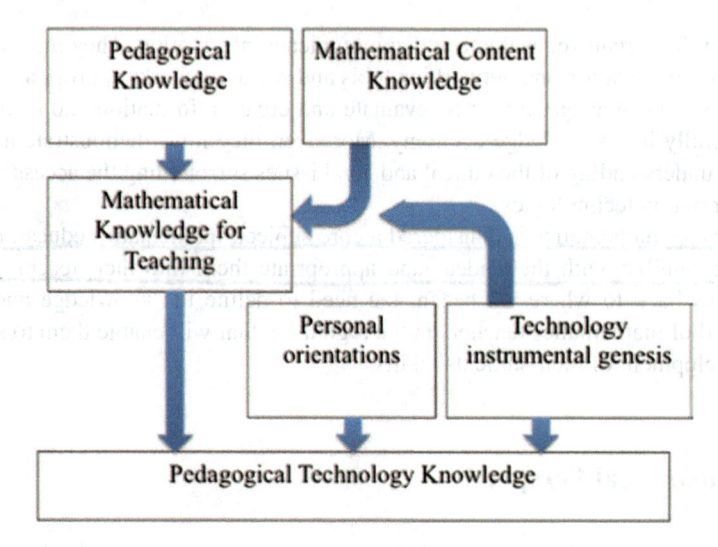

Fig. 2 A model of the PTK framework (Thomas and Palmer, 2014)

3.2 Pedagogical Technological Knowledge

Thomas and Hong (2005) defined the term Pedagogical Technology Knowledge (PTK) as "… knowing how to teach mathematics with the technology" (p. 256). Thomas and Palmer (2014) further developed this concept by interweaving a number of intrinsic teacher factors to produce PTK, among them teachers' instrumental genesis with respect to technology; mathematical knowledge for teaching (Ball et al., 2008); teacher orientations and goals (Schoenfeld, 2011), especially beliefs about the value of technology and the nature of learning mathematical knowledge; and other affective aspects, such as confidence in using technology (Fig. 2).

In our previous studies (Tabach & Trgalová, 2017; Trgalová & Tabach, 2018), we used the construct of double instrumental genesis and the TPACK frameworks to analyse documents outlining teacher ICT competency standards. We found several differences between TPACK and PTK, which were developed around the same period. Yet while PTK was developed based on research specifically related to mathematics teachers, TPACK was developed based on research related to teachers' knowledge in general. PTK is based on the instrumental approach and hence acknowledges the mutual instrumentation and instrumentalization relations between teacher and tools. In contrast, TPACK takes

> knowledge of the existence, components and capabilities of various technologies as they are used in teaching and learning settings, and conversely, knowing how teaching might change as a result of using particular technologies (Mishra & Koehler, 2006, p. 1028).

Moreover, while TPACK makes no reference to affective aspects, PTK explicitly considers teachers' orientation to technology to be an important component in PTK.

Hence, in the current study we chose to retain the lens of double instrumental geneses. In our previous studies this lens was highly relevant and led us to note that teacher ICT competency standards are overemphasized in the professional instrumental genesis. Nevertheless, in this study we decided to use the PTK framework rather than TPACK, as PTK will enable us to consider affective aspects of teachers' work that TPACK cannot capture.

4 Methods

As noted above, in this study we use the PTK framework because it appears to be more relevant for analysing the data sources to gain a better understanding of the specific knowledge and skills needed by mathematics teachers to use ICT efficiently in the mathematics classroom. Indeed, this model acknowledges that teachers must appropriate technology through instrumental genesis processes before using it in the classroom. Before applying the PTK framework in our analysis, we first elaborate each of its three main pillars: instrumental genesis; mathematical knowledge for teaching (Ball et al., 2008); and teacher orientations and goals (Schoenfeld, 2011).

Instrumental genesis: In this research we used the double instrumental genesis proposed by Haspekian (2011). According to this genesis, teachers must first acquire basic skills to master the specific technology they intend to use and develop utilization schemes related to this technology (personal instrumental genesis). They must also develop their understanding of how to support students' mathematics learning in a digital environment (professional instrumental genesis). While the personal genesis may be seen as common to any teacher (though tool specific), the professional genesis is unique to mathematics teachers.

Mathematics knowledge for teaching (*MKT*): Ball et al. (2008) identified six knowledge areas. We have adapted these areas to technology (see Fig. 3) in a modified framework we refer to as Mathematics Digital Knowledge for Teaching (MDKT). Among the six knowledge areas identified in MKT, we focused on the following:

- Teachers' specialized digital content knowledge (SDCK) with respect to the mathematics to be taught;
- knowledge of content and students, which in a technological environment includes additional aspects that may be formulated as knowledge of digital content and students (KDCS);
- knowledge of content and teaching, which in a technological environment may be interpreted as knowledge of digital-content and teaching (KDCT);
- knowledge of content and curriculum in a digital environment, e.g., knowledge of prescribed use of ICT (KDCC).

We believe that SDCK is closely linked to a teacher's personal instrumental genesis. The other three knowledge areas are linked to the professional instrumental genesis, the students' instrumental geneses and the genesis of learning mathematics with digital technology.

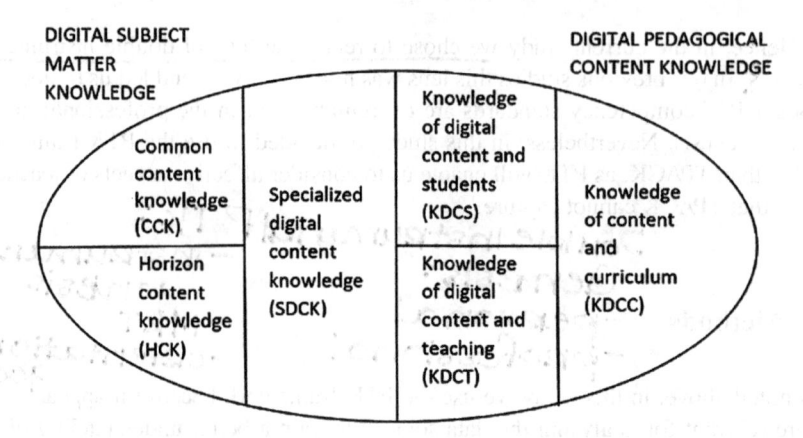

Fig. 3 Mathematical Digital Knowledge for Teaching (MDKT), adapted to a technological environment from Ball et al.'s (2008) Mathematical Knowledge for Teaching (MKT)

Teacher orientations and goals: These are related to affective aspects and teachers' beliefs regarding mathematics, teaching mathematics and technology. What kind of discipline is mathematics and what are its norms and values? What is the teacher's role in teaching mathematics, and how can this role be accomplished? What added value will technology bring to learning mathematics, what affordances and constraints does it involve, and how confident do teachers feel with respect to its use?

5 Findings

In this section, we use the lens of the MDKT theoretical framework to analyse various documents that discuss ICT competencies for teachers. These documents include both international and national documents. At the international level, we were interested in the standards stipulated by UNESCO and by the P21 organisation in collaboration with the National Council of Teachers of Mathematics (NCTM) and the Mathematical Association of America (MAA). These standards have the potential to influence policymakers worldwide. At the national level, we found that Australia and France were among the few countries to define the professional competencies required to integrate ICT in education. Thus, their standards are worth analysing. Finally, we consider a framework developed by US researchers that draws on TPACK (Niess et al., 2009) that will enable us to compare research and institutional points of view on ICT standards.

5.1 Teacher ICT Competencies in UNESCO Documents

UNESCO addresses the issue of ICT in teacher education in its *Planning Guide* (Khvilon & Patru, 2002), which emphasizes the following four competencies (p. 42):

- *Pedagogy* focuses on "teachers' instructional practices and knowledge of the curriculum and requires that they develop applications within their disciplines that make effective use of ICTs to support and extend teaching and learning";
- *Collaboration and Networking* focus on teachers' awareness of "the communicative potential of ICTs to extend learning beyond the classroom walls and the implications for teachers' development of new knowledge and skills";
- *Social Issues* focus on teachers' acknowledgement that "technology brings with it new rights and responsibilities, including equitable access to technology resources, care for individual health, and respect for intellectual property";
- *Technical Issues* are "an aspect of the Lifelong Learning theme through which teachers update skills with hardware and software as new generations of technology emerge".

Whereas the first competency refers to the mathematical knowledge for teaching with technology and implies teachers' professional instrumental genesis, the second and fourth category of competencies clearly indicate the need for an ongoing personal instrumental genesis.

The competencies outlined in the UNESCO *Planning Guide* were further developed into the UNESCO (2011) *ICT Competency Framework for Teachers* (ICT-CFT), which delineates "the competencies required to teach effectively with ICT" (p. 3). This framework stresses that

> it is not enough for teachers to have ICT competencies and be able to teach them to their students. Teachers need to be able to help the students become collaborative, problem solving, creative learners through using ICT so they will be effective citizens and members of the workforce (ibid.).

Hence, the framework is organized into three different approaches to teaching that correspond to three stages of ICT integration. The first is Technology Literacy, which enables "students to use ICT in order to learn more efficiently". The second is Knowledge Deepening, which enables "students to acquire in-depth knowledge of their school subjects and apply it to complex, real-world problems". The third is Knowledge Creation, which enables "students, citizens and the workforce they become, to create the new knowledge required for more harmonious, fulfilling and prosperous societies" (p. 3). It is interesting to note that these stages are formulated in terms of students' abilities to exploit the potential inherent in ICT as a result of how teachers use ICT. These three stages address all aspects of teachers' work, namely understanding ICT in education, curriculum and assessment, pedagogy, ICT, organization and administration, and teacher professional learning (Fig. 4).

The authors of the UNESCO framework claim that

> [t]he successful integration of ICT into the classroom will depend on the ability of teachers to structure the learning environment in new ways, to merge new technology with a new pedagogy, to develop socially active classrooms, encouraging co-operative interaction, collaborative learning and group work. This requires a different set of classroom management

	TECHNOLOGY LITERACY	KNOWLEDGE DEEPENING	KNOWLEDGE CREATION
UNDERSTANDING ICT IN EDUCATION	Policy awareness	Policy understanding	Policy Innovation
CURRICULUM AND ASSESSMENT	Basic knowledge	Knowledge application	Knowledge society skills
PEDAGOGY	Integrate technology	Complex problem solving	Self management
ICT	Basic tools	Complex tools	Pervasive tools
ORGANIZATION AND ADMINISTRATION	Standard classroom	Collaborative groups	Learning organizations
TEACHER PROFESSIONAL LEARNING	Digital literacy	Manage and guide	Teacher as model learner

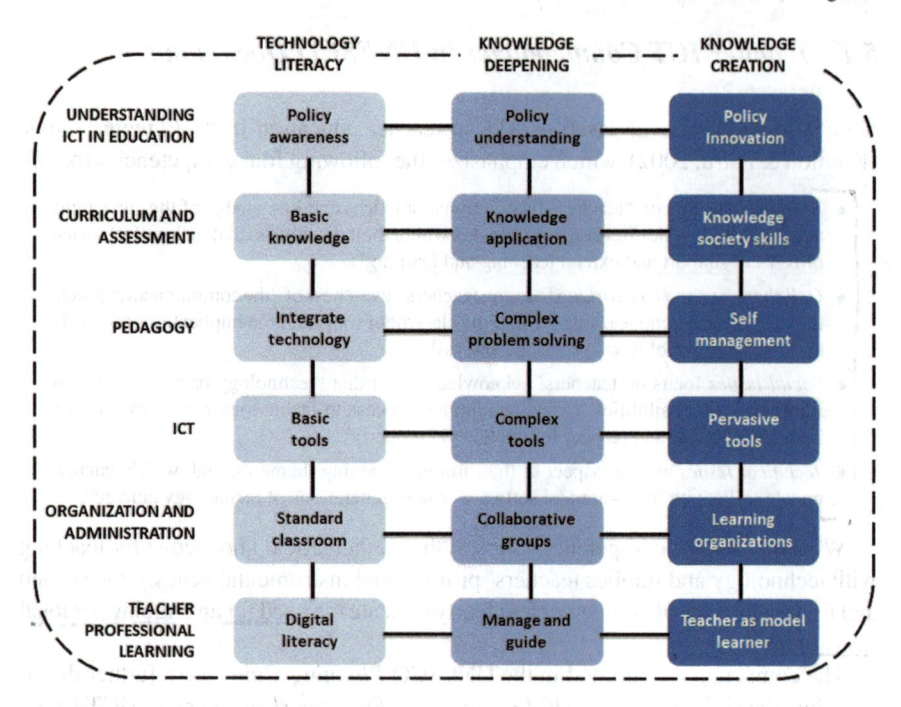

Fig. 4 The UNESCO ICT competency framework for teachers (UNESCO, 2011, p. 13)

skills. The teaching skills of the future will include the ability to develop innovative ways of using technology to enhance the learning environment, and to encourage technology literacy, knowledge deepening and knowledge creation (ibid., p. 8).

The framework specifies the competencies needed by teachers in all aspects of their work. In the Technology Literacy stage, teacher competences […] include basic digital literacy skills and digital citizenship, along with the ability to select and use appropriate off-the-shelf educational tutorials, games, drill-and-practice software, and web content in computer laboratories or with limited classroom facilities to complement standard curriculum objectives, assessment approaches, unit plans, and didactic teaching methods. Teachers must also be able to use ICT to manage classroom data and support their own professional learning (ibid., p. 10).

In the context of the PTK model, "basic digital literacy" can be thought of as resulting from the teacher's personal instrumental genesis and ability to select appropriate resources to "complement […] standard didactic teaching methods" as part of the mathematical knowledge for teaching requiring professional instrumental genesis. Moreover, the framework alludes to teachers' ability to "use ICT to… support their own professional learning". This statement implies the continuous instrumental genesis of teachers, both personal and professional.

The next stage, Knowledge Deepening, refers more directly to teachers' mathematical knowledge for teaching with technology (in square brackets we highlight references to the constellation shown in Fig. 3):

> teacher competences... include the ability to manage information, structure problem tasks, and integrate open-ended software tools and subject-specific applications [SDCK] with student-centred teaching methods and collaborative projects in support of students' in-depth understanding of key concepts and their application to complex, real-world problems [KDCS]. To support collaborative projects, teachers should use networked and web-based resources to help students collaborate, access information, and communicate with external experts to analyze and solve their selected problems [KDCS]. Teachers should also be able to use ICT to create and monitor individual and group student project plans [KDCT], as well as to access information and experts and collaborate with other teachers to support their own professional learning (ibid., p. 11).

knowledge deepening

Finally, in the Knowledge Creation stage, teachers

> will be able to design ICT-based learning resources and environments [KDCT]; use ICT to support the development of knowledge creation and the critical thinking skills of students [KDCS]; support students' continuous, reflective learning [KDCS]; and create knowledge communities for students and colleagues (ibid., p. 14).

knowledge creation

The UNESCO document provides sample syllabi for teacher education that demonstrate ways of operationalizing the ICT competency framework. Table 1 summarizes a few examples of tasks suggested in the syllabi at the three levels: technology literacy (TL), knowledge deepening (KD) and knowledge creation (KC). In the table, we organized the teachers' competencies according to the PTK model and noted the relations to personal and professional instrumental genesis.

These examples of teacher competencies show that the UNESCO ICT framework takes into account teachers' personal and professional ICT knowledge and skills at the TL and KD levels, while at the KC level teachers are thought to have sufficient personal mastery of technology. All technology-related categories of the PTK model are present (except for personal orientations), although the content knowledge itself is not specific to any particular subject matter.

5.2 The P21 Partnership

The Partnership for 21st Century Learning[2] (formerly the Partnership for 21st Century Skills) was founded in 2002 as a coalition that brought together the business community, education leaders, and policymakers to position 21st century readiness at the centre of K-12 education in the US and to kick-start national discourse on the importance of 21st century skills for all students. The organization began by defining a framework for learning that includes students' outcomes and support systems. Currently the framework provides indications of what students should achieve from kindergarten to grade 12. The framework considers mathematics to be one of its nine

---

[2] www.P21.org.

Table 1 Examples of teacher competencies mentioned in the UNESCO ICT framework

	Personal instrumental genesis	Professional instrumental genesis
Teachers should be able to…	TL—Describe the purpose and basic function of graphics software and use a graphics software package to create a simple graphic display (SDCK)	TL—Identify the appropriate and inappropriate social arrangements for using various technologies (technology instrumental genesis)
	TL—Use common communication and collaboration technologies, such as text messaging, video conferencing, and web-based collaboration and social environments (technology instrumental genesis)	TL—Match specific curriculum standards to particular software packages and computer applications and describe how these standards are supported by these applications (KDCC, SDCK)
	TL—Use ICT resources to support their own acquisition of subject matter and pedagogical knowledge (SDCK, KDCT)	TL—Incorporate appropriate ICT activities into lesson plans so as to support students' acquisition of school subject matter knowledge (KDCC, KDCT, KDCS)
	KD—Identify or design complex, real-world problems and structure them in a way that incorporates key subject matter concepts and serves as the basis for student projects (KDCC, KDCT)	KD—Structure unit plans and classroom activities so that open-ended tools and subject-specific applications will support students in their reasoning with, talking about, and use of key subject matter concepts and processes while they collaborate to solve complex problems (KDCT, KDCS)
	KD—Operate various open-ended software packages appropriate to their subject matter area, such as visualization, data analysis, role-play simulations, and online references (SDCK)	KC—Help students reflect on their own learning (KDCS)

Table 2 Analysis of instructional pillars from the P21 framework

Learning and instructional pillars	Analysis
Understanding applications of learning theory; innovative uses of digital tools to support learning and the importance of incorporating global contexts and perspectives into classroom instruction	Professional instrumental genesis; KDCT, KDCC
Investigating, designing and synthesizing innovative curriculum, technology tools and best practices from diverse sources to implement and integrate global content into classroom instruction	Professional instrumental genesis; KDCT, KDCC
Connecting and collaborating with peers in professional learning communities to advance the field of global education	Professional instrumental genesis
Integrating global attitudes, skills and knowledge into curriculum, instruction and assessment	Professional instrumental genesis; Orientation

core content knowledge areas and published an ictmap_math[3] that lists specific mathematics competencies for grades 4, 8 and 12. In addition, the organisation also set *a global-ready teacher competency framework with standards and indicators*, based on the understanding that a document setting out indications for students requires a parallel document for teachers.

The P21 standards for teachers enumerate three instructional practices: "Continually developing understanding of and applications for inquiry-based pedagogical approaches; integrating global content into curriculum; and utilizing next-generation technologies in curriculum practices". The first of these practices reflects the professional instrumental genesis, the second is related to SDCK from the MDKT framework, and the third reflects KDCC.

In addition to enumerating instructional practices, the P21 framework also defines global competence characteristics organized according to four learning and instructional pillars (Table 2). The righthand side of the table includes our classification of each competence characteristics.

The P21 framework for teachers is divided into three domains of practice: pedagogy, content and technology. Each domain includes two standards that outline expertise and leadership characteristics required by global-ready teachers. Each standard is defined by indicators that identify the attitudes, skills and knowledge needed for global competence. Due to space limitations we cannot elaborate the entire framework. We note, however, that the emphasis on attitudes in each domain of practice points to the importance of teachers' orientation towards technology, as mentioned by Thomas and Palmer (2014).

---

[3] http://www.p21.org/storage/documents/P21_Math_Map.pdf.

5.3 The Australian Digital Competencies Standards for Teachers

The Australian Institute for Teaching and School Leadership (AITSL) published two core frameworks regarding teaching and teacher education—*Australian Professional Standards for Teachers* (AITSL, 2011) and *Accreditation of Initial Teacher Education Program in Australia: Standards and Procedures* (AITSL, 2012). The second document was specifically aimed at providing institutions with teacher education programs some program planning guidelines. Three points in the guidelines explicitly mention ICT: 2.6—"implement teaching strategies for using ICT to expand curriculum learning opportunities for students"; 3.4—"Demonstrate knowledge of a range of resources, including ICT, that engage students in their learning"; and 4.5—"Demonstrate an understanding of the relevant issues and the strategies available to support the safe, responsible, and ethical use of ICT in learning and teaching". The guidelines do not refer directly to the personal instrumental genesis and assume that the technological knowledge was developed prior to participating in the teacher education programs. The three points address only the professional instrumental genesis. Examining the guidelines through the lens of the MDKT framework, we can associate point 2.6 with KDCS and point 3.4 with KDCC and KDCT. The last point (4.5) is related to the ethical implications of ICT use and is not represented in the MDKT framework.

The Australian teacher education system has already begun changing its teacher education programs to meet these guidelines. Lloyd (2014) identified four different models for such programs: (1) independent, in which ICT is taught as a separate course; (2) embedded, in which ICT becomes embedded in all courses taught; (3) hybrid, in which ICT is partially separate yet has some modules connected to other subjects; and (4) modified hybrid, which is a variant of the hybrid model. The programs are motivated by the guidelines yet diverge from them, seemingly acknowledging the need for supporting teachers' personal ICT use in the future, thus referring to their students' personal instrumental genesis.

5.4 The French Digital Competencies Standards for Teachers

Until 2014, France was one of the European countries that required a certificate of digital skills known as a "certificate of computer science and Internet"[4] to become a primary or secondary teacher. Since 2014, this certification has been integrated into pre-service teacher education. The certification requirement was instituted in 2010 to guarantee that teachers had the professional skills in pedagogical use of digital technologies necessary for exercising their profession.

---

[4]https://c2i.enseignementsup-recherche.gouv.fr/enseignant/quelles-competences-pour-le-c2i2e.

The national standards for digital competencies needed for certification include two main parts: (A) general skills related to exercising the profession, and B) skills needed for integrating ICT into teaching practice. The general skills (part A) are organized into three domains: (A1) mastery of professional digital environment (e.g., "select and use the most appropriate tools to communicate with the actors and users of the education system"); (A2) development of skills for lifelong learning (e.g., "use online resources or distance learning devices for self-training"); For self-training"); and (A3) professional responsibility in the education system (The national standards for digital competencies needed for certification include two main parts: (A) general skills related to exercising the profession, and (B) skills needed for integrating ICT into teaching practice. The general skills (part A) are organized into three domains: (A1) mastery of professional digital environment (e.g., "select and use the most appropriate tools to communicate with the actors and users of the education system"); (A2) development of skills for lifelong learning (e.g., "use online resources or distance learning devices e.g., "take into account the laws and requirements for professional use of ICT"). Development of these competencies contributes to fostering values and attitudes towards ICT use (personal orientations) and supports teachers' personal instrumental genesis of technology (e.g., A2 sub-domain).

The skills for ICT integration (part B) are classified in four domains: (B1) networking while using collaborative tools (e.g., "search, produce, index, and share documents, information, resources in a digital environment"); (B2) design and preparation of teaching content and learning situations (e.g., "design learning and assessment situations using software that is general or specific to the subject matter, field and school level"); (B3) pedagogical enactment (e.g., "manage diverse learning situations by taking advantage of the potential of ICT (group, individual, small groups work)"); and (B4) implementation of assessment techniques (e.g., "use assessment and pedagogical monitoring tools"). These skills are clearly related to MDKT, especially KDCC (B2 sub-domain) and KDCT (B3 and B4 sub-domains) and require professional instrumental genesis of technology that is specific to subject matter.

These standards thus cover various aspects of the teaching profession. Teaching is not limited to classroom activity. Rather, it takes into account both personal and professional mastery of ICT. The standards apply to all teachers, regardless of school level and subject matter taught.

5.5 Mathematics Teacher TPACK Standards

Niess et al. (2009) contend there is a need for "content-specific ideas that address what students or teachers should know about using technology for learning mathematics". The authors elaborated standards developed around the TPACK model that aim to provide a "framework for guiding professional practice that supports the improvement of mathematics teaching and learning". These standards are organized into four themes that encompass the knowledge and beliefs teachers demonstrate when incorporating technology into mathematics teaching and learning:

I. ***Designing and developing digital-age learning environments and experiences.*** *Teachers design and develop authentic learning environments and experiences while incorporating appropriate digital-age tools and resources to maximize mathematical learning in context.*
II. ***Teaching, learning and the mathematics curriculum.*** *Teachers implement curriculum plans that include methods and strategies for applying appropriate technologies to maximize student learning and creativity in mathematics.*
III. ***Assessment and evaluation.*** *Teachers apply technology to facilitate a variety of effective assessment and evaluation strategies.*
IV. ***Productivity and professional practice.*** *Teachers use technology to enhance their productivity and professional practice.*

In terms of the PTK framework, these themes are related primarily to the mathematical knowledge required for teaching with technology (MDKT). The indicators refer to:

- SDCK: "facilitate technology-enhanced mathematical experiences that foster creativity"—Theme II.
- KDCS: "design appropriate mathematical learning opportunities that incorporate worthwhile mathematical tasks, based on current research and that apply appropriate technologies to support the diverse needs of all students in learning mathematics"—Theme I; incorporate knowledge of all students' understandings, thinking, and learning of mathematics with technology"—Theme II.
- KDCC: "identify, locate, and evaluate mathematical environments, tasks, and experiences in the curriculum to integrate digital technology tools for supporting students' individual and collaborative mathematical learning and creativity"—Theme II.
- KDCT: "use technology to support learner-centered strategies that address the diverse needs of all students in learning mathematics as these strategies help students become responsible for and reflect on their own learning"—Theme II; "use formative assessment of technology-enhanced student learning to evaluate students' mathematics learning and to adjust instructional strategies"—Theme III.

Theme IV addresses teachers' attitudes as reflective practitioners (e.g., "evaluate and reflect on the effective use of existing and emerging technologies to enhance all students' mathematical learning"), which can be linked to their personal orientations.

The authors further question how teachers can develop this TPACK knowledge and suggest the following five-stage developmental process inferred from their observation of a number of teachers who use spreadsheets in their mathematic classes:

1. *Recognizing (knowledge), where teachers are able to use the technology and recognize the alignment of the technology with mathematics content yet do not integrate the technology in teaching and learning of mathematics.*
2. *Accepting (persuasion), where teachers form a favorable or unfavorable attitude toward teaching and learning mathematics with an appropriate technology.*

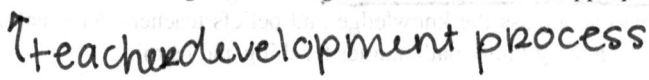

3. *Adapting (decision), where teachers engage in activities that lead to a choice to adopt or reject teaching and learning mathematics with an appropriate technology.*
4. *Exploring (implementation), where teachers actively integrate teaching and learning of mathematics with an appropriate technology.*
5. *Advancing (confirmation), where teachers evaluate the results of the decision to integrate teaching and learning mathematics with an appropriate technology.*

These stages seem to acknowledge that teachers must first express an interest in using technology for teaching and learning mathematics in order to develop a positive attitude (personal orientation) toward it and use it for personal purposes (personal instrumental genesis leading to the development of SDCK). Only after that can they integrate technology in their professional practice (professional instrumental genesis).

6 Summary and Conclusion

At the outset of this chapter, we stated that our aim in the current study was to better understand what specific knowledge and skills mathematics teachers need to use ICT efficiently in the mathematics classroom. We examined documents developed at the international level as well as the national level. In this section we summarise our findings according to the three main components of the PTK framework: personal orientation, teachers' personal and professional instrumental genesis, and MDKT. We then conclude by proposing what we believe is a better way to formulate the knowledge and skills teachers need to teach mathematics in a digital environment.

Not all of the examined documents consider teachers' **personal orientation** towards technology integration. The UNESCO framework did not mention any affective aspects, nor did the Australian documents. Though this absence is surprising, it is in line with Thomas and Palmer's (2014) claim:

> We believe that this latter aspect of teacher orientations and their effect on confidence in using technology has been given less attention in research and development than it deserves. (p. 76).

Indeed, affective aspects must be taken into account in examining how learning takes place in mathematics. These aspects can also help explain teachers' decision-making in class (Schoenfeld, 2011). Clearly, a positive orientation towards technology can serve as an important driving force for technology integration. On the other hand, two of the international documents—the P21 organization document and the one by Niess et al. (2009)—did refer to orientation. P21 explicitly referred to the importance of teachers' positive attitudes in each of its three domains of practice: pedagogy, content and technology. Niess et al. (2009) cite the need to develop positive attitudes at the first two or three levels of teacher technology integration. Consideration of affective issues may appear surprising in the Niess framework as it is explicitly based on the knowledge-based TPACK model. Hence, this inclusion

Table 3 References to elements from the MDKT framework

	UNESCO	P21.org	Niess et al.	Australia	France
SDCK	+	+	+		
KDCK	+		+	+	
KDCT	+	+	+	+	+
KDCC	+	+	+	+	+

underscores the importance of considering more than cognitive aspects in teachers' digital competencies. At the national level, in France the "general skills related to the exercise of the profession" in sub-domain A2 also refer to positive orientation as a necessary aspect. We believe that the importance of personal orientation is perhaps underestimated as a major factor in teacher competencies.

The **double instrumental genesis** expected of teachers refers both to their personal genesis in using digital tools in general and to their professional genesis in using digital tools for teaching mathematics to their students. While all the documents acknowledge the professional genesis, the P21 and the Australian frameworks both take the personal instrumental genesis for granted and do not mention it at all. We contend that disregarding the personal instrumental genesis is a mistake for two reasons. First, the technological skills of the current teacher population vary considerably. Hence, developing teachers' personal use of digital mathematical software is a prerequisite for developing their professional genesis in using the same tools as instruments for mathematics teaching. Second, because of the rapid evolution of digital tools, we believe that even two decades from now, when the workforce of teachers is based on those who are currently K-12 students, organizations cannot assume that personal genesis will emerge on its own.

With respect to **Mathematics Digital Knowledge for Teaching (MDKT)**, various documents refer to different components of this framework, as summarized in Table 3. Clearly, there are variations in the ways these documents refer to the various components of knowledge. The research-based framework suggested by Niess et al. (2009) addresses four knowledge domains. Of the four institutional documents we analysed, only the UNESCO framework also refers to the four knowledge domains. All five documents refer to knowledge of content and teaching in a digital environment (KDCT) and to knowledge of digital content and curriculum (KDCC). Reference to KDCT is understandable, as teaching competencies and skills are essential in all the frameworks. Moreover, all of them place major emphasis on content.

Even those organizations that do refer to the various aspects we have deemed essential in such frameworks still need to operationalize these standards, first for teacher education and professional development programs and second for building support systems for teachers in their immediate working environments. Indeed, the documents we analysed seem to converge toward general ICT competencies, such as teachers' ability to design relevant technology-mediated tasks and learning environments. Nevertheless, specific goals leading to the development of such general competencies need to be identified to help teacher educators devise professional

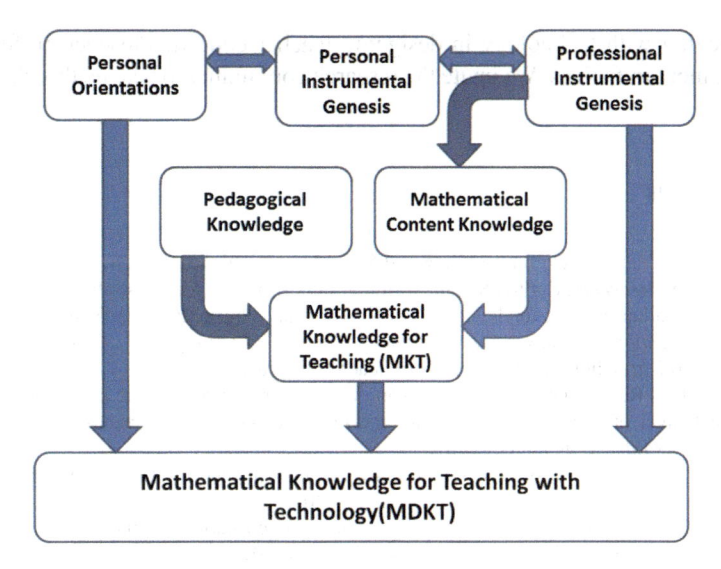

Fig. 5 Mathematical knowledge for teaching with technology (MDKT) framework

development programmes. It is interesting to note that both the UNESCO framework and the one elaborated by Niess et al. (2009) also converge to suggest that teacher professional development toward ICT integration emerges in several stages. Teachers must first perceive the potential of technology and its possible contribution to teaching and learning mathematics in order to adopt it and start using it personally as a mathematical instrument. This stage must take place before they consider using technology in the classroom. These stages should contribute to the development of specific knowledge and skills that need to be identified and formalized. We believe that such a goal may serve as a research agenda for the mathematics education research community.

To conclude, let us return to our theoretical framework, and in particular to our choice of the pedagogical technological knowledge (PTK) model. This choice turned out to be highly relevant as it enabled us to showcase issues we would have not been able to capture with other frameworks, such as the TPACK framework used in our previous studies. Indeed, the PTK model enabled us to understand the importance of teachers' personal orientation component as a crucial aspect of ICT integration. Similarly, personal instrumental genesis appears to be a prerequisite for professional genesis. Therefore, we propose adapting the PTK framework to emphasize the decisive role played by those components of mathematical knowledge for teaching with technology that are related to teacher orientations and to personal and professional instrumental genesis (Fig. 5).

Finally, we believe that the next challenge facing us as a research community is to inform policymakers at both the international and national levels about our findings. Publishing our findings in this book may be a first step towards alerting the education system to the importance of using explicit mathematics knowledge for teaching with technology in designing teacher education/teacher professional development initiatives. We invite the research community to take up this challenge.

References

AITSL (Australian Institute for Teaching and School Leadership). (2011). *National professional standards for teachers*. Carlton South, Australia: Education Services Australia.

AITSL (Australian Institute for Teaching and School Leadership). (2012). *Accreditation of initial teacher education programs in Australia: Guide to the accreditation process April 2012*. Retrieved August 2017 from https://www.aitsl.edu.au/teach/standards.

Ball, L. (2004). Researchers and teachers working together to deal with the issues, opportunities and challenges of implementing CAS into the senior secondary mathematics classroom. *ZDM Mathematics Education, 36*(1), 27–31.

Ball, L. D., Thames, M. H., & Phelps, G. (2008). Content knowledge for teaching: What makes it special? *Journal of teacher education, 59*(5), 389–407.

Borba, M. C. (2012). Humans-with-media and continuing education for mathematics teachers in online environments. *ZDM Mathematics Education, 44*, 801–814.

Bowers, J. S., & Stephens, B. (2011). Using technology to explore mathematical relationships: A framework for orienting mathematics courses for prospective teachers. *J. of Math Teacher Education, 14*, 285–304.

Clay, E., Silverman, J., & Fischer, D. J. (2012). Unpacking online asynchronous collaboration in mathematics teacher education. *ZDM Mathematics Education, 44*, 761–773.

Emprin, F. (2010). A didactic engineering for teachers education courses in mathematics using ICT. In V. Durand-Guerrier et al. (Eds.), *Proceedings of the Sixth Congress of the European Society for Research in Mathematics Education (CERME6)* (pp. 1290–1299). Lyon: INRP.

EU. (2006). Recommendation 2006/962/EC of the European Parliament and of the Council of 18 December 2006 on key competences for lifelong learning. *Official Journal of the European Union, L 394*, 10–18.

Haspekian, M. (2014). Teachers' Instrumental Geneses When Integrating Spreadsheet Software. In A. Clark-Wilson, O. Robutti, & N. Sinclair (Eds.), *The mathematics teacher in the digital era: An international perspective on technology focused professional development* (pp. 241–276). Dordrecht: Springer.

Haspekian, M. (2011). The co-construction of a mathematical and a didactical instrument. In M. Pytlak, T. Rowland, & E. Swoboda (Eds.) *Proceedings of the 7th Congress of the European Society for Research in Mathematics Education (CERME7)* (pp. 2298–2307). Rzeszów: University of Rzeszów.

Hegedus, S., Laborde, C., Brady, C., Dalton, S., Siller, H.-S., Tabach, M., Trgalová, J., & Moreno-Armella, L. (2017). *Uses of technology in upper secondary mathematics education*. ICME13 Topical Surveys. Springer. Retrieved December 14th 2017 at https://link.springer.com/content/pdf/10.10072F978-3-319-42611-2.pdf.

Jones. (2004). *A review of the research literature on barriers to the uptake of ICT by teachers*. British Educational Communications and Technology Agency (BECTA). Retrieved December 14, 2017 at http://dera.ioe.ac.uk/1603/1/becta_2004_barrierstouptake_litrev.pdf.

Khvilon, E., & Patru, M. (2002). *Information and communication technologies in teacher education. A planning guide*. Paris: UNESCO. Retrieved December 14, 2017 from http://www.unesco.org/new/fr/communication-and-information/resources/publications-and-communication-materials/publications/full-list/information-and-communication-technologies-in-teacher-education-a-planning-guide/.

Lee, H. S. (2005). Facilitating students' problem solving in a technological context: Prospective teachers' learning trajectory. *Journal for Mathematics Teacher Education, 8*, 223–254.

Lloyd, M. (2014). ICT in Teacher Education in the Age of AITSL. In *Paper presented at the ADELAIDE conference*.
Mishra, P., & Koehler, M. J. (2006). Technological pedagogical content knowledge: A new framework for teacher knowledge. *Teachers College Record, 108*(6), 1017–1054.
Niess, M. L., Ronau, R. N., Shafer, K. G., Driskell, S. O., Harper, S. R., Johnston, C., ... & Kersaint, G. (2009). Mathematics teacher TPACK standards and development model. *Contemporary Issues in Technology and Teacher Education, 9*(1), 4–24.
P21.org (Nb). Retrieved November 2017 from www.P21.org.
Rabardel, P. (2002). *People and technology—a cognitive approach to contemporary instruments*. Université Paris 8. Retrieved October 15, 2013 https://hal-univ-paris8.archives-ouvertes.fr/file/index/docid/1020705/filename/people_and_technology.pdf.
Robová, J. (2013). Specific skills necessary to work with some ICT tools in mathematics effectively. *Acta Didactica Mathematicae, 35,* 71–104.
Robová, J., & Vondrová, N. (2015). Developing future mathematics teachers' ability to identify specific skills needed for work in GeoGebra. In K. Krainer & N. Vondrová (Eds.) *Proceedings of the 9th Congress of the European Society for Research in Mathematics Education (CERME9)* (pp. 2396–2402). Prague: Charles University.
Schoenfeld, A. H. (2011). Toward professional development for teachers grounded in a theory of decision making. *ZDM Mathematics Education, 43*(4), 457–469.
Shulman, L. S. (1986). Those who understand: Knowledge growth in teaching. *Educational researcher, 15*(2), 4–14.
Tabach, M., & Trgalová, J. (2017). In search for standards: Teaching mathematics in a technological environment. In G. Aldon, & J. Trgalová (Eds.), *Proceedings of the 13th International Conference on Technology in Mathematics Teaching* (ICTMT13) (pp. 293–300). Lyon, France. Retrieved Nov. 2017 from https://ictmt13.sciencesconf.org/data/pages/proceedings_compressed.pdf.
Thomas, M. O. J., & Hong, Y. Y. (2005). Teacher factors in integration of graphic calculators into mathematics learning. In H. L. Chick & J. L. Vincent (Eds.), *Proceedings of the 29th conference of the international group for the psychology of mathematics education* (Vol. 4, pp. 257–264). Melbourne: University of Melbourne.
Thomas, M. O. J., & Palmer, J. M. (2014). Teaching with digital technology: Obstacles and opportunities. In A. Clark-Wilson et al. (Eds.), *The mathematics teacher in the digital era: An international perspective on technology focused professional development* (pp. 71–89). https://doi.org/10.1007/978-94-007-4638-1_4.
Trgalová, J., & Jahn, A. P. (2013). Quality issue in the design and use of resources by mathematics teachers. *ZDM Mathematics Education, 45,* 973–986.
Trgalová, J., & Tabach, M. (2018). In search for standards: Teaching mathematics in technological environment. In L., Ball, P., Drijvers, S., Ladel, H-S., Siller, M., Tabach, & C., Vale, (Eds.), *Uses of technology in primary and secondary mathematics education: Tools, topics and trends* (pp. 387–397). Cham: Springer.
UNESCO. (2011). *ICT competency framework for teachers*. Paris: United Nations Educational, Scientific and Cultural Organization and Microsoft. Retrieved December 14, 2017 from http://unesdoc.unesco.org/images/0021/002134/213475e.pdf.
Voogt, J., Fisser, P., Pareja Roblin, N., Tondeur, J., & van Braak, J. (2012). Technological pedagogical content knowledge—A review of the literature. *Journal of Computer Assisted learning, 29*(2), 109–206. https://doi.org/10.1111/j.1365-2729.2012.00487.x.
Zehavi, N., & Mann, G. (2011). Development process of a praxeology for supporting the teaching of proofs in a CAS environment based on teachers' experience in a professional development course. *Technology, Knowledge, Learning, 16,* 153–181.
Zuccheri. (2003). Problems arising in teachers' education in the use of didactical tools. In M. A. Mariotti (Ed.) *Proceedings of the 3rd Conference of the European Society for Research in Mathematics Education*. Bellaria: University of Pisa.

Co-Design and Use of Open Online Materials for Mathematics and Science Didactics Courses in Teacher Education: Product and Process

Theo van den Bogaart, Paul Drijvers and Jos Tolboom

1 Introduction

The design and use of online materials for learning have been in the spotlight of educational development over the last decade. Notions of blended learning (Bonk, & Graham, 2006) and flipping classrooms (Nwosisi, Ferreira, Rosenberg, & Walsh, 2016; O'Flaherty, & Phillips, 2015; Tucker, 2012) have given rise to an immense growth of online educational resources, that in many cases are the product of processes of co-design in teams of teachers, designers or researchers. These resources facilitate online learning, which is claimed to provide opportunities for increased educational quality, and for more flexible and effective learning (Garrison, & Kanuka, 2004; O'Flaherty, & Phillips, 2015).

To our experience, online learning is particularly gaining momentum with respect to courses that concern subject knowledge, such as courses on calculus in applied mathematics curricula, or on statistics for social science studies. With respect to didactical courses, however, we consider the potential of online and blended learning to be underused. This is probably the case because the transfer between didactical theory and teaching practice, so crucial in didactics courses, makes the design of such a course more complex and subtle. Also, the learning goals of didactical courses often include a mixture of skills, knowledge and attitudes, a mixture that is difficult to address in an online setting. A first challenge of the study presented here, therefore, is to address this complexity and subtlety through designing online learning units that facilitate the transfer from theory to practice, in this case for pre-service mathematics and science teachers.

A second challenge when designing online learning units is the time and energy needed to really produce them, particularly if the target group is relatively small and the budget and time for creating materials are limited. For courses in mathematics and science didactics in the Netherlands, teacher education is relatively small scale, and educators in many cases work in isolation and deal with high time constraints. Despite the existence of a successful cooperative network of Dutch STEM teacher education centres (ELWIeR-ECENT[1]), we observe that the development of the education of STEM didactics is under pressure and that new initiatives in this field are more than welcome (Verhoef, Drijvers, Bakker, & Konings, 2014). As a consequence, it seems logical to try to collaborate with different institutions when it comes to the design of online learning units. The second challenge addressed in the study, therefore, is how to enhance the co-design of online learning units for STEM teacher education across different teacher education institutes.

In this chapter, we address these two issues. First, how do we cope with the challenge of designing online learning units on mathematics and science didactics for pre-service teacher education? How are online learning units for pre-service teacher education for secondary mathematics in a blended learning context designed, implemented and evaluated? Secondly, besides these product-oriented questions, we are also interested in the ways collaboration took place: How can the process of co-design between teacher educators from different institutes be enhanced? To address these questions, we will describe the design, use, and evaluation of two online learning units for pre-service teacher mathematics education, one on mathematical thinking and the other on statistics didactics, as well as the co-design process. As a result, we identify transferable design heuristics and process model characteristics.

2 Theoretical Framework

The theoretical framework that guided this study includes two main lenses, one on online and blended learning, and the other on the co-design of learning units. We will outline them now, and next phrase the study's research questions.

2.1 Online and Blended Learning

Obviously, it is the responsibility of teacher education institutes to ensure that their students, being prospective mathematics and science teachers, not only master the domain knowledge, but also have the skills to adequately teach it. For instance, prospective teachers should be able to exploit the potential of information and com-

---

[1] See https://elbd.sites.uu.nl/ (in Dutch).

munication technology (Hegedus et al., 2017). In teacher education, the possibilities of online learning and blended learning in the domain of didactics nevertheless remain largely unexplored.

When addressing this responsibility, blended learning comes into play. Roughly speaking, blended learning means blending face-to-face education with online learning activities. Nowadays, more than twenty-five years after the introduction of the worldwide web as part of the internet (Berners-Lee, 1989), a staggering amount of digital resources for the teaching and learning of mathematics is available online. This leads educational designers and teachers to selecting, re-designing and arranging resources to orchestrate their students' learning (Drijvers, Doorman, Boon, Reed, & Gravemeijer, 2010). For the case of teacher education, however, and for courses on domain-specific didactics in particular, the affordances of blended learning remain largely unexplored.

In higher education, blended learning has been on the rise since the early 2000's. With respect to terminology, quite a few buzz words came along. In fact, one might wonder if educational goals have fundamentally changed since researchers from the University of Illinois in 1960 utilized a mainframe computer with work stations for their students for computer assisted learning, which they called Programmed Logic for Automatic Teaching Operations [PLATO, see Woolley (1994)]. Terminology evolved from computer-assisted (or-based or-supported) learning to intelligent tutoring systems (Anderson, Corbett, Koedinger, & Pelletier, 1995), E-learning (Clark, & Mayer, 2008), with blended learning as a popular teaching approach nowadays (Bonk, & Graham, 2006). In retrospective, all terminology boils down to roughly the same issue, i.e., how to arrange the educational resources—including information and communication technology—into an educational design that optimizes learning? What we appreciate in the term 'blended learning' is that it explicitly points at the fact that there is more than one medium to be addressed when designing instruction.

From the perspective of learning theory, scientific insights have evolved as well: from the behaviourist view on human learning (Skinner, 1954), suitable for computer assisted mastery learning (Skinner, 1958), to the nowadays accepted social constructivist view, as initiated by Vygotsky (1962), which can be supported by a more open learning environment. Blended learning is a technological paradigm that suits this view on learning and teaching.

A major didactical issue with respect to blended learning is how to arrange the interplay between online, web-based activities (Tolboom, 2004) and face-to-face activities, and how to design such arrangements. In the case of small-scale courses in mathematics and science didactics, it is important to keep in mind that position of such courses, content, size, and approach differ between the teacher education institutes. Also, each educator wants to be able to add a particular focus or flavour to it. Therefore, the online parts of the blended courses should be very flexible and offer opportunities to function as building blocks for adaptation to a particular course in a particular institute.

2.2 Co-design of Online Learning Units

Pre-service teacher education in mathematics and science takes place in different teacher education institutes in the Netherlands, and in many cases have a limited number of students. Besides this, the national curricula and pedagogical culture complicate the use of international materials. For these reasons, is seems beneficial that educators from the different institutes engage in a process of co-design to develop online learning units. Some researchers report persistent tensions in co-design teams (Kvan, 2000; Penuel, Roschelle, & Shechtman, 2007), but others point at good practices in other fields than education and formulate design guidelines for successful teams (Coburn, & Penuel, 2016). As is more often the case with new phenomena, there is some terminological confusion about what precisely co-design, or co-creation, or research-practice partnerships consist of. In this study, we are pragmatic in choosing the term 'co-design', and read it as 'a collaborative effort of a team of mathematics teacher educators in designing and developing learning units'. Some Dutch experiences with the co-design approach have turned out to be effective, such as the co-design of the handbooks of mathematics didactics (Drijvers, Streun, & Zwaneveld, 2012) and science didactics (Kortland, Mooldijk, & Poorthuis, 2017) and a series for bachelor teacher education (van den Bogaart, Daemen, & Konings, 2017). Also, a limited collection of online materials was designed and stored, and made available online as the Knowledgebase Mathematics (Staal, 2006). Co-design of online learning units, however, seems to become more common in higher education in the Netherlands (Baas et al., 2017), and connects to the phenomenon of co-creation in vocational education (Butter, & Schamhart, 2017).

The above experiences have shown that the co-design of educational materials can overcome its challenges and indeed may lead to high-quality didactical materials. As an important side-effect of engaging in a co-design process, we would like to point out the professional development reported by the participating teacher educators. The constructive, in-depth discussion of educational content and didactics, that is inherent in the co-design process, leads to increasing knowledge and skills among the participants, and to more coherent views on teacher education across the different institutes. As such, a co-design team may act as a community of practice (Wenger, 1998), in which knowledge and experiences are co-created and shared.

2.3 Research Questions

The challenges identified in the introduction and the above theoretical lenses lead to three research questions that the study presented here would like to answer.

1. Which features can be identified in the online learning units on mathematics and science didactics produced for teacher education?
2. How can a process of co-design, in which teacher educators design such online learning units, be organized?

3. How do educators and students experience the use of the online learning units that result from the design process?

3 Methods

3.1 Research Context

In 2013, new curricula in upper secondary education (grades 9–12) were implemented in the Netherlands for the natural sciences, and the mathematics curricula followed in 2015. These revised curricula included some new overall perspectives: for science, micro-meso-macro thinking was highlighted, whereas mathematical thinking was an overarching new element in the mathematics curricula. More specifically, in the mathematics curricula for pre-social science students, new approaches to statistics education were introduced, based on large data sets made available through the use of ICT. The crucial factor in curriculum innovation, however, is to make these innovations impact on classroom practice (Anderson, 1997; Fullan, 2007) and teachers play an important role in it. Therefore, teacher education institutes needed to reconsider their curricula as well. Also, most institutes for higher education in the Netherlands were considering forms of blended and online learning. From these perspectives, the study was the right thing to do at the right moment. It was a small, fourteen month project granted by the Dutch ministry of education and supervised by SURFnet, the collaborative organisation for ICT in Dutch education and research.[2]

3.2 Research Design

To address the three research questions phrased in the previous section in the available time frame, the project had the character of a design study with one cycle, consisting of three phases: an initial design phase, a field test phase, and a revision and conclusion phase.

In the initial design phase, participants were twelve teacher educators, six in mathematics and six in science teacher educations. Four design teams were set up. Each design team consisted of three teacher educators: one from the HU University of Applied Sciences, one from Utrecht University, and one from another teacher training institute in the Netherlands. The latter would facilitate dissemination and bring in a wider view. Most of the designers were experienced teacher educators, who had only limited experience with (the design of) blended learning resources. Within the design teams, some colleagues knew each other and others didn't.

---

[2] See https://www.surf.nl/en/innovationprojects/customised education.html.

At the start of the project, it was decided to focus on two themes in the didactics of mathematics in secondary education that were relevant in the light of the curriculum reform: a more generic one on mathematical thinking and a more specific one on the didactics of statistics. Something similar was done with respect to science teacher education: as a general theme, we chose for micro-meso-macro thinking, needed to understand and use the relations between the observed scientific phenomena at the macro level, the models of the invisible particles at the micro level, and the intermediate meso level. As a specific theme in science, we chose the concept of warmth. Experiences with these co-design trajectories are out of the scope of this chapter.

As each of the designers had limited time for the project (about 40 h over the whole one-year period), the coordinating team—this paper's authors—decided to organize short, intensive collaborative "boot camp" design sessions. During the fall of 2016, three of such one-day boot camps were organized, during which the design teams engaged in their co-design, but informal exchange between teams was also possible. Camera teams were available, as well as tools such as light boards for the production of video clips. During the design process within the design teams, the educators brought in the materials they used in their own teacher education and collected freely available materials, as to build up a shared body of resources.

During these boot camp days, the different teams discussed overarching topics, such as learning unit layout and structure, and possible guidelines for use by teacher educators. During the design process, design heuristics and decisions were monitored. To address the first research question, design heuristics and decisions were observed, and the design process was monitored by this chapter's authors, as to evaluate the process of co-design and its organization. These experiences form the basis for answering the second research question. To facilitate ongoing collaboration and co-design in between the boot camp design meetings, a collaborative online design environment was set up.

Based on these criteria, we chose to use the Dutch online platform Wikiwijs,[3] an open platform for educational resources. Wikiwijs also offers extensive search options based on standardised metadata, which is expected to support the dissemination and use of the designed learning units. This ICT environment was hosted by Kennisnet, a Dutch semi-governmental organisation for ICT in education. In this way, a blended design approach was made possible.

Altogether, data in the initial design phase included the first versions of the online learning units, and field notes of the design process made by the researchers.

In the field test phase, the online learning units were field-tested in didactics courses by teacher educators all over the country, including co-designers and educators not involved in the design. Participants included fourteen educators, nine of whom actually field-tested (part of) one or more units, and their students. Out of the fourteen, nine were mathematics educators and five science educators, so mathematics is slightly overrepresented. To monitor these field-tests, the educators filled in an online questionnaire beforehand, to assess their intentions and ideas. After the field

---

[3] See https://www.wikiwijs.nl.

test, they received a second questionnaire to assess their appreciation of the units as well as the ways in which they used them in practice.

The pre-questionnaire focused on the educator's goals, impressions and expectations, whereas the post-questionnaire focused on their experiences and those of their students (see Appendix for the questionnaires). Initially, some more educators reacted to the emails, indicating that they were not able to pilot the learning units. Therefore, they have not been included in the data; in the meantime, such reactions show the educators' interest and the viability of this approach. The responses to the questionnaires were the main data source that were analysed to answer the third research question. To do so, the responses were coded with respect to the categories mentioned in the questionnaire itself, in a bottom-up, open approach. As the number of reactions was limited and the format was rather open, we were unable to carry out a confirmative coding process or to carry out an interrater reliability.

In the third phase, the revision and conclusion phase, the units were revised by the design teams, based on the feedback from the educators who field-tested them, as well as on the input by an external expert committee. Furthermore, the results were disseminated through different means (workshops, journal papers, and online media) and conclusions were formulated.

4 Results

In this section, we will discuss the study's results according to the three research questions.

4.1 Features of the Online Learning Units

The designed learning units for each of the four themes were published online under a *creative commons* license,[4] which implies that they are freely available for use.[5] For the design process, this required some care in using already existing materials or materials featuring persons not directly involved in the project, for example video data in which students are filmed.

A first important design heuristic that emerged during the design process concerned the way in which the learning activities were arranged and elaborated. To enhance their use in teacher education, we felt the learning units should not be stand-alone materials for individual use by the student, but rather should provide the teacher educator with autonomy and opportunity—as is the case when using a textbook—to include them in a teaching arrangement that does justice to the teacher educator's

---

[4] See https://creativecommons.org/licenses.

[5] The learning units are accessible from https://elbd.sites.uu.nl/2017/11/13/open-online-betadidactiek.

didactical expertise and intended role. This implies that the units should offer the possibility to easily incorporate (parts of) the materials into the learning management system used in the teacher educators' own institution. Also, the materials should allow for use in the arrangements the type of interactions preferred by the educator, such as in blended, face-to-face or online teaching formats.

As a consequence, the designed learning units do not provide ready-to-use and fixed learning trajectories, but instead suggest activities that teacher educators can use as building blocks for activities to be carried out with or by their students. As such, the online units serve two target groups: the pre- and in-service mathematics teachers, but in particular their educators, who have their own ideas for their courses but still need input to further improve them. The online available video materials and literature primarily aim at the former target audience, whereas the suggested activities are meant to serve the educators' needs.

For each of the four themes, the learning units share the same structure. For example, each unit contains a part entitled "For the educator", in which suggestions are provided for the use of the materials in a teacher training context, and a part called "Further reading", in which main literature resources on the topic of the unit are collected and made accessible to students through some annotations and reading guidelines.

Apart from the two overarching features of the learning units, namely the building block character and the shared overall structure, we wanted to provide the four design teams with as much freedom as possible to make their own design choices, also in the light of the project's explorative character. To give an impression of the resulting learning units, we will now briefly describe the two mathematics units.

4.2 Unit 1: Mathematical Thinking

The first case we describe concerns an open online unit about didactics for fostering mathematical thinking. Attention to this topic is evident in the international research community (Devlin, 2012; Schoenfeld, & Grouws, 1992) and was invigorated in the Netherlands by recent curriculum developments in Dutch secondary education. As one of the design team members also developed and taught a course on mathematical thinking as in-service training for teachers, there were already some materials and experiences that could serve as points of departure. As a result, the outlines of the online unit were quickly decided on. The unit was planned to consist of several self-contained student activities divided into three topics: (i) designing classroom tasks that stimulate mathematical thinking, (ii) supporting such classroom tasks in the classroom, (iii) assessing proficiency in mathematical thinking.

For the first topic's inspiration was sought in a key article by Swan, Van 't Hooft, Kratcoski and Unger (2005). This resulted in a set of materials, including a video clip, and a guide for teacher educators how the material could be used. An example from this set is a 'speed date activity' were students are asked to discuss in class differences

Fig. 1 An impression of the learning materials on mathematical thinking

between standard school book exercises and exercises specially designed to stimulate mathematical thinking and then to reflect on this activity in an online message board.

The second topic featured three series of three video clips, labelled A, B and C. Clip A showed two team members discussing the exercise before it was used in practice (see Fig. 1). They tried to predict what kind of thought processes the question would evoke in pupils. Clip B was filmed inside a school building. A pupil was asked to work on the set question, and was then interviewed about the strategies he or she had used. Clip C showed the team members again, but now they reflected on their experiences with the pupils. The film projects were placed on the website together with suggestions for use in teacher education. The suggestions involved a choice for the teacher educator. He could either just use the clips B together with digital copies of the exercises, or use the whole series of clips modelling how to discuss potential thought provoking questions. In the former case, his students can predict and reflect on the quality of the exercises in a whole-class discussion. In the latter case, students can be given the task to try it out themselves with other (e.g., self-designed) exercises in their own classrooms.

Besides these series of video materials, the second topic contained other resources such as several interviews with teachers and an expert about mathematical thinking in the classroom. The third topic centered around authentic pupil's materials, taken from high school assignment.

In retrospective, the most salient feature of this unit is the way in which the secondary school students played a role in the video recordings: for pre-service teachers, it is very important to acquire insight into the way students think, in contrast to their own thinking. Video materials and hand-written student work can be very useful for that, combined with additional analysis and design tasks for the teacher-students. This provides us with a third important feature of online units for teacher education.

4.3 Unit 2: Statistics Didactics

Based on general ideas on exploratory data analysis (Tukey, 1977) and the analysis of large data sets through the use of ICT, the Dutch statistics curricula have been reformed recently. Therefore, statistics didactics is an issue in teacher education and this explains the choice for this topic.

It was noticed that many mathematics teachers, due to their education, only have limited knowledge about statistics and the new approach to it. Therefore, content knowledge should be added to the learning units, intertwined with pedagogical and didactical lenses. Similar considerations were acknowledged in other design teams, and have led to the fourth feature of the learning units: it is important to take into consideration the specific content knowledge that is a prerequisite for a didactical approach, and to include opportunities for teacher students to extend their knowledge by adding knowledge components to the learning units that are essentially not didactical in character.

With this characteristic as point of departure, the design team decided to focus on two key aspects of statistics education, which on the one hand are expected to be beneficial to teachers' content knowledge, and on the other hand involve didactical challenges while teaching. The first focus is called Describing data and concerns data visualization, measurement levels and statistical literacy. The second focus is called Beyond data and concerns answering questions about a population based on a sample. Topics addressed here include correlation and causality, the interpretation of significance, and the meaning of confidence intervals.

In the design process, a mix was made of existing resources such as video clips, text books, research papers, and newly designed resources such as tasks for teacher-students and guidelines for the teacher educator, and dedicated video clips. On the one hand, it made sense to make use as much as possible from existing resources. On the other hand, the need was felt to have dedicated resources that fit well to the specific Dutch situation and curriculum. Figure 2 shows a still from a new clip on measurement levels made with light board technology. Figure 3 shows an extract of a dialog between Dutch mathematics teachers' Facebook group on a particular problem, which is used in the online learning unit to enhance discussion between students during the face-to-face part of the blended course. As an overall approach, misconceptions and confusion with respect to statistical and probabilistic issues served as interesting contexts to address content knowledge and didactics in this domain.

Co-Design and Use of Open Online Materials for Mathematics ... 215

Fig. 2 Still from a video made with light board technology

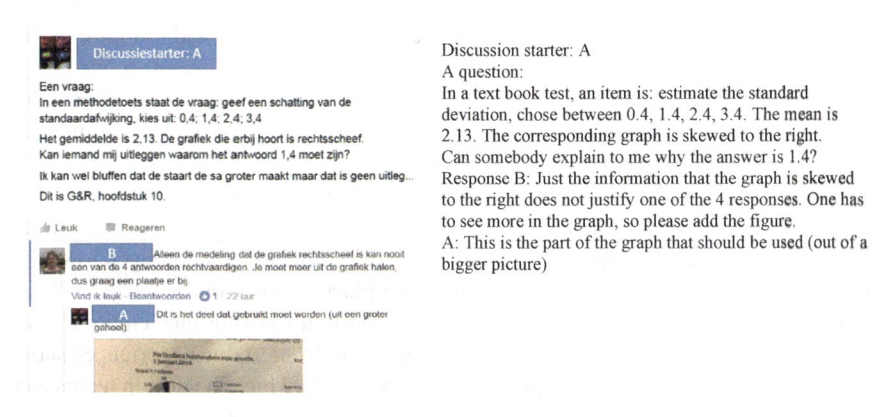

Discussion starter: A
A question:
In a text book test, an item is: estimate the standard deviation, chose between 0.4, 1.4, 2.4, 3.4. The mean is 2.13. The corresponding graph is skewed to the right. Can somebody explain to me why the answer is 1.4?
Response B: Just the information that the graph is skewed to the right does not justify one of the 4 responses. One has to see more in the graph, so please add the figure.
A: This is the part of the graph that should be used (out of a bigger picture)

Fig. 3 Screen dump of a dialog on the Dutch mathematics teachers' Facebook group

4.4 The Process of Co-design

The members of the design teams were acquired through an invitation letter to teacher training institutes. Teacher educators who reacted were contacted to align the purpose and goals of the project and the practical arrangements. As a next step, the authors formed design teams for each of the four topics, each consisting of four designers from different institutes.

To facilitate the co-design beyond face-to-face meetings, and to prepare for the online publication of the learning units, an important choice needed to be made with respect to the online platform to use (Tolboom, 2004). Different requirements played a role for the different target groups. A first requirement for the platform with respect to the end users was that it would make the content freely available without any obstacles. In addition, it should present the multimedia content in an accessible and user-friendly way. Also, it should allow for online collaboration by the design teams. For optimal use for teacher educators, it should allow easy export to specific web-based learning environments (WLO) used in the different teacher training institutes, as well as adaptation within these WLOs. From the financial perspective, finally, we wanted to have a service without any costs, as to increase shared ownership, also beyond the participating teacher training institutes.

In an oral debriefing meeting, the educators indicated that taking part in the design teams had been a personal learning experience, both with respect to their knowledge of the subject matter and the didactics, as to the skills needed to design online learning units for blended learning, including the design of video materials. A limitation of the composition of design teams with members from different institutions, however, was the time needed to get to know each other and to develop a shared view on the topic of and the approach to the learning units. In short, the experiences show that the organization of the design process in small-size design teams of experienced educators enabled them to design rich online learning units, and, through their participation, to engage in a process of professional development. A pitfall may be that much time needs to be spent to developing an overall approach and too little to the actual design.

An important element in the design process was its organization in the boot camp days. During these days, the design teams intensively collaborated, with some plenary, cross-design team meetings to synchronize approaches. The design teams were themselves responsible for their style of working and were technically supported by video technicians. In this way, the design teams on the one hand were quite autonomous, which they appreciated, and in the meantime were encouraged to spend three full days outside of their regular working place to work on the project. Even though it was difficult to schedule these days in this extra-institutional environment, they seemed to be an important organizational factor. In short, the experiences show that the organization of the design process in sessions in which the design teams can collaborate intensively with full attention for the learning units is an efficient and fruitful way to design online units. The attendance of technical support lowers the barriers for the production of video materials.

4.5 *Experiences from Teacher Education Practice*

The main sources for the experiences with the learning units in teacher education practice are the educators' reactions to the pre- and post-field test questionnaires.

The reactions to the pre-test questionnaire show that there was greater interest in the learning units for mathematics than for science, which may be explained by the higher response from mathematics educators. The educators' first impressions of the learning units were positive: the subjects were considered relevant and the presentation was perceived as attractive. The video resources seemed to be the most interesting content. It was appreciated that the units were flexible in that they could also be partially integrated in existing courses. As critical notes, some educators found the units too extensive, both in terms of content and of study load for students. Also, questions were raised on how to really "make a course out of the building blocks", and on the usefulness of the materials for teacher education for lower secondary level. Furthermore, even if the set-up of the units was appreciated, the materials still were not completed and in some cases looked somewhat provisional, which is not a surprise given the stage of the design process when the pre-questionnaires were send out. In the eyes of some of the educators, the learning units might have been more exciting and engaging.

Before the actual field-test, the educators described their goals to do so as to improve the mathematical and didactical content of their course, but also to bring in new dynamics, inspiration and examples that would be applicable in teaching practice. Beforehand, some educators expected to just use the learning units directly in the Wikiwijs platform, whereas others considered inclusion in their institute's WLO. In short, the pre-questionnaires show that the responding educators were very open to the ideas of the project and to using (parts of) the learning units in their didactics courses.

The post-field test questionnaire shows the actual use of the learning units in the educators' courses (see Table 1). Some educators used (parts of) two learning units. Again, Table 1 shows a dominance of mathematics didactics units, compared to the science didactics materials. Most of the units have been used in upper-secondary teacher education. This may be because educators found them more suitable for that than for use in lower secondary education. Our conjecture, however, is that this is mainly caused because of an over-representation of upper-secondary educators in the sample.

The educators' opinions after use were not very different from their impressions beforehand, and overall were positive. Even if improvements on a detailed level were possible, and suggestions for that were provided, and the comments depended on the different units, the educators found them useful for their teacher training practice. Layout, global approach, and accessibility were the suggestions that were most frequent. The learning units could be studied by the students independently. This being said, the educators did struggle with finding ways to embed the online learning units in their courses for different reasons: face-to-face teaching time was limited and it was not easy to decide what to do in the meetings and what to leave over to the online activities. Also, there were existing course materials, and the fine-

Table 1 Number of field tests per learning unit

Learning unit	Number of courses
Mathematical thinking	4
Statistics didactics	4
Micro-meso-macro thinking	2
Warmth didactics	1

tuning between different resources was not always straightforward. Therefore, the actual way to use the learning units in most cases concerned using (part of) it to in the course meetings and leave other parts as online take-home tasks, the results of which for example needed to be uploaded in the students' portfolios.

Most educators were happy with achieving their initial learning goals. This satisfaction not only concerns the learning units, but also the way in which they were used in the frame of the courses, and the suitability for the target group of students. Some of the educators also asked their students to react to the learning units and the results were positive, in particular with respect to the online video resources and the options for variation in activities that the online units allowed for. Students appreciated the freedom to explore the content of the units. Concerning the technical aspect of the integration, most educators provided their students with hyperlinks to the units in the Wikiwijs platform and didn't feel the need to include them in their institute's WLO, even if some educators chose the latter options without any technical problems. Some educators also visited the units in whole-class sessions, for example on the interactive white board.

In short, the educators' responses to the questionnaires and the input by their students suggest that the experiences in using the online units in the institute courses are positive. Probably the most important success factor is the availability of new types of resources and activities that are suitable for students to work on online as part of self-study or homework.

5 Conclusion and Reflection

To address the issue of the co-design of online learning units for mathematics and science teacher education, three research questions were phrased, which we will now revisit. After that, we will reflect on the findings and on possible future steps.

The first research question concerns the features of the online learning units on mathematics and science didactics produced for teacher education. An important finding is that the online units cannot and should not consist of ready-to-use materials, but rather can only contain building blocks for courses that will be further tailored to the educator's ideas. Indeed, teacher educators are used to design their courses in relative autonomy, and want to be able to fine-tune their courses to the target group at stake. Furthermore, some general design heuristics are identified. One is to use the power of video recordings of students working on tasks, and to use them to make teacher-students reflect on possible didactical interventions. A second heuristic is to consider the subject knowledge that is a prerequisite for didactical analysis and intertwine content knowledge and didactical knowledge in the learning units, as to avoid the hindrance of content knowledge deficiencies. Third, it was important to keep in mind the two different target groups: the student teachers and the teacher educators, and to produce learning units that fit both. To summarize, the building block approach was fruitful, the presence of students was an important feature, and the different target groups deserved attention.

The second research question was how a process of co-design, in which teacher educators collaboratively design such online learning units, can be organized. The blended approach of on the one hand intensive joint design meetings, the so-called boot camp days, and on the other hand the distant co-design, made possible by the digital platform, has shown to be a fruitful one. Scheduling design sessions during which teams can collaborate for several hours with full focus on producing materials made it feasible to construct digital blended learning units in a short time span. Readily available technical assistance during these sessions lowered the barrier for producing film clips. It resulted in both rich learning units and processes of professional development within the design teams. The composition of these teams, including different levels of expertise, worked out well. The technical facilities, both for distant collaboration and for the production of video resources, facilitated the co-design process. A drawback of using mixed teams is that people need time to getting to know each other and to form a joint vision on the subject at hand. Although this is important for a fruitful collaboration, care must be taken that teams dwell too long in this phase. To summarize, small design teams of experienced teacher educators from different institutes leads to boundary crossing between institutes, resulting in (i) rich material and (ii) professional development of the educators themselves, although a pitfall is that (iii) too much time may be spent on discussion rather than on the actual design.

The third research question was how educators and students experience the use of the online units that result from the design process. The pilot field tests in the different teacher education institutions have shown that both educators and students appreciated the online learning units that resulted from the co-design as interesting and useful. Even if the units have clear limitations, which are no surprise in the light of the design conditions, they overall were perceived as inspiring. Educators noticed that, as a result from the design heuristic to design building blocks rather than ready-to-use courseware, the actual use of the materials in their courses required considerable time and effort, and that the overall study load of the units for students was high. To summarize the findings on this question, we conclude that the experiences are encouraging, but that more time might be needed for designers to finalize the design and for educators to prepare their incorporation in their courses.

Of course, these conclusions need to be considered in the light of the limitations of this small-scale and short-term project, which covered a period of 14 months. In spite of these limitations, we can extract some suggestions for future work. A first step is to further disseminate the results and to take care of their sustainability, for example, through the website of the mathematics and science educators community represented in ELWIeR/Ecent, and by setting up an editorial board to deal with new submissions. Furthermore, a next step might be to further investigate how teacher educators can continue to engage in the co-design of teaching materials, based on these and newly developed resources. In this way, they can develop professional expertise in the field of online learning and contribute to the community.

Appendix: Pre- and Post-field Test Questionnaires

Pre-field Test Questionnaire

1. Which learning unit do you intend to use in a course mathematics didactics?
2. Do you already have an impression of this learning unit and if so, could you describe it? Relevance, consistency, usability? Content, design, appearance?
3. In which subject and for which target group will you use the learning unit? Size in ECTS[6]? In what period?
4. What are the goals you hope to realize with the deployment of the learning unit? What expectations do you have? In what need the learning unit can hopefully provide?
5. How are you going to tackle this?
   a. Technically: refer to the online learning unit, or import parts in your own web-based learning environment?
   b. Practical: replace parts of the existing course, as additional material, as part of homework, as part of assessment? In what way do the students will work with the learning unit?
   c. Content: which content parts of the learning unit do you intend to use?

Post-Field Test Questionnaire

1. Which of the learning units did you use in your course didactics of mathematics?
2. What is, looking back, your opinion about this learning unit? Relevance, consistency, usability? Content, design, appearance?
3. To what extent have the goals you hoped to achieve with the deployment of the learning unit actually been achieved?
4. How did you use the learning unit:
   a. Technically: refer to the online learning unit, or import parts in your own web-based learning environment?
   b. Practical: replace parts of the existing course, as additional material, as part of homework, as part of assessment? In what way do the students will work with the learning unit?
   c. Content: which content parts of the learning unit do you intend to use?
5. How did the learning unit please the students? Were there any positive or negative reactions?

---

[6]European Credits Transfer System "is a credit system designed to make it easier for students to move between different countries". See https://ec.europa.eu/education/resources-and-tools/european-credit-transfer-and-accumulation-system-ects_en.

6. Do you have assignments to share with us that were given to students regarding the learning unit? Do you have students' work with respect to these assignments to share with us?

References

Anderson, J. R., Corbett, A. T., Koedinger, K. R., & Pelletier, R. (1995). Cognitive tutors: Lessons learned. *The Journal of the Learning Sciences, 4*(2), 167–207.

Anderson, S. E. (1997). Understanding teacher change: Revisiting the concerns based adoption model. *Curriculum Inquiry, 27*(3), 331–367.

Baas, M., Van Hees, J., Houwen, G., Ouwehand, M., Van der Spek, E., & Veelo, K. (2017). *Good practices. Open leermaterialen binnen vakcommunities.* Utrecht: SURFnet.

Berners-Lee, T. (1989). *Information management: A proposal.* Geneva: CERN.

Bonk, C. J., & Graham, C. R. (2006). *The handbook of blended learning: Global perspectives, local designs.* Hoboken, NY: Wiley.

Butter, R., & Schamhart, R. (2017). *Cocreatie als duurzame energiebron voor de toekomst.* Utrecht: Hogeschool Utrecht.

Clark, R. C., & Mayer, R. E. (2008). *e-learning and the science of instruction: proven guidelines for consumers and designers of multimedia learning.* San Fransisco, CA, USA: Pfeiffer.

Coburn, C. E., & Penuel, W. (2016). Research-practice partnerships in education: Outcomes, dynamics, and open questions. *Educational Researcher, 45*(1), 48–54.

Devlin, K. (2012). *Introduction to mathematical thinking.* Palo Alto, CA: Keith Devlin.

Drijvers, P., Doorman, M., Boon, P., Reed, H., & Gravemeijer, K. (2010). The teacher and the tool: Instrumental orchestrations in the technology-rich mathematics classroom. *Educational Studies in Mathematics, 75*(2), 213–234.

Drijvers, P., Streun, A. V., & Zwaneveld, B. (Eds.). (2012). *Handboek wiskundedidactiek.* Utrecht: Epsilon.

Fullan, M. (2007). *The new meaning of educational change* (4th ed.). New York: Teachers College Press.

Garrison, D. R., & Kanuka, H. (2004). Blended learning: Uncovering its transformative potential in higher education. *The Internet and Higher Education, 7*(2), 95–105.

Hegedus, S., Laborde, C., Brady, C., Dalton, S., Siller, H.-S., Tabach, M., et al. (2017). Uses of technology in upper secondary mathematics education. *Uses of technology in upper secondary mathematics education* (pp. 1–36). Cham: Springer International Publishing.

Kortland, K., Mooldijk, A., & Poorthuis, H. (Eds.). (2017). *Handboek natuurkundedidactiek.* Utrecht: Epsilon.

Kvan, T. (2000). Collaborative design: What is it? *Automation in Construction, 9*(4), 409–415.

Nwosisi, C., Ferreira, A., Rosenberg, W., & Walsh, K. (2016). A study of the flipped classroom and its effectiveness in flipping thirty percent of the course content. *International Journal of Information and Education Technology, 6*(5), 348–351.

O'Flaherty, J., & Phillips, C. (2015). The use of flipped classrooms in higher education: A scoping review. *The Internet and Higher Education, 25* (Supplement C), 85–95.

Penuel, W., Roschelle, J., & Shechtman, N. (2007). Designing formative assessment software with teachers: an analysis of the co-design process. *Research and Practice in Technology Enhanced Learning, 2*(1), 51–74.

Schoenfeld, A. H., & Grouws, D. A. (1992). Learning to think mathematically: Problem solving, metacognition, and sense making in mathematics. *Handbook of research on mathematics teaching* (pp. 224–270). New York: McMillan Publishing.

Skinner, B. F. (1954). The science of learning and the art of teaching. *Harvard Educational Review, 24,* 86–97.

Skinner, B. F. (1958). Teaching machines. *Science, 128*, 969–977.
Staal, H. (2006). *De Kennisbank Wiskunde en competentiegericht opleiden van wiskundeleraren. Verslag van een samenwerking tussen de Educatieve Hogeschool van Amsterdam en het Ruud de Moor Centrum*. Heerlen: Open Universiteit.
Swan, K., Van 't Hooft, M., Kratcoski, A., & Unger, D. (2005). Uses and effects of mobile computing devices in K-8 classrooms. *Journal of Research on Technology in Education, 38*(1), 99–112.
Tolboom, J. L. J. (2004). An organisational model for a digital learning environment, based on a hierarchical decomposition. *International Journal of Continuing Engineering Education and Life Long Learning, 14*(1–2), 68–78.
Tucker, B. (2012). The flipped classroom. *Education Next, 12*(1), 82–83.
Tukey, J. W. (1977). *Exploratory data analysis*. Reading, MA: Addison-Wesley.
van den Bogaart, T., Daemen, J., & Konings, T. (2017). Katern 3: inhoud en vakdidactiek van de lerarenopleidingen. In G. Geerdink & I. Pauw (Eds.), *Kennisbasis lerarenopleiders wiskunde* (pp. 283–291). Eindhoven: Velon.
Verhoef, N., Drijvers, P., Bakker, A., & Konings, T. (2014). *Tussen val en schip*. Rapport Onderwijsonderzoekscommissie: Wiskundig-didactisch onderwijsonderzoek in Nederland.
Vygotsky, L. S. (1962). *Thought and language* (E. Hanfmann & G. Vakar, Trans.). Cambridge, MA: MIT press.
Wenger, E. (1998). *Communities of practice: Learning, meaning, and identity*. Cambridge: Cambridge University Press.
Woolley, D. R. (1994). The emergence of on-line community. *Computer-Mediated Communication Magazine, 1*(3), 5–5.

MOOCs for Mathematics Teacher Education to Foster Professional Development: Design Principles and Assessment

Gilles Aldon, Ferdinando Arzarello, Monica Panero, Ornella Robutti, Eugenia Taranto and Jana Trgalová

1 Introduction

The emergence of Massive Open Online Courses (MOOCs) in 2008, enabled by technology and social networking, has opened new educational possibilities. McAuley, Stewart, Siemens, and Cormier (2010, p. 10) define a MOOC as "an online course with the option of free and open registration, a publicly shared curriculum, and open-ended outcomes". The authors put forward that "a MOOC builds on the active engagement of several hundred to several thousand students who self-organize their participation according to learning goals, prior knowledge and skills, and common interests" (ibid.).

M. Panero (✉)
Department of Education and Learning, SUSPI, Manno, Switzerland
e-mail: monica.panero@supsi.ch

G. Aldon
S2HEP-EducTice, Institut Français de L'Education, ENS de Lyon, Lyon, France
e-mail: gilles.aldon@ens-lyon.fr

F. Arzarello · O. Robutti · E. Taranto
Department of Mathematics, University of Turin, Turin, Italy
e-mail: ferdinando.arzarello@unito.it

O. Robutti
e-mail: ornella.robutti@unito.it

J. Trgalová
S2HEP, Université Claude Bernard Lyon 1, Villeurbanne, France
e-mail: jana.trgalova@univ-lyon1.fr

E. Taranto
Department of Educational Science, University of Catania, Catania, Italy
e-mail: eugenia.taranto@unict.it

© Springer Nature Switzerland AG 2019
G. Aldon and J. Trgalová (eds.), *Technology in Mathematics Teaching*, Mathematics Education in the Digital Era 13,
https://doi.org/10.1007/978-3-030-19741-4_10

These distinguishing features, which some actors of education consider as revolutionizing and transforming education (Fidalgo-Blanco, Sein-Echaluce, & García-Peñalvo, 2015), challenge a number of aspects such as pedagogical design, roles of teacher/trainer and student/trainee in these massive courses, monitoring learners' activity and performance, validation and accreditation etc.

According to the pedagogical model, MOOCs can be classified into three main categories: xMOOCs, e.g., Coursera[1] or Udacity,[2] that are designed on a pedagogical model "dominated by the 'drill and grill' instructional methods with video presentations, short quizzes and testing" (Yuan & Powell, 2013, p. 7); cMOOCs or connectivist MOOCs, e.g., CCK11,[3] that are "based on a connectivism theory of learning with networks developed informally" (ibid.) and a participative pedagogical model; and quasi-MOOCs, e.g., Khan academy[4] or Didattica della Matematica,[5] which provide online open educational resources aiming at supporting learning-specific tasks and do not offer social interaction of cMOOCs or a structured course of xMOOCs (Siemens, 2013, p. 8).

There is a growing interest in designing technology-mediated teacher professional development programs (Brooks & Gibson, 2012). These initiatives highlight the importance of combining instruction with peer community learning, which can be well fitted in MOOCs, according to Laurillard (2016) who claims that "if the MOOC format is to be an effective mechanism for promoting educational innovation it must be able to support a co-learning model of professional development for the community of teachers" (p. 13).

In this chapter we report about the experience of three MOOCs for mathematics teachers' professional development: the eFAN Maths MOOC, designed in France and aimed at supporting teachers to integrate digital resources and technology in their classes; and two MOOCs designed in Italy, Geometria and Numeri, aimed at fostering mathematics teachers' professional development with the use of innovative practice for teaching geometry, arithmetic, and algebra, while also using interactive learning environments, e.g., GeoGebra. These MOOCs aspire to create collaborative contexts for teachers' work, where they can learn through sharing their practices and working collaboratively on joint tasks. Taking into account the necessity for teachers to be supported in exploiting technological affordances, the objectives of such MOOCs are: accompanying teachers in the production of teaching resources, by examples of activities and reflection on their ongoing resource design; fostering a sound use of technology through encouraging teachers to choose proper digital tools for the classroom; sharing innovative didactical practices. Such aims are related to the interest in the design and the implementation of teacher professional development programs to include the role of teachers working and learning in communities (Jaworski & Goodchild, 2006; Wenger, 1998).

---

[1] https://about.coursera.org/.

[2] https://www.udacity.com/.

[3] http://cck11.mooc.ca/.

[4] https://www.khanacademy.org/.

[5] https://www.youtube.com/user/DIFIMARobutti.

The originality of our research resides in those design principles that are relevant and useful to mediate teachers' professional development courses with technology and in the assessment of the impact of such distance courses on mathematics teachers' engagement. Specific attention is paid to trainers and their role in supporting interactions and learning communities that emerged during the MOOC. Trainers' techniques and their evolution are presented and analyzed in order to highlight and discuss their methodological and theoretical justifications.

2 Description of the MOOCs

2.1 Four Dimensions of Collaborative Work

For the description of the MOOCs, we use the frame introduced in the recent ICME survey (Robutti et al., 2016), focused on learning that occurs when mathematics teachers work together collaboratively. It is based on three themes: (i) contexts and features of mathematics teachers working collaboratively; (ii) theories and methodologies; and (iii) outcomes. The first theme is the one that we will use in this chapter to introduce the experiences of the different MOOCs. This theme is particularly useful in framing the educational initiative, because it is spread out in different dimensions that give specific information on MOOCs' birth, structure, and participants, which are:

(1) The initiation, foci, and aims of collaborations;
(2) The scale of collaborations (numbers of teachers and timeline);
(3) The composition of collaborative groups and the roles of the participants;
(4) Collaborative ways of working and their conception.

In what follows, we present the experiences of MOOCs according to this frame, in order to contextualize them in a general perspective. The reason is that the (numerous) teachers involved in these MOOCs collaborated intensely, although not in a face-to-face modality, but rather through a platform in order to accomplish the tasks presented by the trainers. In addition, the teachers spontaneously collaborated, according to their professional needs, roles, and competencies, and worked together in various ways, such as following the structure of the MOOCs and using the available technological tools. Working together they learnt, because they were involved in various kinds of knowledge (content, pedagogical, technological, and institutional).

2.2 Geometria and Numeri, the Italian MOOCs

2.2.1 Initiation, Foci and Aims of Collaboration

From spring 2015, at the Mathematics Department of Turin University, the Math MOOC UniTo project has come to life. It is the result of a long development process over many years by the researchers of the Department and characterized by many previous experiences of teacher education projects in which the team has been involved (e.g., the m@t.abel project[6]) alongside years of research on teacher education. The Math MOOC UniTo project consists of the design and delivery of MOOCs for education and professional development of in-service mathematics teachers of secondary schools (both lower and higher). In particular, the Italian MOOCs are for teachers and designed by teachers, in collaboration with university researchers. The teacher-designers previously attended a Master's Programme in Mathematics Education and Innovation, based on the didactical material from the m@t.abel project. After this experience, fresh needs emerged from those who had concluded the Master's: awareness of the need to support teaching activities with teacher education; willingness to develop best practices of innovation using software; reconsidering in terms of learning social media mostly used by the students. Hence, it was decided to offer the opportunity of an authentic professional development experience designed for a larger group of teachers that could become a community of practice (Wenger, 1998).

2.2.2 Scale of Collaboration (Number of Teachers and Timeline)

The first two MOOCs (others are following them: Relations and Functions in 2018) delivered are open, free, and available online on the Moodle platform DI.FI.MA. (Didactics of Physics and Mathematics[7]) and make use of open source tools (e.g., GeoGebra), enabling teachers to easily adapt to them in their teaching practices.

The Geometria MOOC was focused on geometry and its 424 participants were secondary school teachers from all over Italy. It lasted 8 weeks, plus a further 2 weeks for completing the final tasks, from October 2015 to January 2016. The Numeri MOOC was focused on arithmetic and algebra, with 278 participants, secondary school teachers from all over Italy. It was delivered over 6 weeks, with 4 further weeks for completing the final tasks, from November 2016 to January 2017. 36 and 42% of teachers completed all of the Geometria and Numeri MOOCs activities respectively, which is quite a high completion rate compared to the current average completion rate for MOOCs, approximately 15%.[8]

---

[6]http://www.scuolavalore.indire.it/superguida/matabel/.

[7]http://difima.i-learn.unito.it/.

[8]Retrieved October 25, 2017 from http://www.katyjordan.com/MOOCproject.html.

2.2.3 Composition of Collaborative Groups and Participants' Roles

Within the Italian MOOCs, and in general within MOOCs for teacher education, two communities can be distinguished. On the one hand, there are trainers (two university professors, about 15 in-service teachers enrolled in the Master's in Mathematics Education and a Ph.D. student); all of them were involved in the design of the course, its delivery, and monitoring of its evolution in terms of interaction among participants and educational resources made available. In particular, the 15 teachers also created the activities delivered in the MOOCs, adapted from m@tabel project and revised by the university professors. Moreover, the trainers helped trainees to solve technical problems, made tutorials, and recalled the tasks to be done week by week with weekly emails. On the other hand, there are trainees (the teachers enrolled in the MOOC); they have an active role in learning not only the activities and methodologies proposed but also in using technological tools for interaction. In fact, every week the trainees are assigned an individual workload and use environments and methodologies at different levels, in order to collect their weekly badges. These activities include: watching a video where an expert introduced the conceptual knot of the week; watching a "cartoon video" with some guidelines to carry out the units; reading the activities based on a mathematics laboratory (and the option to experiment with these in their classroom); answering multiple choice questions (MCQs) on the themes of the week. Moreover, they have to use suitable communication message boards (Forum, Padlet, Tricider) to express opinions about the content of the course, exchange experiences with colleagues, and benefit from other participants' ways of thinking. In both Italian MOOCs there was a collaborative climate and, surprisingly (for the trainers), some of the participants started voluntarily sharing material they created and were using in their lessons. The team of trainers chose to limit their own interventions in these message boards to a minimum in order to support the birth of a trainees-only online community of practice (Wenger, 1998). The trainers were more active within the webinars, educational online events for trainees.

Each Italian MOOC design included as a final module two production activities: designing a teaching activity (or Project Work, hereafter PW) using specific software and reviewing (or Peer Review, hereafter PR) a project designed by a colleague. For all those who took part in all MOOC stages (that is, completing all tasks for collecting all weekly badges and completing the PW and PR), a participation certificate was issued by the Mathematics Department of the University of Turin.

2.2.4 Collaborative Ways of Working and Their Conception

Collaboration among the trainees (both small groups and the whole group) is made possible through the platform and the communication boards provided (for more details see Taranto et al., 2017). Trainees are engaged in discussions on didactics, activities, classroom experiments, and formative assessment (the PR that each trainee has to do). On the other hand, the collaboration among the trainers takes place both during the design and during the monitoring stages of the MOOC. The experience of monitoring trainees' discussions on the communication boards and the feedback received through questionnaires filled in by trainees were certainly taken into account to make courses evolve from one season to another.

2.3 eFAN Maths, the French MOOC (Seasons 2 and 3)

2.3.1 Initiation, Foci, and Aims of Collaboration

In autumn 2015, at the French Institute of Education (Ifé) of the Ecole Normale Supérieure in Lyon, a team of researchers and teachers had the idea to take up and readapt some contents of the first season of the eFAN Maths MOOC[9] delivered in 2014. It was hence possible to preserve the aim of the MOOC, namely fostering mathematics teachers' professional development with the use of technology for teaching and training mathematics, especially at secondary school level. The MOOC is in line with the *Stratégie Mathématiques* program of the French Ministry of Education, which stresses the relationship of mathematics with other sciences and with the world, and aims at training teachers in this perspective in order to give students a refreshed view of mathematics. More specifically, the second season of the MOOC was designed with a double institutional aim: to support teachers and teacher educators in understanding and implementing new aspects of the French curriculum[10] applicable since September 2016 in all French primary and secondary schools, namely computational reasoning and interdisciplinary work, and to promote collaboration within the French-speaking community all over the world. The MOOC experience offered teachers a possibility to collaborate in small groups within a wider online community, with a goal of making these groups evolve into lasting communities of practice (Wenger, 1998).

2.3.2 Scale of Collaboration (Number of Teachers and Timeline)

Three seasons of the eFAN Maths MOOC were delivered in 2015, 2016, and 2017 respectively. In this chapter, we draw on data from the latter two experiences. The MOOC is delivered on a French national platform (FUN, France Universités Numérique[11]) and only free open source tools are suggested, so that enrolled teachers could easily find and appropriate them. The second and the third seasons of the MOOC delivered essentially the same content, with some differences in the dynamics that we will discuss further in this chapter. The number of participants was 2,572 in the second season and 2,690 in the third, mostly French-speaking mathematics

---

[9]Enseigner et Former avec le Numérique en Mathématiques (Teaching and Training Mathematics with Technology), https://www.fun-mooc.fr/courses/ENSDeLyon/14003S03/session03/about.

[10]French curriculum and supporting material are available at http://eduscol.education.fr/.

[11]https://www.fun-mooc.fr/courses/ENSDeLyon/14003S03/session03/about.

teachers and teacher educators interested in the use of technology. The second season was organized to take place over five weeks, from February to mid-March 2016, while the third one, running from February to the end of March 2017, added two central weeks for facilitating group work. The percentage of participants who completed all of the proposed activities was about 12% in the second season, which is comparable with the current average completion rate reported in literature (15%, see above Sect. 2.2.2); this percentage decreased in the third season of the MOOC to 6% (see comments to Table 2 and conclusion for possible interpretations).

2.3.3 Composition of Collaborative Groups and Participants' Roles

The trainers' team was composed of researchers in Mathematics Education from France as well as Senegal and Cameroon, and mathematics secondary teachers involved in research at the Ifé. Their role was to design and deliver courses and to monitor trainees' activities. Since the delivered content was the same, the trainees were generally different from one season to another. They were encouraged to play an active role in designing and analyzing a mathematical task integrating the use of a digital tool and, week after week, the proposed activities aimed to support trainees in their work. Each of the five weeks of the MOOC proposed three video lessons on key concepts related to technology in mathematics education, MCQs, an activity related to the theme of the week, and a few articles for an in-depth study. The examples discussed in the video lessons were selected and adapted from different European research projects (e.g., FaSMEd,[12] MC Squared[13]) with a focus on the use of technology supporting formative assessment and enhancing creative mathematical thinking.

The designers provided an open environment to encourage trainees' participation. Some trainers worked as community managers: they helped trainees to solve technical problems; made tutorials; created and regularly updated a list with all the trainees' ongoing projects to help teachers to find a project to join; they recalled the tasks to be done week by week. Furthermore, every week began with a short video titled "From one week to another" in order to bridge two consecutive weeks of the MOOC.

2.3.4 Collaborative Ways of Working and Their Conception

To encourage collaboration, trainees were invited to work on the proposed activities in a collaborative way, by forming groups around common interests in a mathematical theme on Viaéduc,[14] a professional social network for teachers that essentially allows

---

[12]Formative Assessment in Science and Mathematics Education (fp7/2007-2013 grant agreement n.612337).
[13]Mathematical Creativity Squared (ICT-2013.8.1 "A Computational Environment to Stimulate and Enhance Creative Designs for Mathematical Creativity", Project 610467).
[14]http://www.viaeduc.fr.

members to post comments, to create groups, to create and publish documents, and to comment or recommend or share them. Group members can work collaboratively either asynchronously, being authors of the same online document, or synchronously, writing on the same online collaborative board (padlet). To cultivate and trigger the formation of trainees' groups as communities of practice, one trainer per group followed the development of the group project from within, intervening to encourage and trigger collaborative work (Panero, Aldon, Trgalová, & Trouche, 2017). The project, elaborated collaboratively, went through two phases of assessment: a peer assessment with the possibility of improving the work taking into account the received feedback, and a trainers' assessment (by the trainer who followed the group all along the MOOC).

In what follows we focus on the analysis of the design principles and the assessment choices made by the trainers.

2.4 Research Questions

The research questions that guided our study were:

(i) *What design issues emerge when trainers aim at mediating distance teachers' professional development courses with technology?*
(ii) *How to assess the impact of such courses on mathematics teachers' engagement?*

While searching for answers to these questions, we aim not only to highlight relevant designs for teacher professional development programs on the one hand, and their impact on teachers' engagement on the other, but also to gain insights into possible links and consequences of one over the other. The collaboration of the teachers involved in such MOOCs is a consequence of the design of activities that encourage participants to collaborate at large-scale and with different modalities. Collaborative work is also a step towards learning different kinds of knowledge in different contexts (Robutti et al., 2016). Thus, in what follows, we focus on the design principles and assessment strategies that—as trainers—we use in our MOOCs. We hope that the reader will benefit from our expertise with online educational environments such as MOOCs.

3 Theoretical Framework

The theoretical elements used independently by both trainer-researchers teams in Italy and France are the notion of communities of practice (Wenger, 1998) and the theory of the Meta-Didactical Transposition (MDT; Arzarello et al., 2014). Therefore, when the two teams exchanged details of their respective experiences with MOOCs for mathematics teacher education, these theoretical elements emerged as a common global frame. They allow us to describe and analyze practices of the two MOOC communities (trainers and trainees) and their evolution over time when the two communities interact.

The MDT model is grounded in the Anthropological Theory of the Didactics (ATD, Chevallard, 1999), borrowing and extending the notions of praxeology and of didactical transposition. More specifically, trainers and trainees develop their own praxeologies to solve specific types of task. Given a type of task, we can identify the related praxeology, which is composed of a practical part (techniques to solve tasks of the given type) and theoretical part (the *logos* justifying the used techniques). The MDT model distinguishes between didactical and meta-didactical praxeologies. The didactical praxeologies aim to model the mathematical activity when solving a didactical task, such as "to teach a particular mathematical topic". The meta-didactical praxeologies concern meta-didactical tasks, such as "to reflect on possible praxeologies for teaching that particular concept". Concretely, trainers' praxeologies are meta-didactical in the sense that they deal with a discourse about the didactical issues given as tasks to the trainees, who, from their side, have their own didactical praxeologies. The two types of praxeologies, namely those of trainers and trainees, can initially differ: some of their components can be internal to one community but external to another. Thanks to the interactions of the two communities they can evolve from external to internal (internalization process, Arzarello et al., 2014, pp. 9–10).

More specifically, this chapter focuses on the trainers' meta-didactical praxeologies related to crucial tasks of teacher education within MOOCs.

Adapting the MDT model to MOOCs, Taranto et al. (2017) notice that in these online environments trainers and trainees are led to solve tasks using multiple procedures or multi-procedures, which we call—with the intent of extending the ATD language—"multi-techniques". They are multiple procedures because if one considers only one of them individually, the task cannot be solved in a satisfactory manner. Instead, as we will see in the analysis section, a fair number of them need to be considered. Note, however, that it may be inaccurate to talk about techniques since what follows will be a list of suggested procedures that we want to share with other potential MOOC trainers in teacher education. The procedures will in fact become techniques once they are universally shared and institutionally recognized (Chevallard, 1999) by the research community.

In our study, we analyze the trainers' praxeologies that can be considered as meta-didactical in the sense that they deal with a discourse about didactical issues: hence, we identify the meta-didactical praxeologies by selecting the tasks that are essential for the design of a MOOC. These tasks concern both the design principles and the assessment strategies. Through the analysis of the praxeologies associated with these tasks, we will catch several essential topics regarding MOOCs: (i) the relationship between design principles and professional development that can be grasped through the audience of each of the MOOC; (ii) the theme(s) of the MOOC, which is(are) essential from an institutional point of view; (iii) the delicate question of the relationships between trainees and trainers; (iv) the assessment strategies included in the MOOC design, which gives important clues to assess trainees' engagement.

4 Methodology

Drawing on the MDT, we pointed out some essential meta-didactical types of tasks that, according to our experiences, any trainer of a MOOC for mathematics teacher education should address. We consider four topics related to the design principles: Target; Theme; Trainers' interaction with trainees; Collaboration among trainees. Moreover, we take into account three topics related to the assessment strategies: Test; Project Work; Peer Review. For each topic, we describe the Italian and the French meta-didactical praxeologies. In fact, we identify the related meta-didactical types of tasks, the techniques adopted by trainers in the Italian Geometria MOOC and in the French eFAN Maths MOOC respectively to solve such tasks, as well as the related justifications (logos). For the logos, we particularly wondered how the chosen techniques were justified and supported by theories in Mathematics Education or more generally in the educational field. The Italian and the French praxeologies may not coincide because of the different nature of the MOOCs (e.g., institutional context) but they will be similar in their purpose.

The identification of these meta-didactical praxeologies has been possible by reflecting on the design phases in which we were involved both in the first and the second season of our MOOCs, but also on the massive assessment phases. In particular, we analyzed the evolution of the trainers' meta-didactical praxeologies to design the subsequent season of the respective MOOCs (Numeri MOOC and season 3 of the eFAN Maths MOOC). The reasons for this evolution (intended as an improvement of the professional development program) came from the trainers' self-analysis of the respective experiences but also from some trainees' comments (via questionnaires or posts in communication message boards). In the following section, we focus on these aspects, also highlighting similarities and differences between our online educational experiences.

5 Analysis

5.1 Design Principles

The analysis reported in this section is driven by our first research question:

What design issues emerge when trainers aim at mediating distance teachers' professional development courses with technology?

Both the Italian and the French MOOCs aim at professional learning and raising awareness of the possibilities for technology use in schools. Given this aim, it is important to identify a hypothetical target trainee: who could be the teachers that can benefit from this educational massive open online course? However, MOOC designers cannot know in advance the teachers who will decide to enroll in the MOOC and they will never meet them in person. For these reasons, as trainers, "you

Table 1 The meta-didactical praxeology related to "Target"

(1)	Target	
	Italian MOOC	French MOOC
Task	To identify a hypothetical target trainee (lower and higher secondary school teachers)	To identify a hypothetical target trainee (lower secondary school teachers)
Technique	To choose activities of a specific school level (according to the target) related to specific mathematics topics	To design activities for this specific target, related to important themes of the curricula
Logos	To hypothesize a mean ZPD of the future trainees	
Evolution	None	

are forced to hypothesize a *mean* Zone of Proximal Development (ZPD) of your future trainees" (Taranto, 2018). The ZPD (Vygotsky, 1978) concerns an internal level and comes into play when the trainers think about the ideal didactical praxeologies that they want to transpose to trainee teachers who will follow the MOOC. Therefore, as Taranto (ibid.) notices, trainers assume a certain level of prior knowledge (ZPD) of the trainees' community (not of the individual teacher since they are forced to consider mean values). They prepare and administer certain activities in order to help the trainees' community move from the current level (their present didactical praxeologies) to the potential level (the ideal didactical praxeologies).

The current level of the trainees' community can be recognized in what Goos (2013) lists to describe the teacher's ZPD. Precisely, it includes the professional development level of the trainee-teachers in terms of: mathematical knowledge; pedagogical content knowledge; skill/experience in working with technology; beliefs about mathematics, teaching and learning (Goos, 2013, p. 524). Such current level of professional development could evolve thanks to the contents the trainees find in the MOOC. The MOOC contents are carefully designed and implemented by trainers and they are related to specific mathematics topics or important themes of the curricula.

Regarding the topic "Target" (Table 1) there has not been any evolution from one MOOC season to another. The target was clearly stated and, since the enrolled participants proved to be in line with our expectations, no changes were needed.

Another essential aspect of a MOOC design for mathematics teacher education is the "Theme" (Table 2). To this purpose, the trainers face two types of tasks and for each they can adopt different multi-techniques.

The choice of the theme is naturally related to the identified target and to institutional purposes of the professional development program. Both MOOCs aim to respond to specific teachers' needs identified in the institutional and social contexts, referring to national plans for teacher professional development and to crucial (or even new, in the French case) aspects of the national curriculum.

Designers have to evaluate essentially two possibilities, according to their long-term educational aim: to change the MOOC theme from season to season, trying to cover, one by one, different crucial aspects and educational objectives; or to keep the same theme and deliver the same content, considered as crucial in professional development, in every season. Such a decision influences the potential MOOC audience. Indeed, with the former choice, as in the Italian case, the same group of teachers can enroll into every season of the MOOC to pursue their professional development; with the latter choice, as in the French case, the opportunity of professional development is offered to an increasing group of teachers (including those who have not completed the previous season).

Once the theme and its possible evolution from season to season are decided, designers have to consider the time variable. The Italian and French experiences highlight again two possible approaches: decide how much time has to be devoted to each module of the MOOC or how much material is possible to be read and worked on in a module that has a fixed duration (e.g., one week). The Italian team chose the first approach and, according to the theme, decided to devote one week or two to the same content or methodology because of its complexity or the material profusion. The French team chose the second approach and, given the fixed module duration of a week, designed and adapted the material in such a way that trainees could manage appropriating it.

In both cases, designers need to make an average of the estimated learning times of the target audience (Carroll, 1963). After the first season of the Geometria MOOC, the Italian team decided to reduce the quantity of the provided material and to pay greater attention to differencing the material for various school levels. The French team, in contrast, has not changed its praxeology. Indeed, although trainees complained about the amount of time needed to complete reading of the provided material on FUN, the reason was not the quantity of the material but rather managing the work between courses on FUN projects on Viaéduc.

To make the online interaction with the trainees possible, the trainers are called to put some multi-techniques into action (Table 3). The first kind of interaction is reading of available materials and didactical resources. Digital resources replace the voice and explanations of the trainers that usually feature in face-to-face courses: the trainees instead interact through videos, images, interactive texts. In this way trainers are able to communicate their training intentions at a distance, and share research results, methodologies, and teaching strategies that can be used in class with students. In the Italian MOOCs the activities have been transposed into a digital format according to the E-tivity framework (Salmon, 2013). The E-tivities are designed before opening the MOOC to participants. They support learners in achieving the learning outcomes: in fact, they promote a learner-centered task and problem-based approach to online learning.

Based on the 7Cs of learning design (Conole, 2014), and in particular "Capture[15]" and "Communication[16]", as well as on the pillars of the "pedagogical contract" and

---

[15] In terms of capturing resources to be used: What resources are being used and what other resources need to be developed? (Conole, 2014, pp. 1, 3).

[16] Mechanisms to foster communication: How are the learners interacting with each other and their tutors? (Conole, 2014, pp. 3–4).

Table 2 The meta-didactical praxeology related to "Theme"

(2)	Theme	
	Italian MOOC	French MOOC
Task 1	To identify the main theme to address in the MOOC	
Techniques	To focus every season on a different core part of the curriculum (Geometry, Numbers) and to choose activities around specific topics according to the theme	To focus every season on the same theme (teaching and training in mathematics with technology) and to choose activities around specific topics according to the theme
Logos	To innovate methodology and strategies of teaching mathematics as highlighted in the Piano nazionale per la formazione docenti and the Italian curriculum (Indicazioni nazionali[a])	To renew the vision of mathematics as highlighted in the Ministry plan of Stratégie Mathématiques and the French curriculum (Programmes: see Footnote 11)
Evolution	To cover Geometry (first season), followed by numbers (second season). Once a topic is covered (see Sect. 2.2.2), the professional development program moves on to another one, with the long-term aim of deepening the professional development of the same group of teachers. 50% of teachers enrolled in the second season came from the previous one	None
	Time	
Task 2	To decide how much time is devoted to each module of the MOOC	To decide how much material can be read/worked on in a week (fixed duration of a module)
Techniques	To estimate the time necessary To complete the treated topic, taking into account an estimated engagement of 4 h per week – If necessary, to divide theoretical and practical parts – If the material is too dense, to devote two weeks to the same topic	To create materials (three videos, related quiz, activity) in such a way that 4 h (estimate per week) Are enough to appropriate and make use of them
Logos	To calculate average of estimated learning times of the target	
Evolution	To reduce the quantity of the material provided; greater attention to differentiating the material for different school levels	None

[a]Link to the Italian curriculum: http://www.indire.it/lucabas/lkmw_file/licei2010/indicazioni_nuovo_impaginato/_decreto_indicazioni_nazionali.pdf

of the "trainer as a facilitator" in the accompanied auto-education (Carré, 2003), the Italian and French trainers created institutional mail addresses to send e-mails periodically in order to have moments of direct contact with the group and/or with the individual trainees. Weekly mails were sent to all members to remember the content and required activities, and private emails or specific forums were set up for technical issues. The French team also created video summaries of weekly activities and performance.

The French team decided that one trainer per group of trainees would follow the development of the work from within, by becoming a group member, encouraging collaborative work and helping the group turn into a community of practice. The Italian team, instead, preferred to alternate the platform control moments, managed by groups of teacher-designers per module, with moments of synchronous contact through webinars. These are online meetings in which an expert (seen through a camera) shares with the trainees (who can only interact via chat) some issues about the research in mathematics education and focuses on some questions that could be raised during the previous weeks in the MOOC.

Questionnaires were administered for feedback on the degree of satisfaction for the educational offer. The Italian team administered 3 questionnaires (at the beginning, halfway through the course, and at the end); the French team one at the end. Meetings with some of the French trainees were organized at the end of the MOOC, which allowed the French trainers to gather more explanatory answers. From the feedback they received, both teams understood how to better refine some questions to get clearer information. Moreover, the French team decided to announce the questionnaire from the beginning of the MOOC in order to reach trainees who would drop out of the course before its end.

Fostering collaboration among trainees (Table 4) is not a peculiarity of all MOOCs, but we stated from the beginning that it is a fundamental aspect common to the Italian and the French MOOCs, distinguishing them from other kinds of online courses where the trainee alone has to watch videos and complete activities. We conceive our MOOCs as authentic collaborative experiences and for this reason we described them above (Sect. 2) in terms of the four dimensions of the collaborative work. However, collaboration cannot be considered as a spontaneous way of work, especially within such remote contexts; designers have to make it possible through specific multi-techniques. The Italian and the French praxeologies related to this task constitute some effective examples of how to solve this issue. In both cases, the trainers' teams provide an "open environment" (Carré, 2003) and ground their choices on "Collaborate" in the sense of Conole (2014). The French team opted for a professional social network with integrated collaborative tools: a careful coordination of the two platforms (FUN for the courses and Viaéduc for the collaborative projects) is extremely important to support the trainees in finding their bearings between the two online workspaces. The Italian team used forums provided by DI.FI.MA. platform, where the courses were delivered, and decided to add some collaborative tools such as Padlet and Tricider from the outside. In both cases, the trainers felt the need to augment the "official" platform with additional tools to properly foster collaboration. This fact is relevant for us and can be interpreted as the current lack of

Table 3 The meta-didactical praxeology related to "Trainers' interaction with trainees"

(3)	Trainers' interaction with trainees	
	Italian MOOC	French MOOC
Task	To make the interaction WITH the trainees possible	
Techniques	– To transpose in a digital format materials and didactical resources for teacher education – To create institutional e-mail addresses for sending periodic e-mails (e.g., weekly e-mails, private e-mails for technical problems) – To open forums for technical and didactical issues – To organize webinars for creating occasions of synchronous contact – To prepare and administrate questionnaires	– To transpose in a digital format materials and didactical resources for teacher education – To send an e-mail at the beginning of the week to all members as a reminder of the activities to be completed, and private emails or specific forums on FUN for technical problems – To create videos "from one week to another" as the first content of a new module – To prepare and administrate the final questionnaire – To evaluate the MOOC face-to-face with some trainees (the most active ones) at the end of the experience – To follow the development of group work on Viaéduc (one trainer per group)
Logos	E-tivity framework for digital transposition "Capture" and "Communicate" from the 7Cs	"Pedagogical contract" between trainers and trainees and the role of the "trainer as a facilitator" as pillars of the accompanied auto-education
Evolution	Some questions in the questionnaires have been changed	Some questions in the questionnaires have been changed and a questionnaire has been announced in the first weeks of the MOOC

remote platforms for online courses, which can fully support collaboration among participants. We will discuss this point further in the conclusion (Sect. 6).

In the forums, both teams adopted a technique to initiate discussions with a prompting question in order to accompany trainees in reading the materials and identifying their focus. As a difference between the meta-didactical praxeologies of the trainers of the two MOOCs, we can identify the influence of a technique used for interacting with the trainees, that is how and how much to intervene in the trainees' work. It turned out that the Italian team is focused on global collaboration, fostering it within the entire community of the MOOC and aiming at the creation of a global community of practice made only by trainees (Taranto et al., 2017). The French team, instead, is focused on local collaboration, fostering it within small groups of the MOOC community and aiming at the creation of small local communities of practice around a common project, where the trainer intervenes and acts as a tutor before and as an assessor in the end (Panero et al., 2017).

Table 4 The meta-didactical praxeology related to "Collaboration among trainees"

(4)	Collaboration among trainees	
	Italian MOOC	French MOOC
Task	To make the interaction among trainees possible	
Techniques	– To provide a suitable space for making remote communication possible (communication message boards such as forum, padlet, tricider) – To initiate discussions on forums with a prompting question (in order to accompany trainees in reading the materials and identifying their focus) – To reduce trainers' interventions, monitoring behind the scenes	– To open a collaborative workspace on Viaéduc for making remote communication possible – To initiate discussions on forums with a prompting question (in order to accompany trainees in reading the materials and identifying their focus) – To foster collaboration from the inside (one trainer per group)
Logos	"Collaborate" in the 7Cs; to foster the birth of a community of practice	The presence of an "open environment" among the pillars of Carré's model; to foster the birth of small communities of practice
Evolution	To provide more tutorials to allow trainees to move autonomously in the collaborative space and to use collaborative tools as efficiently as possible	None

5.2 Methodological Choices Based on Design Principles for Assessment

In this section, we address the second research question:

How to assess the impact of such courses on mathematics teachers' engagement?

The large numbers of teachers participating makes it very difficult to personally follow every participant. In Table 3 we stated that we were always vigilant, with private emails, through a group trainer in the French team or more trainers per module in the Italian team, to follow the development of the work from within. But how one might get an immediate understanding of the progress of each trainee?

Both teams introduced weekly tests to understand the trainees' appropriation of the video content and module activities. The tests consisted of MCQs (Table 5) allowing up to 2 (in the French case) or 3 (in the Italian version) attempts: we also gave feedback about the given answers (Velan, McNeil, Jones, & Kumar, 2008). The trainees could review the resources and try to find the correct answer. Correct answers indicated that the resources had been explored in depth and not superficially. Additionally, granting multiple attempts was a guarantee of success for trainees.

In the Geometria MOOC, the trainees did not share the same opinion about the tests: they saw them as an overload of work besides the commitment already required

Fig. 1 Badge of Module 4 in the Geometria MOOC

Table 5 The meta-didactical praxeology related to "Test"

(5)	Test	
	Italian MOOC	French MOOC
Task	To assess the trainees' degree of participation weekly	
Techniques	– Multiple Choice Questions with up to 3 attempts related to the video content and module activities – Release of the badge (the test was a necessary and sufficient condition for its release)	– Multiple Choice Questions with up to 2 attempts related to the video content and module activities
Logos	Choosing MCQs because MOOCs are massive	
Evolution	Test was present in the first season, but removed from the second one	Modifications of some questions

by the MOOC on a weekly basis. Therefore, tests were removed from the second season (Numeri MOOC). The French trainees, in contrast, did not complain about the tests; therefore, there is no change regarding the use of MCQs because this is required by the institution (FUN) in order to provide a certificate of attendance.

Another technique that remained unchanged in the Italian MOOCs is the end module badge. It was obtained if the trainee self-declared to have seen some specific resources, if he/she wrote on the communication message boards when required and if he/she uploaded specific materials when asked. Once all the module requests were accomplished (test included), the platform released the badge (Fig. 1). In this way, it was quite easy for the Italian team to monitor the progress of the trainees, knowing the amount of badges they had collected.

Both the Italian and the French teams chose a project-based methodology (Bender, 2012) to assess the trainees' engagement, but articulated it in different ways, and both turned out to be efficient (Table 6).

The project consisted in designing a classroom activity: by describing and analyzing a priori its potential for the learning of mathematics, trainees had to demonstrate acquired teaching competencies and expertise. The project-based methodology was chosen to give trainees the opportunity to get involved in the MOOC activities in terms of methodology, creativity, and with the aim of sharing and discussing them in the community: the entire MOOC community in the Italian case, or their own small collaborative group in the French version. Italian trainees had to produce an individual Project Work (PW), while French trainees a collective one. They were free to choose the theme of their project: in the Italian case, a geometric theme, while in the French case, any mathematical theme involving technology. A big difference between the trainers' praxeologies can be found in the instructions given to trainees for carrying out the PW. The Italian team gave a lot of freedom to trainees: trainers did not want either to influence them or to restrain their creativity. Trainees had to use a web-based tool, the Learning Designer (hereafter LD) designed by Laurillard (2016). LD is a software that guides and encourages planning of a lesson: it is characterized by a standard format that allows the teacher to integrate technologies, to have an overview of the teaching/learning dynamics centered on the student, and to share online what the teacher has produced. The French group projects had to be collaboratively written on Viaéduc, using its integrated tools. French trainers gave trainees clear instructions and guidelines to carry out their PW: for each phase of the PW, corresponding to each week's activity, description, and analysis, grids were provided. They consisted in guiding questions grounded in the instrumental approach (Rabardel, 1995) and in the instrumental orchestration (Trouche, 2004). Moreover, a fictive group was created among the real ones in order to give trainees possible examples of the expected activities.

Both trainers' techniques include the creation of video and pdf tutorials in order to familiarize the trainees with LD on the one hand, and with collaborative tools of Viaéduc on the other hand. In the Geometria MOOC, these tutorials were available two weeks before the opening of the last module while in the eFAN Maths MOOC they were available from the beginning of the MOOC because the PW was carried out throughout the MOOC experience. Furthermore, deadlines for accomplishing the PW were announced promptly because trainers wanted to allow everyone to complete a Peer Review (see Table 7). However, some trainees voiced the need to have more time to complete their PW; as a result, in both subsequent seasons of the MOOCs the deadline was extended by 2 weeks. Moreover, the French team decided to make PWs of the previous season available on Viaéduc.

To stimulate collaboration among trainees and to foster formative assessment among peers (Black & Wiliam, 2009), both teams proposed a Peer Review (PR) activity (Table 7). As for the PW, for the PR the trainers have to complete two tasks and for each they can adopt different multi-techniques.

In the Italian case, it was a one on one peer review: each trainee had to review a colleague's PW from an educational point of view, without any marking intention. The teachers were divided, thanks to an Excel table, taking into account their school level. In the Excel table each trainee found the PW's title and links to LD to facilitate identifying of the PW to review. The instructions for the PR were given in a more specific way compared with the PW. In the week dedicated to the PR, a revision grid containing the review criteria was given: attention to the main aspects of each educational intervention and to a conscious use of digital software. The grid provides 5 categories: Connections to the real world; Creativity; Collaboration; Use of technology; General considerations. For each of these categories, some features are indicated. They are to be evaluated by using a scale from 1 (=little present aspect) to 5 (=highly present aspect). The final request was to leave a comment highlighting

MOOCs for Mathematics Teacher Education … 241

Table 6 The meta-didactical praxeology related to "Project work"

(6)	Project work	
	Italian MOOC	French MOOC
Task 1	To assess the competences acquired through the MOOC	
Techniques	– Trainees are asked to carry out an individual project – Recommendation to use the LD software – Trainees can choose the content to address in their project according to the theme of the MOOC	– Trainees are asked to carry out a group project – Description and analysis grids are provided – Trainees can choose the content to address in their project according to the theme of the MOOC – To provide a visual organization of what is done and what is to be done (framaboard) for each group and update it frequently
Logos	Project-based learning	
	Time	
Task 2	To decide how much time is devoted to the individual project work	To decide how much time is devoted to the group project work
Techniques	To estimate the time necessary to carry out the individual project (one week); to give instructions/tutorials about LD starting from the week before	To estimate the time necessary for sharing ideas in the group and carrying out the project (throughout the MOOC); to give instructions/tutorials on Viaéduc starting from week 0
Logos	The time for appropriation of an artefact as LD	Cultivation of small communities of practice requires time; time for appropriation of the artefacts of Viaéduc
Evolution	The deadline to carry out the project work was extended by 2 weeks	The deadline was extended by 2 weeks; PW of the previous season were left as examples on Viaéduc

the strengths of the project, the parts that could be improved, and possible curiosities of the reviewer. The Italian team gave teachers one week to complete this task, considering this a suitable time for internalizing (Arzarello et al., 2014, pp. 9–10) the criteria of assessment.

Also in this case, in the forum dedicated to technical problems, some trainees voiced the need for more time to complete their PR and also to receive in advance the criteria to better complete the design task of the PW. In the subsequent season of the MOOC, the deadline for completing the PR was extended from one to two weeks. Moreover, the revision grid was given at the beginning of the two weeks of PW (i.e., two weeks before the start of the revision process). In addition, the project to be reviewed was assigned by the trainers to each trainee taking into account the school level. This was done because in the previous season more than one trainee selected the same PW and some PWs remained without a reviewer. In both seasons the PRs were delivered on the platform and made available to each trainee.

Table 7 The meta-didactical praxeology related to "Peer Review"

(7)	Peer review	
	Italian MOOC	French MOOC
Task 1	To review the PW	To review the group project
Techniques	– Trainees are asked to complete a peer review (one on one) of a project they choose at the same school level – An Excel table is provided to organize the finalized PWs (with links to LD) to facilitate the choice of the potential reviewers – Revision grid – PRs (sent as a task on Moodle) shared with all the participants on the platform	– Trainees are asked to complete an individual peer review of version 0 of a project of one or more groups of their choice – An Excel table is provided to present version 0 of the group projects (with links to pdf and Viaéduc) – Evaluation grid – Collection of feedback via a questionnaire (google form) and sharing of the resulting table with the entire group on Viaéduc – The tutor of each group assesses version 1, revised by the group
Logos	Stimulate collaboration, peer assessment (formative assessment), deal with the massive nature of MOOC	Stimulate collaboration, peer assessment (formative assessment), criteria shared in the trainers' group to assess version 1
	Time	
Task 2	To decide how much time is devoted to the peer review	
Techniques	– To estimate the time necessary for reviewing one colleague's project (one week) – Provide the revision grid in the week of the PR (last module)	– To estimate the time necessary for reviewing one group's project (one week) – Provide the evaluation grid in the week of the peer review (last week)
Logos	Time for internalizing the criteria of assessment (MDT)	
Evolution	The deadline to accomplish the PR was extended and the revision grid was given at the beginning of the two weeks of PW. The project to be reviewed was assigned by the trainers to each trainee taking into account the school level	None

In the French case, the projects, written collaboratively, went through two phases of evaluation: a peer evaluation with the possibility of improving the work based on the received feedback, and a trainer's evaluation (by the trainer who followed the group from within). For the first phase, trainees were asked to complete an individual peer review of version 0 of the project of one or more groups of their choice. An Excel table was provided to present version 0 of the group projects (with links to pdf and Viaéduc). The evaluation grid was elaborated by the trainers to encompass all the phases of the project developed in the MOOC week after week. This grid was structured around the following four criteria: (1) Accuracy of the definition and description of the project; (2) Relevance of the mobilised digital tools and resources with respect to the educational goals of the designed mathematical task; (3) Relevance of the analysis of the students' expected mathematical activity; (4) Relevance of the analysis of the teacher's role. For each criterion, some guiding questions were proposed with a double objective: to foster the production of justified feedback and to deepen the reflection carried out in the previous weeks of the MOOC. The grid finally asked for brief global feedback on the project and some suggestions to improve the work. Each trainee was invited to use the grid individually to evaluate a project of another group, by answering each guiding question with an evaluation: very good; satisfactory; fragile; or insufficient, accompanied by a justification. The community managers gradually collected feedback and comments in a table and shared it in a specific space on Viaéduc.

6 Discussion and Conclusion

In this chapter, we analyzed two seasons of MOOCs aimed at mathematics teachers' professional development, designed in France (Lyon) and in Italy (Turin). As pointed out in the Introduction, MOOCs with this aim are rare. Some authors claim that they are a promising tool for such a use, but they generally do not directly address the issue of mathematics teachers' professional development: the researches on effective teachers' learning processes within such new environments are not so diffuse, and those concerning mathematics teachers are de facto missing. Hence, our analysis humbly tries to open a new road on this terrain.

A major starting issue for us was to define a proper theoretical frame for our analysis, which could satisfy two constraints:

– The literature is rich in papers concerning the way mathematics teachers can improve their professional knowledge: e.g., Robutti et al. (2016) present a wide survey of this topic, but the related literature generally describes situations where technology concerns mainly how to improve students' learning (at school or at home), not teachers' learning in MOOC courses.
– In MOOCs where teachers are the direct addressees of such online courses, the philosophy behind MOOCs focuses primarily on the transformative impact of technology in teaching and learning that occurred during the last decade. This focus must be shaped specifically by MOOCs designed for (mathematics) teachers' professional development.

The two instances are not contradictory but require a sophisticated analysis tool that can combine these two sides in a complementary, coherent, and productive way. Our task can be expressed through a winegrowers' maxim: "put the new wine in old barrels". Namely, we had to analyze a new way of designing and assessing courses for teachers through MOOCs, using (at least a part of) the old and powerful theoretical lens of the MDT (Arzarello et al., 2014). Our chapter accomplished this task, making it possible to produce a "good wine", apt to the analysis we faced; moreover, it made possible a comparison of pros and cons with respect to the two approaches (the Italian and the French ones).

Out of the metaphor, the praxeological analysis of the selected types of tasks, as detailed in Tables 1, 2, 3, 4, 5, 6 and 7 of Sect. 5, gives significant answers to the research questions. In particular, regarding the usefulness of design principles, the great importance of the institutional context, which has been taken into account by the two teams, is highlighted as an essential issue of a MOOC design. Also, as pointed out by Bozkurt, Akgün-Özbek, and Zawacki-Richter (2017): "Findings of this research revealed that the least explored research areas are learner support services; management and organization; access, equity, and ethics" (p. 12). Our methodological approach tends to give information about the learner support services through the design choices made by the teams regarding the trainers' interaction with trainees and the assessment of the trainees' work through crossed analysis.

One main similarity between the two MOOCs experiences lies on the methodological choice of the project-based assessment. The model itself of MOOC does not allow researchers to directly observe the effect of the training courses proposed by the MOOCs and to gather feedback from observations in classes. For this reason, both teams considered the PW as a suitable way to assess the competencies acquired by the trainees. The PW was individual in the Italian case, collaborative in the French one, but in both cases a PR was proposed to evaluate the work. The evidence is that the connection between trainees does not go without saying and that the role of tutors as well as their scope of activities must be included in the design principles of MOOCs. Moreover, time to devote to these tasks is an important issue to consider and it was increased in both experiences. On the contrary, leaving the PW carried out in the previous season as inheritance for the next one may inhibit trainees' creativity, as did occur in the French case (this could be a reason why the completion rate decreased).

The main differences between the two MOOCs are underlined: often, they are of minor relevance, but sometimes not so. For example, see the final part of the comments to Table 4, where an important difference between the forms of trainees' collaboration is made apparent through the analysis of the trainers' techniques: global collaboration in the Italian MOOC versus local collaboration within the French one. While this result puts forward a sort of inner difference between the two MOOCs experiences, an external difference between our MOOCs and other forms of experiences through the use of platforms (even in different experiences of MOOCs) is made apparent in Table 4: the four dimensions of the collaborative work show the specific involvement of trainees in our MOOC and the form of their active involvement. A main result is that collaboration cannot be considered as a spontaneous way of working, especially within such remote contexts; designers have to make this possible through specific techniques. The Italian and the French praxeologies related to this task constitute some effective examples of how to solve this issue. The French team opted for a professional social network Viaéduc to integrate collaborative tools that were missing on the FUN platform. The same necessity was felt by the Italian team using forums provided by DI.FI.MA. platform, as well as some collaborative tools such as Padlet and Tricider from the outside. Our analysis shows that real involvement of trainees in collaborative work needs to be triggered and supported by suitable tools added to the platform. The availability in the platform of tools consonant with the social networks used in everyday life increases the triggering of what Manlove, Lazonder, and de Jong (2007) call co-regulated learning, in the sense that the trainees themselves regulate their tasks and collaboration. Our analysis leaves open the question of which devices are best for improving active collaboration among the trainees: further research and concrete experimentations could give a more definitive contribution to this crucial issue. What is interesting here is that our analysis, centered on collaboration processes through the adaptation of the meta-didactical lens, has made it possible to grasp this important problem in a clear way. This suggests that the way of research we have undertaken is promising and fruitful for further results along this stream.

Acknowledgements We thank very much the entire Italian and French MOOCs teams for their help and their crucial contribution to the design and management of the project. Without them neither the MOOCs nor this chapter would exist.

References

Arzarello, F., Robutti, O., Sabena, C., Cusi, A., Garuti, R., Malara, N., et al. (2014). Meta-didactical transposition: A theoretical model for teacher education programs. In A. Clark-Wilson, O. Robutti, & N. Sinclair (Eds.), *The mathematics teacher in the digital era: An international perspective on technology focused professional development* (Vol. 2, pp. 347–372). Dordrecht, The Netherlands: Springer.

Bender, W. N. (2012). *Project-based learning: Differentiating instruction for the 21st century*. Corwin Press.

Black, P., & Wiliam, D. (2009). Developing the theory of formative assessment. *Educational Assessment, Evaluation and Accountability, 21*(1), 5.

Bozkurt, A., Akgün-Özbek, E., & Zawacki-Richter, O. (2017). Trends and patterns in Massive Open Online Courses: Review and content analysis of research on MOOCs (2008–2015). *The International Review of Research in Open and Distributed Learning, 18*(5). http://dx.doi.org/10.19173/irrodl.v18i5.3080.

Brooks, C., & Gibson, S. (2012). Professional learning in a digital age. *Canadian Journal of Learning and Technology, 38*(2).

Carré, P. (2003). L'autoformation accompagnée en APP ou les sept piliers revisités. In P. Carré & M. Tétart (dir.), *Les ateliers de pédagogie personnalisée ou l'autoformation en actes*. Paris: L'Harmattan.

Carroll, J. B. (1963). A model of school learning. *Teachers College Record, 64*(8), 723–733.

Chevallard, Y. (1999). L'analyse des pratiques enseignantes en théorie anthropologique du didactique. *Recherches en Didactique des Mathématiques, 19*(2), 221–266.

Conole, G. (2014). The 7Cs of learning design a new approach to rethinking design practice. In *Proceedings of the 9th International Conference on Networked Learning* (pp. 502–509).

Fidalgo-Blanco, A., Sein-Echaluce, M. L., & García-Peñalvo, F. J. (2015). Methodological approach and technological framework to break the current limitations of MOOC Mode. *Journal of Universal Computer Science, 21*, 712–734.

Goos, M. (2013). Sociocultural perspectives in research on and with mathematics teachers: A zone theory approach. *ZDM-Mathematics Education, 45*(4), 521–533.

Jaworski, B., & Goodchild, S. (2006). Inquiry community in an activity theory frame. In J. Novotná, H. Moraová, M. Krátká & N. Stehlíková (Eds.), *Proceedings of the 30th Conference of the International Group for the Psychology of Mathematics Education* (Vol. 3, pp. 353–360).

Laurillard, D. (2016). The educational problem that MOOCs could solve: Professional development for teachers of disadvantaged students. *Research in Learning Technology, 24*(1), 29369. https://doi.org/10.3402/rlt.v24.29369.

Manlove, S., Lazonder, A. W., & de Jong, T. (2007). Software scaffolds to promote regulation during scientific inquiry learning. *Metacognition and Learning, 2*, 141–155.

McAuley, A., Stewart, B., Siemens, G., & Cormier, D. (2010). *The MOOC model for digital practice*. University of Prince Edward Island.

Panero, M., Aldon, G., Trgalová, J., & Trouche, L. (2017). *Analysing MOOCs in terms of their potential for teacher collaboration: The French experience*. Presented to TWG15 of the 10th Conference of European Research on Mathematics Education (CERME). Dublin, Ireland.

Rabardel, P. (1995). *Les hommes et les technologies; Approche cognitive des instruments contemporains*. Armand Colin.

Robutti, O., Cusi, A., Clark-Wilson, A., Jaworski, B., Chapman, O., Esteley, C., et al. (2016). ICME international survey on teachers working and learning through collaboration. *ZDM Mathematics Education, 48*(5), 651–690.

Salmon, G. (2013). *E-tivities: The key to active online learning*. Abingdon, UK: Kogan Page.

Siemens, G. (2013). Massive Open Online Courses: Innovation in education? In R. McGreal, W. Kinuthia, & S. Marshal (Eds.), *Open educational resources: Innovation, research and practice* (pp. 5–16). Vancouver: Commonwealth of Learning and Athabasca University.

Taranto, E., Arzarello, F., Robutti, O., Alberti, V., Labasin, S., & Gaido, S. (2017). Analysing MOOCs in terms of their potential for teacher collaboration: The Italian experience. In T. Dooley & G. Gueudet (Eds.), *Proceedings of the 10th Congress of European Research on Mathematics Education* (pp. 2478–2485). Dublin, Ireland.

Taranto, E. (2018). *MOOC's zone theory: Creating a MOOC environment for professional learning in mathematics teaching education*. Ph.D. Dissertation, Turin University.

Trouche, L. (2004). Managing complexity of human/machine interactions in computerized learning environments: Guiding students' command process through instrumental orchestrations. *International Journal of Computers for Mathematical Learning, 9*, 281–307.

Velan, G. M., McNeil, H. P., Jones, P., & Kumar, R. K. (2008). Integrated online formative assessments in the biomedical sciences for medical students: Benefits for learning. *BMC Medical Education, 8*(1), 52.

Vygotsky, L. (1978). *Mind in society*. Cambridge, MA: Harvard University Press.

Wenger, E. (1998). *Communities of practice: Learning, meaning, and identity*. Cambridge University Press.

Yuan, L., & Powell, S. (2013). *MOOCs and open education: Implications for higher education*. White paper. JISC CETIS.

Part IV
Teaching and Learning Experiences with Digital Technologies

Chronis Kynigos
National and Kapodistian University of Athens, Athens, Greece;
kynigos@ppp.uoa.gr

Introduction

In the introduction to this section, a brief synthesis of the three studies in this section focused on the notion of experience referring to Dewey's call for developing genuine and at the same time educative experiences for learners (ibid.). There is a diversity in these studies regarding the kinds of focus on experience, the technologies used and the theoretical constructs employed to study students' meanings. Coming up with a synthesis between them thus makes an interesting challenge. Uygan and Turgut focus on the kinds of dependencies a student's reasoning had on different representations of a geometrical object in 3D space. They employed a multimodal, embodied cognition approach, addressing the issue of students' grappling with diverse semiotic systems including gesture and sketching. They looked for meanings the students generated to understand properties and techniques for constructing a 3D model. Lisraelli focused on meanings generated with the use of a specifically designed representation to elucidate function as co-variation of two dependent quantities. She used the notion of microworlds developing into instruments of semiotic mediation and discussed the semiotic potential of a digital artefact. In this case, meanings emerged from manipulations of a representation embedding focus on a particular aspect of a mathematical construct, function. Jedtke and Greefrath very differently looked to the potential effects on mathematical performance of a particular kind of computer feedback on the students' answering of questions and providing solutions to tasks, that of explanation and hint. Their focus was on the type of computer feedback and its effect on student performance and self-assessment capability. Here, meaning is understood as emerging from students' responses to tasks embedding conventional concepts and representations of quadratic function in line with an established curriculum. It is interesting to read the papers with respect to the notion of experience with a digital medium for expressing mathematical meaning and I do this here with respect to three aspects of this process.

Mathematical Experiences with Representations

The first is to do with the interactions between meaning making and the use of one or more representations. I focus on the idea of a dependency on a representation and ways in which this dependency may gradually fade out allowing students to understand and use a mathematical idea irrespective of the representation it may be embodied in (Piaget & Beth, 1966). Noss and Hoyles (1996) talk about situated abstractions, a significant aspect of a situation being the particular representations at hand. Morgan and Kynigos (2014) discuss the interconnectivity of diverse representations as a particular aspect of students' representational repertoires. All three papers in this section discuss the use of representations as semiotic systems to express and communicate mathematical meaning (Edwards, 1998; Janvier, 1987). They address the potentials but also the difficulties to understand the meanings embedded in conventional or innovative representations especially in the case where they are inter-dependent as in the case of the first two papers. Lisraelli argues that 'the computer presents an opportunity for students to see change directly and to call into play all the intuitions that they have developed about movement, time and speed' but also concludes that 'the Cartesian representation of function is extremely rich in meaning and useful but, at the same time, the interpretation and manipulation of a graph requires a deep understanding of the relations existing between its elements' (ibid). Jedtke and Greefrath also point out that 'the distinct representations of quadratic functions need to be understood, constructed, interpreted, converted from one to another, and applied to solve problems' (ibid). A particular issue underlying the papers is students' difficulty to focus on a mathematical meaning independently of the particular representation at hand. Uygan and Turgut found that 'Within the context of our study, gestures are limited to the use of specific tools. There did not appear to be any gestures independent of the artefact (mouse and keyboard), such as hand movements, tracing with a finger and so on'. (ibid). All three papers thus use connected representations carrying some innovative aspect (e.g. dynamic manipulation of co-varying numbers) and all come to the joint conclusion that despite the semiotic potential for meaning making there are old and new obstacles to students' ability to gain independency for a particular representation, they tightly connect meaning with the representation it is conveyed through (Latsi & Kynigos, 2011; Falcade et al., 2007). The question thus arises, how can we find ways in which representation dependency may fade out allowing students to make connections between different representation of the same concept or construct and ultimately to convey meaning irrespective of the representation used?

Instrumentalization as a Design Experience and a Mathematical Experience

The papers discuss meaning making through the use of representations, but vary in addressing ways in which students may make changes to the representation or the functionalities of the model or the medium at hand. Digital media, however, offer the potential for meaning making through instrumentalization of a digital artefact, i.e. for the generation of the schemes of use (Verillon & Rabardel, 1995) formed as a reciprocal shaping (to use Noss and Hoyles' terminology here, 1996) of the artefact's functionalities and the embedded meanings.

The affordances, the semantics and the constructs, given to students to work with, all constitute artefacts which are changed through the process of instrumentalization (Kynigos & Psycharis, 2012). As a whole, in the section papers, little attention is given to how students change the artefacts while giving meaning to their experiences with them and reciprocally how the changes in the artefacts shape their new experiences, their reflections and their discussions around them. Lisraelli makes the 'assumption that this kind of dynamic representation can support a dynamic conception of functions because it draws attention to variables' variations and movements and to the relation between these variations. And, we wanted to study how to exploit this potential to introduce students to the idea of functional dependence. Jedke and Greefath discuss how CBLEs offer structured paths through a sequence of related exercises, inviting learners to work independently and autonomously and consider the difficulties students may have to understand aspect of quadratic equations because of the complexity of the ways they are conventionally represented. A fundamental affordance of digital media is that representations and mathematical objects are somehow constructed (Kynigos, 2012) and are represented by means of connected semiotic systems (Morgan et al, 2009; Morgan & Kynigos, 2014). When didactical engineering affords students with access to the objects' construction and the connections between representations, then meaning making draws from the instrumentalization process (Kynigos, 2007). So, a question to ask here would be 'what is the semiotic potential of artefacts purposefully affording instrumentalization' and what is the specific nature of meaning making during that kind of process?

Meaning Making Through Experience with Digital Media
The third, is about the process of meaning making with the use of digital media, how meanings are connected both to the mathematical idea and to the functionalities and modalities of use of the respective medium. Meanings therefore cannot be thought of in terms of pedagogy as silo objectives in a didactical situation but are necessarily at the centre of conceptual fields (Vergnaud, 2009) and are formed in situations where there had been an explicit or implicit restructuration with respect to conventional curriculum structure (Willensky and Papert, 2006). Attention needs to be given to new meanings connected to functionalities, representations and activities specific and relevant to the media used. Lisraelli, for instance, argues that the particular artefact in her study was designed to promote 'a co-variational view of functions, seen as relations between the movements of quantities that are varying in an interval of real numbers. In the same way also some mathematical properties of functions are conceived dynamically, for example Rob identified the domain of the function as a certain range of movement of the independent variable' (ibid). Jedke and Greefath consider how self-assessment ability of students (in mathematics) increases when their learning path for the topic of quadratic functions incorporates feedback that features additional explanations and hints (EF) compared to feedback that is limited to knowledge of the correct solution (Karkalas et al., 2016; Mavrikis et al., 2013). So, meanings are embedded in new kinds of situations and expressed with new representations involving the process of new or alternative

structures of mathematical concepts in relation to conventional curricula. So the question here is 'how can we design for and how can we understand meaning making without taking conventional curricula structures as unquestionable truths and mathematical concepts and objects as silo objectives'?

Conclusion

The main points in this synthesis were about meaning making with digital representations. How their features and uses of them may help students focus on the concept rather than the representation itself, how instrumentalization affects and is affected by meaning making and how concepts necessarily lie in diverse contexts and structures. The papers delightfully provide the arena for questions arising from these issues which have direct relevance to the nature and the uses of digital media. Yet, they also provide a diversity of conceptual fields and of mathematical concepts ranging from spatial awareness to function as co-variation and quadratic function. They further address different aspects and affordances of digital media such as 3D geometrical figures and their properties, special designs for co-variation and quite differently, the notion of the kinds of computer feedback supporting self-assessment with the use of CBLE tools. The papers show just how much more work is needed and the question provided in the papers and in this introduction are just a few of the avenues opening up for further research.

References

Ainsworth, S. E., Bibby, P. A., & Wood, D. J. (1997). Information technology and multiple representations: New opportunities—new problems. *Technology, Pedagogy and Education, 6*(1), 93–105.

Edwards, L. D. (1998). Embodying mathematics and science: Microworlds as representations. *The Journal of Mathematical Behavior, 17*(1), 53–78.

Falcade, R., Laborde, C., & Mariotti, M. (2007). Approaching functions: Cabri tools as instruments of semiotic mediation. *Educational Studies in Mathematics, 66*(3), 317–333.

Janvier, C. (1987). *Problems of representation in the teaching and learning of mathematics.* Hillsdale, NJ: Lawrence Erlbaum Associates.

Karkalas, S., Mavrikis, M., Xenos, M., & Kynigos, C. (2016). Feedback authoring for exploratory activities: The case of a logo-based 3D microworld. In *International Conference on Computer Supported Education* (pp. 259–278). Cham: Springer.

Kynigos, C. (2007). Half-baked microworlds as boundary objects in integrated design. *Informatics in Education, 6*(2), 335–359.

Kynigos, C. (2012). Constructionism: Theory of learning or theory of design? *Proceedings of the 12th International Congress on Mathematics Education.* Seoul, S. Korea.

Kynigos, C., & Psycharis, G. (2013). Designing for instrumentalisation: Constructionist perspectives on instrumental theory. *International Journal for Technology in Mathematics Education, 20*(1), 15–19.

Latsi, M., & Kynigos, C. (2011). Meanings about dynamic aspects of angle while changing perspectives in a simulated 3d space. Proceedings of the 35th Conference of the International Group for the Psychology of Mathematics Education. (Vol. 3, pp. 121–128). Ankara, Turkey: PME.

Mavrikis, M., Noss, R., Hoyles, C., & Geraniou, E. (2013). Sowing the seeds of algebraic generalization: Designing epistemic affordances for an intelligent microworld. *Journal of Computer Assisted Learning, 29*(1), 68–84.

Morgan, C., & Kynigos. C. (2014) Digital artefacts as representations: Forging connections between a constructionist and a social semiotic perspective. Special Issue in Digital representations in mathematics education: Conceptualizing the role of context and networking theories, Educational studies in mathematics, In Lagrange, J. B. and Kynigos, C. (Eds.), Dordrecht: Springer Science + Business Media, 85(3), 357–379.

Morgan, C., Mariotti, M. A., & Maffei, L. (2009). Representation in computational environments: Epistemological and social distance. *International Journal of Computers for Mathematical Learning, 14*(3), 241–263.

Noss, R., & Hoyles, C. (1996). *Windows on mathematical meanings: Learning cultures and computers*. Dordrecht: Kluwer.

Piaget, J., & Beth, E. W. (1966). *Mathematical epistemology and psychology*. Dordrecht: D. Reidel.

Vergnaud, G. (2009). The Theory of Conceptual Fields. *Human Development, 52*, 83–94. doi:10.1159/000202727.

Vérillon, P., & Rabardel, P. (1995). Cognition and artefacts: A contribution to the study of thought in relation to instrumented activity. *European Journal of Psychology of Education, 10*(1), 77–101.

Wilensky U., & Papert S. (2010). Restructurations: Reformulations of knowledge disciplines through new representational forms. In: J. Clayson & I. Kalas (Eds.), *Constructionist approaches to creative learning, thinking and education: Lessons for the 21st century. Proceedings of the Constructionism 2010 Conference* (pp. 97–105). Paris: American University of Paris.

Spatial-Semiotic Analysis of an Eighth Grade Student's Use of 3D Modelling Software

Candas Uygan and Melih Turgut

1 Introduction

Semiotics can be considered as a science of signs and, generally speaking, it can be considered as a perspective that looks for attachments to individuals' meaning-making processes. Because such a perspective has the potential to provide theoretical lens for searching construction, visualization and communication of mathematical concepts (Presmeg, Radford, Roth, & Kadunz, 2016; Sáenz-Ludlow & Kadunz, 2016), it has recently received robust attention from mathematics educators, yet learning mathematics is a kind of complex process and includes different types of reasoning based on mathematical objects (Godino, Font, Wilhelmi, & Lurduy, 2011). However, reasoning on mathematical objects not only includes the acts of thinking, constructing and expressing meaning per se but also those acts interlaced with our gestures, mimics and sometimes with specific sketches. Consequently, the involvement of our sensory-motor functions' productions in the communication and learning processes can be considered to be a *multimodal* process (Arzarello & Robutti, 2008). In other words, while reasoning on mathematical objects, as human beings, we intentionally or unintentionally produce gestures, mimics and sketches through our sensory-motor functions to describe specific mental pictures in our mind. Within a semiotic perspective, as mathematics educators, to understand learners' meaning-making on figures, charts or diagrams, we are interested in *signs* that are attachments to students' mental pictures, because these signs are also descriptors of learners' *internal representations* which are important as their *external representations* (Dreyfus, 1995).

C. Uygan · M. Turgut (✉)
Faculty of Education, Eskisehir Osmangazi University, Eskisehir, Turkey
e-mail: mturgut@ogu.edu.tr

C. Uygan
e-mail: cuygan@ogu.edu.tr

In this work, we acknowledge such a viewpoint aiming to analyse an eighth grader's mental pictures (i.e., mental images or spatial images) through a multimodal paradigm while he uses 3D modelling software to solve spatial tasks including unit cubes. Therefore, we combine the words spatial and semiotic to spatial-semiotic and focus on the following research question: *What kinds of spatial-semiotic resources emerge when an eighth-grade student solves spatial tasks with 3D modelling software?* Consequently, this chapter is divided into six main sections. The next section describes the theoretical constructs that we focus on, such as multimodal, embodied cognition and action, production and communication paradigms (Arzarello, 2008), as well as spatial images and spatial thinking (i.e., visualization). The third section describes a conceptual framework as a tool of data analysis, while the fourth section presents the methodology of the chapter. The fifth section presents a spatial-semiotic analysis of the data, with the chapter ending with a conclusion and discussion section.

2 Background: Theoretical Constructs

We organize this section into three parts to overview the multimodal paradigm, embodied cognition and action, production and communication perspectives and definitions of the mental or spatial images in spatial thinking and/or visualization process that we consider in this chapter.

2.1 Multimodal and Embodied Cognition Paradigm

The multimodal paradigm underlines the role of sensory-motor functions in the development of thinking and communication and its roots are thanks to psychological theories underpinned by an embodied cognition perspective (Maffia & Sabena, 2015). In other words, the embodied cognition hypothesizes 'that cognitive processes are rooted in interactions of the human body and the physical world' (Alibali, Boncoddo, & Hostetter, 2014, p. 150). In the physical world, the development of cognitive productions and therefore concepts, are due to our entire body's sensory-motor functions and, as a result of this, our perception has a multimodal character, such as speech, gestures, touch and so on, which are integral to *learning*. Such a multimodal character of cognition is discussed in learning mathematics by seminal papers in the field (Lakoff & Núñez, 2000; Nemirovsky, 2003; Núñez, Edwards, & Filipe Matos, 1999). Developments of such a perspective appear as a special issue in Educational Studies in Mathematics (vol. 61, 1/2), and also in a collection of recent developments published in an edited book (Edwards, Ferrara, & Moore-Russo, 2014).

Following the multimodal paradigm, Arzarello (2008) proposes a triadic model to consider the development of mathematical phenomena in the classroom. He addresses a *shared environment* for cognition and defines an Action, Production and Communication space (APC-space). This APC-space is constructed of three components; (i) the body, (ii) the physical world, and (iii) the cultural environment. All these together build a complex dynamism and provide several semiotic resources in the classroom, where students' actions, productions and communications with each other, and/or with their teacher, occur. Gradually, in order to analyse such classroom dynamism, Arzarello (2006) introduces the notion of the *semiotic bundle*, which is an extension of existing (e.g., Duval, 2006) semiotic systems, as follows:

> ... A semiotic bundle is a system of signs (with Peirce's comprehensive notion of sign) that is produced by one or more interacting subjects and that evolves over time. Typically, a semiotic bundle is made up of signs that are produced by a student or by a group of students while solving a problem and/or while discussing a mathematical question. Possibly, the teacher also participates in this production, and so the semiotic bundle may also include signs produced by teacher. (Arzarello, Paola, Robutti, & Sabena, 2009, p. 100)

A semiotic bundle is an analysis tool including two independent, but related, components; a synchronic analysis and a diachronic analysis. A synchronic analysis refers to 'the relationships among different semiotic resources simultaneously activated by the subjects at a certain moment', while a diachronic analysis means the 'evolution of signs activated by the subject in successive moments' (Arzarello et al., 2009, p. 109).

2.2 Spatial Thinking and Visualization

Spatial thinking is a core concept in the teaching and learning of mathematics, which can be defined as an amalgam of different sub-skills in relation to the visualization and rotation of 2D and 3D figures and objects in the mind. Inductively, considering such skills associated with geometric reasoning and also the goals of geometric content as stated by National Council of Teachers of Mathematics (2000), *Visualization* is expressed as one of the geometry content goals, along with *Shapes and Properties*, *Transformation* and *Location*. While the visualization content goal focuses on improving skills related to spatial thinking for supporting the study of geometrical figures, it is clearly emphasized that spatial thinking plays a crucial role in geometry content for identifying geometrical figures, understanding the relationships between two dimensional (2D) and three dimensional (3D) objects, visualizing the views of 3D objects from different perspectives and for modelling them within activities using paper-pencil or concrete materials. The need to support spatial thinking in learning geometry has motivated researchers to understand its factors and the psychological processes that students encounter. McGee (1979) defines two factors within spatial thinking; *spatial visualization* and *spatial orientation*. According to McGee, spatial visualization is related to the manipulation of an object in the mind, such as rotating an object as a whole or folding the surfaces of a 3D object. Secondly, McGee explains spatial orientation as a capacity for someone to describe her/his location after certain movements within a spatial configuration and to imagine the appearance of objects in this configuration from different perspectives.

Since spatial thinking plays a crucial role in learning geometry, a number of epistemological analyses have been conducted to elaborate and explain how individuals think spatially when they commence a mathematical task, and specific definitions of mental, visual or spatial images appear in the literature. However, as is clear from the previous paragraph, specific terms are extremely similar and are sometimes used interchangeably. This issue is addressed by Gutiérrez (1996), within a mathematics education perspective, where he notes that several terms, such as mental image, spatial image or visual image, are being used synonymously. Gutiérrez (1996) uses the term visualization for spatial thinking by defining it as, "a kind of *reasoning activity based on the use of visual or spatial elements, either mental or physical*, performed to solve problems or (to) prove properties" (p. 9, emphasis in the original). Here, as Sutherland (1995) points out, the existence and production of a mental (or visual) image play a core role because such a mental image could correspond to cognitive depictions of mathematical objects in the mind.

2.3 Mental and Spatial Images

Presmeg (2006) defines a visual image as "a mental construct depicting visual or spatial information" (p. 207). In her Ph.D. research, Presmeg (1986) proposes five visual (or mental) images when an individual uses one or more of them while studying mathematics (not limited to learning geometry). They are as follows (ibid., pp. 43–44):

- *Concrete, pictorial images*: This kind refers to existing (*already created*) images in the mind, obtained through phenomenological experiences; for instance, an image of a cube.
- *Pattern images*: Such images are in relation to visual mathematics-ready patterns; for instance, a $3k$, $4k$ and $5k$ triangle.
- *Images of formulas*: Remembering the formulas directly from a notebook or a textbook and so on.
- *Kinaesthetic images*: These refer to images that are integral to physical movements; for instance, describing or pointing at something with a finger or explaining something using bodily actions.
- *Dynamic images*: Creating, transforming and manipulating images to other images in the mind; for example, use of spatial visualization and/or spatial orientation skills.

Because our research context is developed through 3D modelling software, SketchUp®, we limit ourselves to the consideration of concrete images, kinaesthetic images and dynamic images. Use of concrete images, kinaesthetic images and dynamic images may be considered as a *visualization process*, and such a process can be fulfilled through two sub-visualization abilities (Bishop, 1983); "the ability for interpreting figural information (IFI), and the ability for visual processing (VP)" (p. 177). He defines these as follows:

> IFI involves knowledge of the visual conventions and spatial 'vocabulary' used in geometric work, graphs, charts, and diagrams of all types. Mathematics abounds with such forms and

> IFI includes the 'reading' and interpretation of these. ... VP, on the other hand, involves the ideas of visualization, the translation of abstract relationships and non-figural data into visual terms, the manipulation and extrapolation of visual imagery, and the transformation of one visual image into another. (Bishop, 1983, p. 177)

Therefore, an individual's use of spatial vocabulary to describe his/her steps or reasoning on a given figure or object in a spatial task can be considered as a part of IFI, while VP includes the creation and manipulation of mental images such as concrete images, kinaesthetic images and dynamic images. This last paragraph may be considered as a summary of our *focus* in this paper.

3 Conceptual Framework

In order to analyse classroom activities through a *spatial-semiotic* lens, Turgut (2017) proposes a conceptual framework based on the hypothesis that thinking spatially in a 3D modelling software environment is also multimodal. Spatial-semiotic lens combines three theoretical constructs; (i) spatial thinking and visualization (Gutiérrez, 1996), (ii) specific mental images (Presmeg, 1986) and interpreting figural information and visual processing processes of spatial thinking (Bishop, 1983), and (iii) APC-space and the notion of the semiotic bundle (Arzarello, 2006, 2008; Arzarello et al., 2009).

These are used to view the emergence of signs linked to spatial thinking. Following the multimodal paradigm, spatial-semiotic lens frames classroom productions that include specific signs, such as words, gestures, sketches and acts and so on, which are attachments to students', as well to the teacher's, spatial thinking processes. To do so, spatial-semiotic lens distinguishes spatial thinking as two major processes: IFI and VP. In our context, we consider and rewrite that IFI includes the emergence of spatial vocabulary and the interpretation of visual images, while VP includes the emergence of concrete images, kinaesthetic images and dynamic images. Concrete images may be considered as pictures in the visual memory, whereas kinaesthetic images refer to physical movements and gestures as an attachment to thinking or reasoning. Dynamic images cover conceiving and manipulating dynamic mental images (Presmeg, 1986; Turgut, 2017). Figure 1 summarizes the spatial-semiotic lens and its components.

In Fig. 1, arrows show complex and dialectic relationships between different kinds of visualization processes and tools and functions of digital tools. Within the context of the present paper, and borrowing research results coming from the literature on the use of strategies in spatial ability (Burin, Delgado, & Prieto, 2000; Janssen & Geiser, 2010; Maeda & Yoon, 2013), we identify two strategies under IFI; a spatial-analytic strategy, meaning focusing on parts of the object, and a spatial-holistic strategy, which refers to comprehending and reasoning on the object as a whole. Spatial-semiotic lens offers analysis on the emergence of signs through the notion of the semiotic bundle (Arzarello, 2006), which constitutes two different, but complementary analysis tools: a synchronic analysis and a diachronic analysis as expressed above.

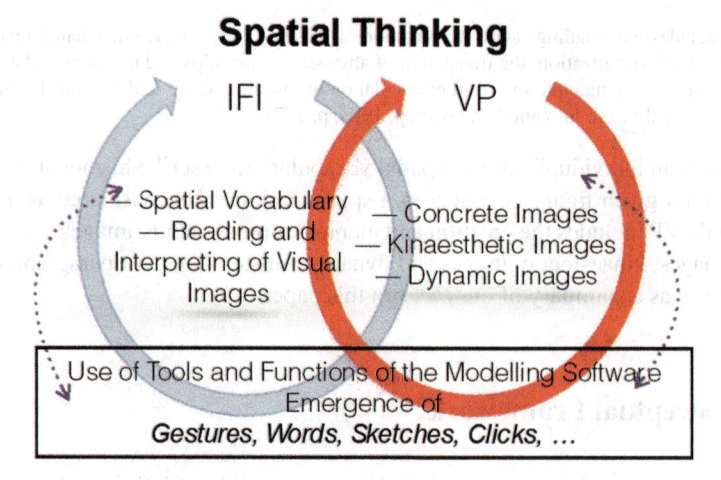

Fig. 1 The spatial-semiotic lens with its components (modified from Turgut, 2017, p. 183)

4 Methodology

A task-based interview was conducted with an eighth-grade student, Atakan (pseudonym), who performs moderately well in mathematics and has experience with the use of various geometrical software that was provided in previous research. In order to research Atakan's spatial reasoning process, we presented two tasks in SketchUp, which is 3D modelling software originally designed for 3D modelling in various areas, such as engineering, architecture and game designing (Murdock, 2009). It should be noted that Atakan has experience in the use of SketchUp since, as part of a larger study he participated in 3D geometry tasks using the same software when he was in 7th grade.

4.1 SketchUp Context and Design of the Tasks

In this chapter, we focus on 3D modelling software chosen based on the results from the literature, where it is stated that students' interaction with SketchUp could reflect spatial thinking processes in depth through analysing functions and tools of the software (La Ferla et al., 2009; Turgut & Uygan, 2015). As a result, SketchUp was used in our study as a digital tool to analyse the spatial thinking processes of an 8th-grade student. While SketchUp has a comprehensive toolbar within its standard surface (see Fig. 2), Atakan was expected to use five basic tools; 'Select', 'Move', 'Orbit', 'Pan' and 'Zoom' as well as 'Views Toolbar'. On the other hand, we designed our tasks using 'Select', 'Shapes', 'Move', 'Push/Pull' and 'Paint'.

While designing the task we used the 'Shapes' and 'Move' tools to sketch squares (Fig. 3a) and combined them to form a table on the floor (Fig. 3b). Then we hid the

Spatial-Semiotic Analysis of an Eighth Grade Student's ...

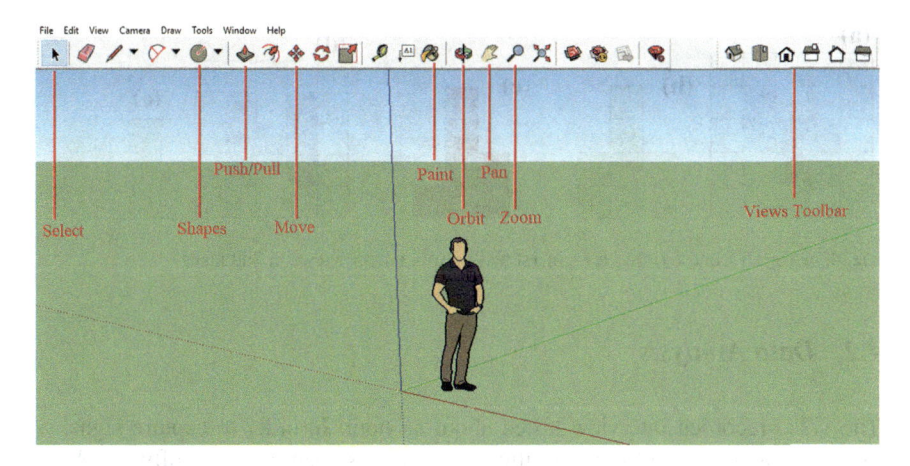

Fig. 2 Standard interface of SketchUp and the tools expected to be used by Atakan

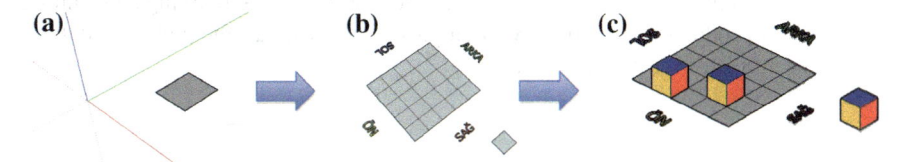

Fig. 3 a–c Steps of designing the SketchUp aided tasks including cube-building

axis on the scene and coded each side of the table with Turkish words as 'Ön', 'Arka', 'Sağ', and 'Sol' (Fig. 3b) with different coloured 3D text; translated into English as 'Front', 'Rear', 'Right' and 'Left', respectively. For the next step, we utilised the 'Push/Pull' tool to build a cube by pulling one of the squares upward. As a final step, we painted each surface of the cube with the same colours as the corresponding direction codes, and also designed the cube as an object that can be selected as a whole by one click with the use of the 'Select' tool. In addition to this, we inactivated a two viewpoints perspective option in the SketchUp and provided isometric perspective views for the tasks, so as to be in accordance with the mathematics curriculum including activities on isometric drawings of the 3D objects (see Fig. 3c).

In the context of acquisition, as described in the Turkish middle school mathematics curriculum, we prepared two 3D building tasks. During the interview, we first proposed three (top, front and right) views of a building (Fig. 4a–c) consisting of unit cubes on paper and asked Atakan to construct the related building. This initial task included two main steps: (i) constructing the building using concrete unit cubes, and (ii) using virtual cubes within SketchUp, which provide a zero-gravity environment with the aim of making alternative solutions with fewer cubes. In the second task, we asked Atakan to complete a building within the SketchUp environment according to the top and front views given on the paper (Fig. 4d, e), and also to find alternative solutions with fewer cubes.

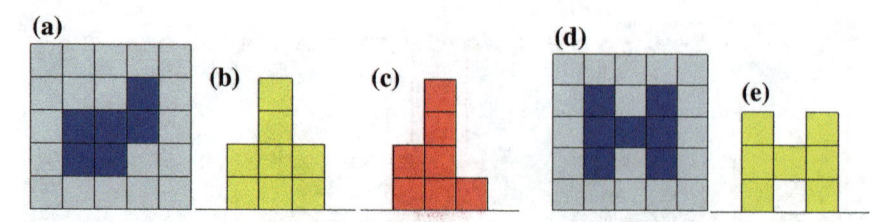

Fig. 4 **a** Top, **b** front, **c** right views in 1st task; **d** top, **e** front views in 2nd task

4.2 Data Analysis

The video-recorded interview lasted about an hour. In order to capture signs, we used two cameras in different positions as well as screen recorder software. A thematic analysis (Braun & Clarke, 2006) (including steps of familiarisation of data, generation of initial codes, search for themes, review of themes, defining and naming themes and producing the report) was employed covering all the collected data to elaborate Atakan's reasoning steps.

5 Spatial-Semiotic Analysis of the Data

In order to present an evolution of the student's reasoning, we first briefly present a macro analysis of the initial part of Task 1. As the first step, Atakan constructed the first floor of the building in such a way as to provide a top view (constructing a block of cubes parallel to the ground) to form a building with concrete cubes in accordance with the views given in the worksheet. For the second step, he built a block of cubes in a vertical position relative to the ground to form a front view without changing the top view. For the third step, he compared the right view of the building (with the right view given in the worksheet) changing his viewpoint by bending his body. Finally, for the fourth step of the initial part, without changing the top and front views, he put a cube in an appropriate place to complete the right view. By the end of the process, Atakan had constructed a building using twelve cubes.

5.1 Synchronic and Diachronic Analyses of the Second Part of Task 1

Several spatial-semiotic resources appeared synchronously when Atakan solved the second part of Task 1 with SketchUp. Table 1 briefly provides a summary of the most frequent spatial-semiotic resources categorized under the interpretation of figural information (IFI) and visual processing (VP) processes.

Spatial-Semiotic Analysis of an Eighth Grade Student's ...

Table 1 An overview of spatial-semiotic resources attached to the reasoning steps of Task 1

IFI		VP		
Spatial-analytic strategy	Spatial-holistic strategy	Concrete images	Kinaesthetic images	Dynamic images
– Exploring an appropriate viewpoint to add or remove a cube – Adding a block of cubes which are parallel to the base to form a top view – Adding a block of cubes which are vertical to the ground to form a front view – Focusing a single view of the object	– Evaluating the object from different viewpoints – The reasoning which cubes can be removed without changing views – Spatial vocabulary: expressing why the completed front view limited his next strategies – Spatial vocabulary: expressing strategies in relation to the top and right views	– Using a mental picture derived from the paper and concrete object – Basing an obtained mental image in the completed (reasoning) step(s)	– Using the *Orbit* tool to complete different steps – Adding, moving or removing the cubes using the *Select* tool – Using the cursor to point out cubes or the object while explaining the situation	– Linking 2D and 3D representations mentally – Mental rotation with respect to given directions – Spatial orientation with respect to different viewpoints

In order to present the emergence of the specific resources expressed in Table 1, we summarize Atakan's reasoning steps for Task 1 in the following statements. At first, he repeated the steps in the initial part of Task 1 to create a representation of the building (formed with twelve concrete unit cubes) in SketchUp. In this process, by making use of the tool 'orbit', he conducted a reasoning process (using the tool slowly) concerning the procedures to be applied (kinaesthetic image, dynamic image). He explored a viewpoint appropriate to cube addition (using the tool fast) (kinaesthetic image) (Fig. 5a), and he evaluated the top, front and right views of the building he had formed (using the tool fast) (kinaesthetic image, concrete image) (Fig. 5b). In the second step, it was seen that without changing the top, front and right views, Atakan deleted a cube from the first floor in the process of transitioning from a 12-cube building to an 11-cube building (dynamic image, kinaesthetic image). He then deleted a cube from the second floor in the process of transitioning to a 10-cube building (dynamic image, kinaesthetic image). Next, he deleted a cube from the first floor in the process of transitioning to a 9-cube building (dynamic image, kinaesthetic image). Finally, he deleted a cube from the second floor in the process of transitioning

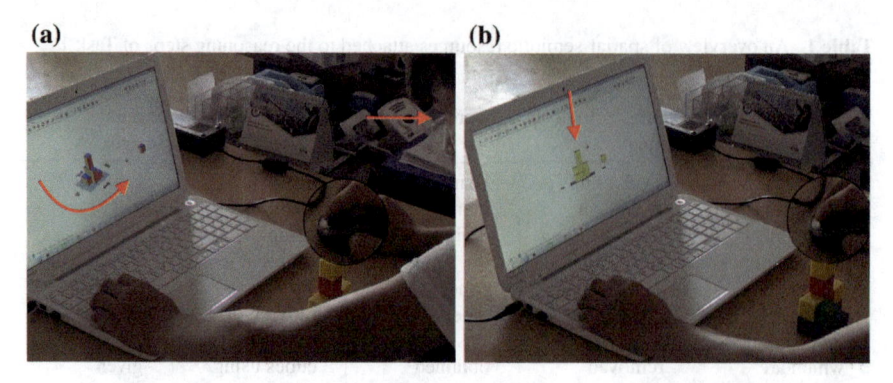

Fig. 5 **a** Atakan's exploration for a viewpoint, **b** evaluating the object from different viewpoints

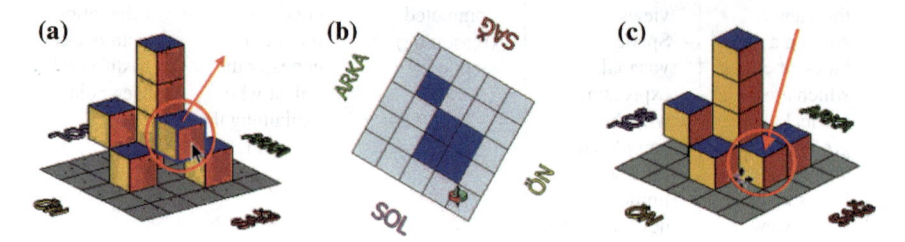

Fig. 6 **a** Deleting the wrong cube, **b** evaluating the top view, **c** replacing the deleted cube wrongly

to an 8-cube building (dynamic image, kinaesthetic image) and evaluated the views of the new building at the end of each step (concrete image, kinaesthetic image).

In the third step, it was seen that without making any changes to the top and right views, Atakan changed the top view by deleting a wrong cube from the second floor in the process of transitioning from an 8-cube building to a 7-cube building (kinaesthetic image) (Fig. 6a). He recognized the wrong strategy in the second step (concrete image) (Fig. 6b) and placed the cube (he had deleted) unintentionally in the first floor rather than in the second floor (Fig. 6c) while trying to cancel this deletion (kinaesthetic image).

In the fourth step, Atakan examined the building he had formed previously with concrete cubes when he failed to develop a strategy for transition from the 8-cube building to the 7-cube building. He then returned to the 11-cube building by adding cubes (kinaesthetic image) (Fig. 7a). In the following steps, the participant used a cube-deletion strategy, respectively, to form an 8-cube building (Fig. 7b), and finally to form the 7-cube building (Fig. 7c) that provided the top and right views, thereby reaching the correct result (dynamic image, kinaesthetic image, concrete image). In the 2nd and 3rd parts of Task 1, Atakan, with the help of 'orbit', did the following: (i) conducted a reasoning process in relation to solution strategies (using the tool slowly) (kinaesthetic image, dynamic image); (ii) evaluated the views of the new buildings formed (using the tool fast) (kinaesthetic image, concrete image); and (iii) searched for a viewpoint appropriate to cube-deletion and cube-addition (using the

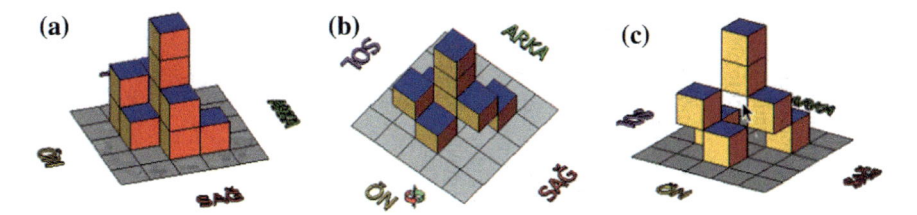

Fig. 7 a 11-cube building, b 8-cube building, c 7-cube building

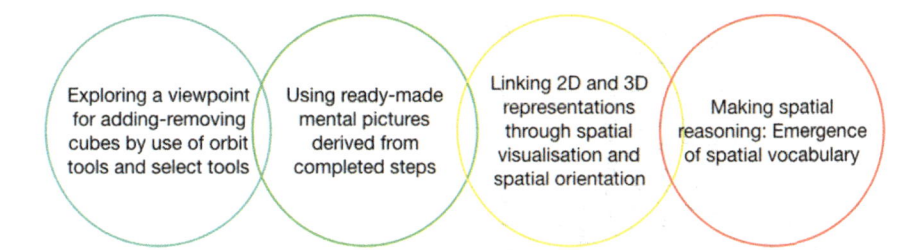

Fig. 8 A semiotic chain showing evolution of Atakan's reasoning for Task 1

[Semiotic chain circles: Exploring a viewpoint for adding-removing cubes by use of orbit tools and select tools → Using ready-made mental pictures derived from completed steps → Linking 2D and 3D representations through spatial visualisation and spatial orientation → Making spatial reasoning: Emergence of spatial vocabulary]

tool quickly) (kinaesthetic image). When the researchers asked Atakan why he had returned to the 11-cube building from the 8-cube building, he replied, "*Well, it didn't work. I had formed it according to the front view... the previous shape*".

In order to summarize a combination of synchronic and diachronic analyses of Task 1, i.e., to articulate specific signs with respect to an evolution of reasoning, we borrow the notion of the semiotic chain to Bartolini Bussi and Mariotti (2008) and express Fig. 8 to overview an evolution of Atakan's reasoning process.

5.2 Synchronic and Diachronic Analyses of Task 2

As the aims of the second step of the first task and the second task are close, the emergence of spatial-semiotic resources is similar to Table 1. However, in the second task, Atakan's strategies differed; in this case, two views of the building were provided on paper. Therefore, he exploited his experience from the first task and, in this way, he developed new insight for exploration of the situation and all of this changed the interpretation of figural information (IFI) and visual processing (VP) columns in Table 1. Another fact is that, in the present case, spatial vocabulary is more apparent compared to Task 2. Table 2 showing our findings summarizes the emergence of specific signs.

Atakan focuses on constructing the first floor of the building to form the top view in the first step and forms the cube-block in a position parallel to the ground. In this process, Atakan changes the viewpoint on the screen with the help of 'orbit'

Table 2 A summary of spatial-semiotic resources attached to the reasoning steps of Task 2

IFI		VP		
Spatial-analytic strategy	Spatial-holistic strategy	Concrete images	Kinaesthetic images	Dynamic images
– Exploring an appropriate viewpoint to add or remove a cube – Adding a block of cubes which are parallel to the ground to form the top view – Adding a block of cubes which are vertical to the ground to form the front view – Focusing single views of the object – Spatial vocabulary: emphasis on a partial solution strategy – Work (temporarily) on the cubes that satisfy a front view while not satisfying a top view – Spatial vocabulary: evaluating the different views part by part – Spatial vocabulary: explaining why he could not develop a strategy for removing cubes with respect to the floors of the building – Focusing only on removal of the cubes and, as a result of this, failing to visualise the object with 7 cubes	– Evaluating the object from different viewpoints – The reasoning which cubes can be removed, but which do not change the views – Spatial vocabulary: explaining and pointing out the cubes that can be moved, but also satisfying the views – Determining symmetrical cubes satisfying two different views when they are deleted – Spatial vocabulary: reasoning on the relationship between the front and rear views – Building the object from the beginning to develop new strategies – Spatial vocabulary: a new strategy for moving cubes on the third floor – Spatial vocabulary: generalizing strategy of removing cubes to satisfy top view	– Using a mental picture derived from the paper – Basing single views (top, front and so on) of the object – Basing obtained mental images in the completed (reasoning) step(s)	– Using the '*Orbit*' tool to complete different steps – Adding, moving or removing the (symmetric) cubes and/or *blocks* using the '*Select*' tool – Using the cursor to point out cubes or the object while explaining the situation – Using the zoom in-zoom out tool – Deleting the whole object	– Linking 2D and 3D representations mentally – 2D mental rotation of top view with respect to given directions – Spatial orientation with respect to different viewpoints – Visualising new views of the object when some cubes are moved – Visualising different views synchronously in the case of removing and/or moving the cubes

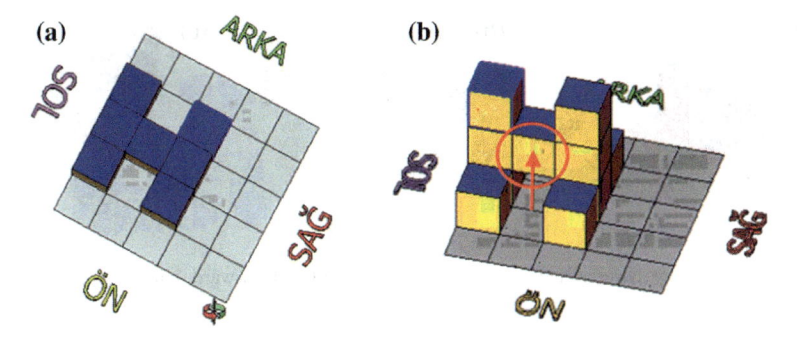

Spatial-Semiotic Analysis of an Eighth Grade Student's ...

Fig. 9 **a** Initial building, **b** moving the cube from the first to the second floor

(fast use) to add the cubes to appropriate places (kinaesthetic image). In addition, the participant considers the direction codes on the screen and rotates the image for the top view in his mind as appropriate to the direction codes while constructing the first floor of the building (dynamic image) (Fig. 9a).

In the second step, Atakan focuses on the second and third floors of the building as appropriate to the front view given in the worksheet and realizes that the first floor he had formed in the first step provides a top view, but not a front view (concrete image). In addition, he carries one of the cubes from the first floor to the second floor (dynamic image, kinaesthetic image) (Fig. 9b).

Following this process, Atakan builds a block of cubes in a vertical position to the ground (kinaesthetic image) and forms a building that provides the front view. In addition, it can be seen that Atakan explores a viewpoint appropriate to cube addition with the help of 'orbit' (fast use) (kinaesthetic image) and evaluates views of the building (concrete image). For the third step, Atakan focuses on constructing a building that provides top and front views using fewer cubes. In this process, Atakan focuses on symmetrical cube pairs on the right and left sides that do not change the top and front views when deleted (dynamic image), and he deletes two symmetrical cubes from the first floor (kinaesthetic image) (Fig. 10a). Following this, while the participant evaluates the top and front views of the new building with the help of 'orbit' (fast use) (concrete image, kinaesthetic image), the researchers ask him whether there was an alternative solution, which includes nine cubes. Within the scope of this question, it can be seen that Atakan initially again replaces the two symmetrical cubes he has deleted (kinaesthetic image) and then simultaneously examines the 11-cube building and the views given in the worksheet to produce new strategies (dynamic image). In such a way, it can also be seen that Atakan examines the building from different viewpoints with the help of 'orbit' (slow use) (kinaesthetic image), searches for the cubes that would not change the views when deleted, and fails to produce solution strategies.

Therefore, the researchers ask Atakan whether he would be able to form an alternative 11-cube building with the same top and front views. Within the scope of this question, to begin with, Atakan simultaneously examines the 11-cube building and

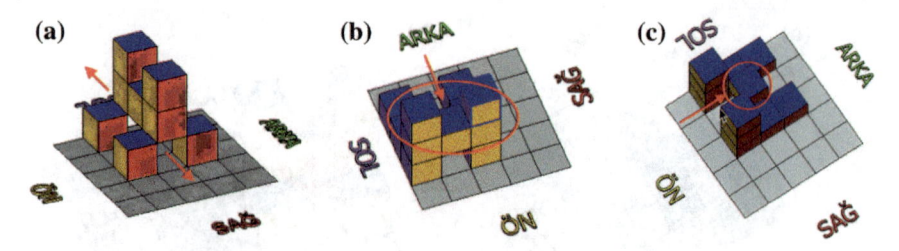

Fig. 10 **a** Deleting symmetrical cubes, **b** moving a cube-block forward, **c** moving the cube backwards

the views given in the worksheet (dynamic image) and then says that the block of cubes which forms the second and third floors could be moved one unit backwards or one unit forward (dynamic image, spatial vocabulary). In the following step, he moves this block one unit forward (kinaesthetic image) (Fig. 10b). Following this, Atakan, with the help of 'orbit' (fast use), evaluates the top and front views of the building (concrete image, kinaesthetic image) and sees that the top view has changed. As a result, he moves one cube on the second floor to provide the top view (kinaesthetic image) (Fig. 10c). Following this step, Atakan evaluates the views with the help of 'orbit' (concrete image, kinaesthetic image), realizes that the top view is wrong again and deletes one cube in the second floor, which changes the top view (kinaesthetic image). Following this strategy, in which the participant does not change the top or front views, he evaluates the views with the help of 'orbit' again (fast use) (kinaesthetic image, concrete image) and says that the alternative 11-cube building is complete (spatial vocabulary).

In the next part of the discourse, the researchers ask Atakan whether he could work on the building and form an alternative 9-cube building. Within the scope of this question, the participant, with the help of 'orbit' (slow use), searches for symmetrical cubes, which would not change the top and front views when deleted (kinaesthetic image, concrete image, dynamic image) and says that he has failed to find a strategy to form such a building (spatial vocabulary). In the following process, Atakan continues his search with 'orbit' (slow use) (kinaesthetic image) and realizes that there would be no change in the top and front views when two symmetrical cubes in the first floor are deleted. The participant deletes the symmetrical cubes he had determined (kinaesthetic image), and then evaluates the top and front views of the new building with the help of 'orbit' (fast use) (concrete image, kinaesthetic image).

Finally, the researchers ask Atakan whether he could form a 7-cube building without changing the top and front views. Within the scope of this question the participant, with the help of 'orbit' (slow use), searches for cubes that would not change the top and front views when deleted (kinaesthetic image, concrete image, dynamic image). As a result, Atakan reasons in relation to the 9-cube building and the views in the worksheet (dynamic image), but fails to develop a strategy to form the 7-cube building at the end of the process. Eventually, he deletes all the cubes on the screen to re-form the 11-cube building (kinaesthetic image, spatial vocabulary).

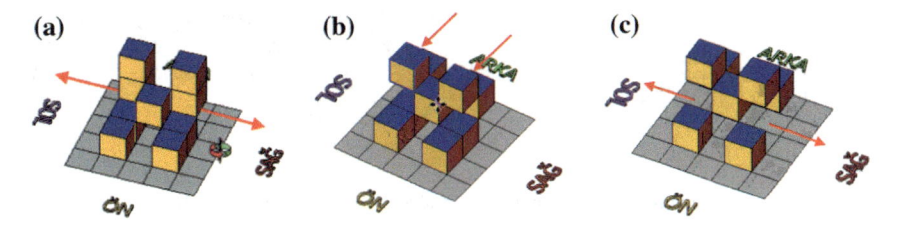

Spatial-Semiotic Analysis of an Eighth Grade Student's ...

Fig. 11 a–c Process of transitioning from a 9-cubes building to a 7-cubes building

This time Atakan, who starts constructing the building again, forms the block of cubes in a vertical position to the ground to complete the front view (concrete image, kinaesthetic image). This block is built in such a way as to form the rear of the building differently from his previous buildings.

In the next part, the participant examines the building from the top with the help of 'orbit' (fast use) (kinaesthetic image) and sees that one of the cubes he has added to the second floor changes the top view (concrete image). Therefore, he moves this cube one unit forward (dynamic image, kinaesthetic image). Following this, Atakan works on the block of cubes in a position parallel to the ground and builds the first floor (kinaesthetic image) in such a way as to complete the top view without changing the front view (concrete image, dynamic image). As a result, he completes the 11-cube building. The participant deletes two symmetrical cubes from the first floor during transitioning to the 9-cube building (kinaesthetic image) (Fig. 11a). Next, he uses the 'zoom' tool to examine the building in more detail (kinaesthetic image). Lastly, with the help of 'orbit' (slow use), he searches for cubes he could delete to make a transition to the 7-cube building (kinaesthetic image). In this process, Atakan reasons in relation to the 9-cube building regarding the views given in the worksheet (concrete image, dynamic image). He says that he did not make a transition to the building with the cube-deletion strategy as required in the question (dynamic image, spatial vocabulary). In this respect, when the researchers ask Atakan whether he has developed his thinking strategy based on a cube-deletion strategy, he responds positively to the question and says that he would think about the building a little longer and move two symmetrical cubes in the third floor one unit forward. He then adds that these cubes would hang in the air at the end of the process without changing the views (dynamic image, spatial vocabulary). Following this, the participant moves the symmetrical cubes in the third floor one unit forward (kinaesthetic image) (Fig. 11b). Next, he adds the cubes to places he wants and examines the building with the 'zoom' tool (kinaesthetic image). After this, he deletes the symmetrical cubes in the first floor, which were under the symmetrical cubes he moved forward (dynamic image, kinaesthetic image) (Fig. 11c). In the last step, Atakan evaluates the top and front views with the help of 'orbit' (fast use) (concrete image, kinaesthetic image).

At the end of the solution process, the researchers ask Atakan to explain his reasoning processes, and he explains that the cubes placed under one cube in the upper floors are not visible from the top view and that deleting the cubes below

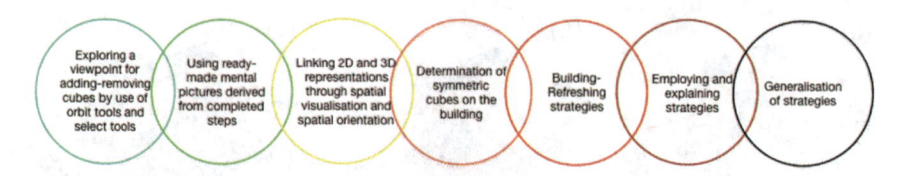

Fig. 12 A semiotic chain showing evolution of Atakan's reasoning regarding Task 2

would not change the top view (spatial vocabulary). In this respect, Atakan reports, *"From the top view, we see the upper cubes, and the ones below are not visible. If we take the ones below, those at the top look the same"*. In addition, Atakan states that he has evaluated how simultaneously the deletion process, which did not change the top view, did not change the front view (spatial vocabulary). In this respect, Atakan says, *"When we did not move to the front, and if I take these* (showing the symmetrical cubes he has deleted from the first floor in the last step), *then these* (coming to the top view rapidly with the help of 'orbit') *would have looked as if they had been removed* (pointing to the procedure that changed the top view). *In addition, if I had taken these* (showing the symmetrical cubes in the second floor at the back) ... *they would have remained at the back* (showing the symmetrical cubes in the third floor he had moved one unit forward) ... *Then they would have looked* ... (taking the front view rapidly with the help of 'orbit' and showing the spaces that would appear in the front view at the end of the process)". As a result of all this progress, Fig. 12 refers to a combination of synchronic and diachronic analyses of Atakan's reasoning processes associated with Task 2.

From a technical point of view, Fig. 12 reflects Atakan's reasoning steps from primitive uses of SketchUp to using his concrete images, spatial visualization and spatial orientation skills, making conjectures and establishing strategies and generalizing strategies; in other words, his reasoning process from artefact type steps to cognitive type steps.

6 Conclusions and Discussion

In this paper, we consider the following research question: 'What kind of spatial-semiotic resources emerge when an eighth-grade student solves spatial tasks using 3D modelling software?' A spatial-semiotic analysis of the data obtained provides us with a detailed understanding of the student's spatial reasoning processes in SketchUp. In the first task, the student easily constructs the building with concrete unit cubes with different views provided on the paper. In the second task, the student's reasoning steps appear with an emphasis on a spatial-analytic strategy based on exploring a viewpoint for adding or removing cubes, using ready-made mental pictures, linking 2D and 3D representations through spatial visualisation and spatial orientation, and an emergence of spatial vocabulary, including his strategies. How-

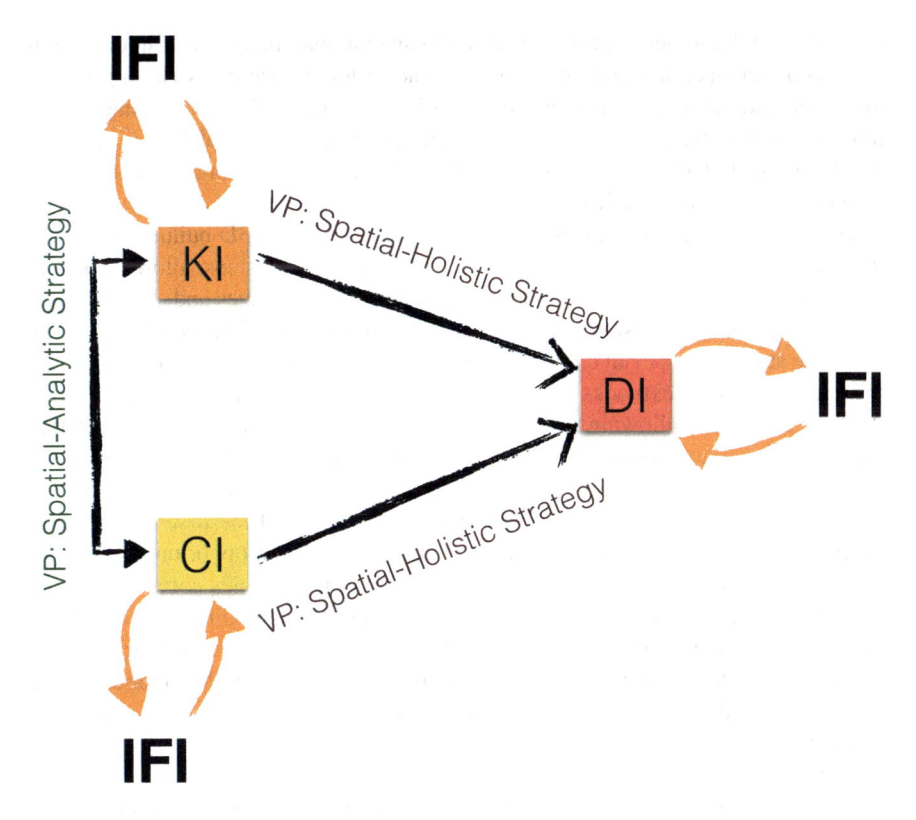

Fig. 13 Synergies between spatial thinking processes in our case (KI: kinaesthetic images, CI: concrete images, DI: dynamic images)

ever, in the second task, certain specific reasoning steps appear as spatial-holistic strategies more than in the previous task, such that focusing on an environment with zero-gravity, symmetric cubes, and constructing and explaining strategies, interlaces into completed steps in the second task. Within the context of our study, gestures are limited to the use of specific tools ('orbit', 'select', mimicking with the cursor, ctrl+v, delete and 'zoom'). There did not appear to be any gestures independent of the artefact (mouse and keyboard), such as hand movements, tracing with a finger and so on.

In terms of the obtained results, we finally summarize the synergies among kinaesthetic images, concrete images, dynamic images and VP and IFI processes through Fig. 13, which are a theoretical contribution and an attachment to Fig. 1.

Figure 13 implies that spatial-analytic and spatial-holistic strategies that we consider in this paper commonly intertwine with the IFI process and the emergence of kinaesthetic images, concrete images and dynamic images. The IFI process always emerges when the student solves spatial tasks and this seems that IFI is *the core element* in spatial thinking and creation of dynamic images. The emergence of signs confirms that the student's initial strategy was spatial-analytic in which student studies to construct a certain part of the building according to one of its views given on the paper, and the specific images are kinaesthetic images and concrete images. The next step is the emergence of dynamic images in terms of a spatial-holistic strategy in which the student constructs the building as a whole according to all the views given in the paper and the IFI process.

According to the results, it is shown that SketchUp based 3D building activities that were conducted using concrete cube models in previous steps, allow Atakan to meet a new challenge, thereby taking his dynamic mental images and strategies one step further. Our conclusion corresponds to the arguments of Gutiérrez (1996) in which he states that special computer software could enable students to form richer mental images than textbooks that provide pencil-paper activities and the use of real concrete models, by allowing them to study 3D objects on a computer screen through various ways and to manipulate them. In this regard, when students handle 3D objects by only using real concrete models, they carry out certain operations, such as rotation unconsciously, so that they cannot reflect on the features of the manipulations. In addition, in our context, it is difficult for students to have an opportunity to reflect on alternative solutions using less real cubes within the buildings corresponding to the views given. At this point, the SketchUp based tasks applied in our study differ from real cube aided activities in terms of providing an isometric perspective and moveable, but un-rotatable cubes as limited representations. On the other hand, the SketchUp based tasks differentiate from real space because the context provides a zero- gravity environment for students in order for them to consider expansive ways for constructing alternative buildings.

As pointed out by Sutherland (1995), students' mental images can be elaborated only by analysing how they use verbal, graphical and gestural supports while solving a task. In the context of our study, due to the fact that spatial thinking tasks are designed and conducted in SketchUp, Atakan's gestures transform to the use of the cursor and orbit tool on the computer screen in ways that differ from real building activity in which he used his fingers or body. At this point, we identify his thinking processes in SketchUp by interpreting which objects they move the cursor on, how fast he uses the orbit tool and which sides of the building are viewed from while studying it. Additively, these gestures enable us to identify Atakan's kinaesthetic images while reasoning. On the other hand, it is shown that there is limited data regarding the spatial vocabulary used by Atakan in the tasks. According to Gutiérrez (1996) in interviews with students who study spatial tasks, researchers should not ask students to explain their mental images because students may not be aware of their own mental images or the questions may distract students from reasoning about the task. Therefore, as stated by Gutiérrez, we mostly prefer to identify Atakan's mental image by analysing his gestural and graphical actions. We use questions that aim to enlighten us on how Atakan reasons after a task is completed, while we are doubtful about the reasons for his actions in SketchUp. The spatial vocabulary that is used by Atakan in the verbal answers figure out certain strategies that he generalizes on while removing or moving a cube within the building according to the positions of other cubes around it. Additionally, certain verbal answers help us understand if he studies a certain part of a building independently from other parts (spatial-analytic strategy) or considers all parts of the building (spatial-holistic strategy).

In our analysis regarding Atakan's spatial thinking strategies, we base the conceptual framework as IFI within which we identify two components; a spatial-analytic strategy that is related to constructing a certain part of a 3D building independently from other parts, and a spatial-holistic strategy which is about constructing a building as a whole by considering the relationships between all the parts within it. At this point, these two concepts are identified as inherent in the context of the SketchUp based cube-building task and are different from the perspective of Burin et al. (2000) that identifies two strategies; an analytic strategy by which a subject examines key features of an object to match them with the target figure, and a holistic strategy in which a subject imagines manipulations of objects, such as rotation, reflection or combination. According to Burin et al. (2000), a basic variant of analytic strategy is verbal labelling of the features and their context which involves comparing an object with the target figure. On the other hand, the context of our tasks concerns constructing 3D buildings using equal and un-rotatable cube models in SketchUp and tasks that do not include comparing features of a cube with the target 3D building. From this point of view, both the spatial-analytic strategy and the spatial-holistic strategy are handled as different approaches regarding the construction of 3D buildings. However, all these results come only from an eighth grader's result. Therefore, it would be meaningful to explore a group of students' results in order to discuss articulation of Figs. 1 and 13 in future research settings.

Acknowledgements The authors would like to thank the Scientific and Technological Research Council of Turkey (TUBITAK) for their financial support during M.Sc. and Ph.D. studies. Special thanks go to the anonymous reviewers and Editors for their careful readings of the paper and making constructive suggestions that improved the presentation of the paper.

References

Alibali, M. W., Boncoddo, R., & Hostetter, A. B. (2014). Gesture in reasoning: An embodied perspective. In L. Shapiro (Ed.), *The Routledge handbook of embodied cognition* (pp. 150–159). NY: Routledge.

Arzarello, F. (2006). *Semiosis as a multimodal process* (pp. 267–299). Especial: RELIME.

Arzarello, F. (2008). Mathematical landscapes and their inhabitants: Perceptions, languages, theories. In E. Emborg & M. Niss (Eds.), *Proceedings of the 10th International Congress of Mathematical Education* (pp. 158–181). Copenhagen: ICMI.

Arzarello, F., Paola, D., Robutti, O., & Sabena, C. (2009). Gestures as semiotic resources in the mathematics classroom. *Educational Studies in Mathematics, 70*(2), 97–109.

Arzarello, F., & Robutti, O. (2008). Framing the embodied mind approach within a multimodal paradigm. In L. English, M. Bartolini Bussi, G. Jones, R. Lesh, & D. Tirosh (Eds.), *Handbook of international research in mathematics education* (2nd ed., pp. 720–749). Mahwah, NJ: Erlbaum.

Bartolini Bussi, M. G., & Mariotti, M. A. (2008). Semiotic mediation in the mathematics classroom: Artifacts and signs after a Vygotskian perspective. In L. English, M. Bartolini Bussi, G. Jones, R. Lesh, & D. Tirosh (Eds.), *Handbook of international research in mathematics education* (2nd ed., pp. 746–783). Mahwah, NJ: Erlbaum.

Bishop, A. (1983). Spatial abilities and mathematics thinking. In M. Zweng, T. Green, J. Kilpatrick, H. Pollak, & M. Suydam (Eds.), *Proceedings of the Fourth International Congress on Mathematical Education* (pp. 176–178). Boston: Birkhäuser.

Braun, V., & Clarke, V. (2006). Using thematic analysis in psychology. *Qualitative Research in Psychology, 3,* 77–101.

Burin, D. I., Delgado, A. R., & Prieto, G. (2000). Solution strategies and gender differences in spatial visualizations tasks. *Psicológica, 21,* 275–286.

Dreyfus, T. (1995). Imagery for diagrams. In R. Sutherland & J. Mason (Eds.), *Exploiting mental imagery with computers in mathematics education* (Vol. 138, pp. 3–19). Berlin Heidelberg: Springer.

Duval, R. (2006). A cognitive analysis of problems of comprehension in a learning of mathematics. *Educational Studies in Mathematics, 61*(1–2), 103–131. https://doi.org/10.1007/s10649-006-0400-z.

Edwards, L. D., Ferrara, F., & Moore-Russo, D. (Eds.). (2014). *Emerging perspectives on gesture and embodiment in mathematics*. NC: Information Age Publishing.

Godino, J. D., Font, V., Wilhelmi, M. R., & Lurduy, O. (2011). Why is the learning of elementary arithmetic concepts difficult? Semiotic tools for understanding the nature of mathematical objects. *Educational Studies in Mathematics, 77*(2), 247–265. https://doi.org/10.1007/s10649-010-9278-x.

Gutiérrez, A. (1996). Visualization in 3-dimensional geometry: In search of a framework. In L. Puig & A. Gutiérrez (Eds.), *Proceedings of the 20th Conference of the International Group for the Psychology of Mathematics Education* (pp. 3–19). Valencia: Universidad de Valencia.

Janssen, A. B., & Geiser, C. (2010). On the relationship between solution strategies in two mental rotation tasks. *Learning and Individual Differences, 20,* 473–478.

La Ferla, V., Olkun, S., Akkurt, Z., Alibeyoglu, M. C., Gonulates, F. O., & Accascina, G. (2009). An international comparison of the effect of using computer manipulatives on middle grades students' understanding of three-dimensional buildings. In C. Bardini, D. Vagost, & A. Oldknow (Eds.), *Proceedings of the 9th International Conference on Technology in Mathematics Teaching—ICTMT9*. France: University of Metz.

Lakoff, G., & Núñez, R. (2000). *Where mathematics come from: How the embodied mind brings mathematics into being*. New York: Basic Books.

Maeda, Y., & Yoon, S. Y. (2013). A meta-analysis on gender differences in mental rotation ability measured by the Purdue spatial visualization tests: Visualization of rotations (PSVT:R). *Educational Psychology Review, 25,* 69–94.

Maffia, A., & Sabena, C. (2015). Networking of theories as resource for classroom activities analysis: The emergence of multimodal semiotic chains. *Quaderni di Ricerca in Didattica (Mathematics), 25*(2), 419–431.

McGee, M. G. (1979). Human spatial abilities: Psychometric studies and environmental, genetic, hormonal, and neurological influences. *Psychological Bulletin, 86*(5), 889–918.

Murdock, K. L. (2009). *Google SketchUp and SketchUp Pro 7 Bible*. Indianapolis, USA: Wiley Publishing Inc.

National Council of Teachers of Mathematics. (2000). *Principles and standards for school mathematics*. Virginia: NCTM.

Nemirovsky, R. (2003). Three conjectures concerning the relationship between body activity and understanding mathematics. In N. A. Pateman, B. Dougherty, & J. Zilliox (Eds.), *Proceedings of the 27th Conference of the International Group for the Psychology of Mathematics Education* (Vol. 1, pp. 105–109). Honolulu, Hawaii: PME.

Núñez, R. E., Edwards, L. D., & Filipe Matos, J. (1999). Embodied cognition as grounding for situatedness and context in mathematics education. *Educational Studies in Mathematics, 39*(1), 45–65. https://doi.org/10.1023/a:1003759711966.

Presmeg, N. C. (1986). Visualization in high school mathematics. *For the Learning of Mathematics, 6*(3), 42–46.

Presmeg, N. C. (2006). Research on visualization in learning and teaching mathematics. In A. Gutiérrez & P. Boero (Eds.), *Handbook of research on the psychology of mathematics education, past, present and future* (pp. 205–235). Rotterdam, The Netherlands: Sense Publishers.

Presmeg, N. C., Radford, L., Roth, W.-M., & Kadunz, G. (2016). *Semiotics in mathematics education*. Switzerland: Springer International Publishing.

Sáenz-Ludlow, A., & Kadunz, G. (Eds.). (2016). *Semiotics as a tool for learning mathematics*. The Netherlands: Sense Publishers.

Sutherland, R. (1995). Mediating mathematical action. In R. Sutherland & J. Mason (Eds.), *Exploiting mental imagery with computers in mathematics education* (pp. 71–81). Berlin, Heidelberg: Springer.

Turgut, M. (2017). A spatial-semiotic framework in the context of information and communication technologies (ICTs). In M. S. Khine (Ed.), *Visual-spatial ability in STEM education: Transforming research into practice* (pp. 173–194). Switzerland: Springer International Publishing.

Turgut, M., & Uygan, C. (2015). Designing spatial visualization tasks for middle school students with 3D modelling software: An instrumental approach. *International Journal for Technology in Mathematics Education, 22*(2), 45–52.

Activities Involving Dynamic Representations of Functions with Parallel Axes: A Study of Different Utilization Schemes

Giulia Lisarelli

1 Introduction and Theoretical Background

The concept of function is very important in both secondary school and university mathematics but it also has a central role in everyday situations. For a long time, this notion has been at the core of several studies in mathematics education, which have revealed students' difficulties in understanding the concept in all its aspects (Vinner & Dreyfus, 1989; Tall, 1991; Dubinsky & Harel, 1992). As Goldenberg, Lewis and O'Keefe (1992) write: "the act of representing functions graphically has as much potential to produce confusion as enlightenment" (p. 240).

Difficulties in interpreting the dependence relation as a dynamic relation between co-varying quantities and also difficulties in manipulating graphs and recognizing functions' properties from graphs are widely reported (Carlson, Jacobs, Coe, Larsen, & Hsu, 2002; Carlson & Oehrtman, 2005). In particular, it is argued that if the graph is seen as representing the picture of a physical situation then it is reasonable that the main aspects characterizing the functional relation remain hidden. However, these are necessary aspects for overcoming possible obstacles in the analysis and the interpretation of all the information presented in a graph.

Indeed, the tendency to think of functions and graphs as static objects, rather than as dynamic processes, may contribute to students' struggles in the learning of Calculus (Ng, 2016). At the same time, the emergence of different available software has fostered a variety of new teaching and learning approaches. A number of studies have investigated the learning of functions from a graphical point of view, using graphing and computer environments, or using technology to manipulate multiple representations of functions, zoom in on parts of graphs and explore the dependence relation between two variables through dragging (Kaput, 1992; Confrey & Smith, 1995; Healy & Sinclair, 2007).

G. Lisarelli (✉)
Department of Mathematics, University of Florence, Florence, Italy
e-mail: giulia.lisarelli@unifi.it

Falcade, Laborde and Mariotti (2007) suggest that the use of a DIMLE (i.e., Dynamic and Interactive Mathematics Learning Environment), as Martinovic and Karadag (2012) call software such as GeoGebra (in this chapter, we adopt this denomination), allows students to experience functions as co-variation and to exploit the functionalities of the dragging and trace tools to communicate co-variance dynamically. This is a crucial aspect of the notion of function (Tall, 1996; Kieran & Yerushalmy, 2004). According to these assumptions, we are interested in studying the potentialities of a particular representation of functions in a dynamic environment and students' cognitive processes involved in working with this representation.

As far as DIMLE is concerned, previous studies (Mariotti & Bartolini Bussi, 1998; Arzarello, 2000; Mariotti & Cerulli, 2001; Mariotti, 2002) have focused on the analysis of specific elements of the microworld as instruments of semiotic mediation that the teacher can use in order to introduce students to different mathematical ideas. In line with these studies, we are interested in analyzing the semiotic potential (Bartolini Bussi & Mariotti, 2008) of the representation of functions with parallel axes, to gain insight into how to exploit it didactically.

The theory of semiotic mediation describes the semiotic potential of an artifact as follows:

> On the one hand, personal meanings are related to the use of the artifact, in particular in relation to the aim of accomplishing the task; on the other hand, mathematical meanings may be related to the artifact and its use. This double semiotic relationship is named the *semiotic potential of an artifact*. (ibid., p. 754)

The educational aim, developed within the didactical theory centered on the notion of semiotic mediation, is to achieve the internalization of a technical tool used by the student to fulfil a task, into a sign, which is able to stand for a certain mathematical meaning. When analysing the semiotic potential of an artifact, a possible approach consists of studying the emergence of different signs: artifact signs, pivot signs and mathematical signs. However, rather than focusing on a possible evolution of signs that is guided by the teacher during the mathematical discussion, we have focussed on students' ways of interacting with the artifact and on their descriptions of these explorations.

For this reason, we referred to Rabardel, who studied the processes through which a subject interacts with an artifact (Rabardel, 1995; Bèguin and Rabardel 2000), and who conceptualized instruments as psychological and social realities when studying instrument-mediated activity. Rabardel (1995) introduced a distinction between *artifacts* and *instruments*, which results in the definition of utilization schemes. According to him, an artifact is just a material or a symbolic object, while an instrument is a mixed entity composed of two parts: an artifact-type component and a schematic component that the author calls a utilization scheme. Therefore, an instrument cannot be reduced to an artifact; it has to be associated with a subjective scheme.

The utilization schemes are progressively elaborated while using the artifact in relation to accomplishing a particular task, thus the instrument is an individual construction, it has a psychological character and it is strictly related to the context within which it originates and its development occurs. Moreover, the utilization schemes may not be consistent with the pragmatic goals for which the artifact was designed.

In this chapter, we describe a particular representation of functions that requires the use of a DIMLE. Although it is not possible to show dynamic visualization on the static medium of paper, we attempt to discuss it in detail in order for the reader to recreate the dynamism by herself/himself.

A study of students' utilization schemes elaborated during the exploration of this artifact (Lisarelli, 2017) highlights different types of dragging, some of which can also be recognized by a technical tool, while others are identified in association to an aim that may be utilized as a particular strategy for accomplishing a task.

In the rest of this chapter, we discuss some results from a pilot study that was conducted in 2016. The analysis aims at exploring the semiotic potential of the representation of functions with parallel axes, as well as at gaining insight into how to exploit it didactically. This study is part of a larger research project, whose focus is investigating how certain aspects of the mathematical concept of function can be introduced to students with such dynamic representations.

2 Dynagraphs

Cartesian graph is a very common representation of a real function and consists of representing a set of points $(x, f(x))$, where the independent variable x belongs to the domain of the function and the dependent one $f(x)$ is its image in the Cartesian plane. Therefore, in the Cartesian plane, a curve representing the graph of a function is made up of points that represent the functional relation between two variables: their coordinates. In particular, these two numbers belong to the same set (the real number set) but they correspond to two points belonging to two different axes. This double representation of the same set makes the construction of the curve possible, but it can also generate difficulties for a student who is approaching functions for the first time. In particular, a common obstacle for students involves recognizing that each point on the graph is a coordinated presentation of two pieces of information, a domain point and its image.

In conclusion, the Cartesian representation of a function is extremely rich in meaning and useful, but at the same time, the interpretation and manipulation of a graph requires a deep understanding of the relations between its elements. As previously mentioned, the reconstruction of these relations is not an easy task from a cognitive point of view. For this reason, some researchers started considering alternative visualizations in which the domain variable can be dynamically varied by the student, and is separately represented by its image. Goldenberg et al. (1992) refer to this class of function-visualizing tools as DynaGraphs, which represent both the x- and y-axes horizontally, in one dimension, as shown in Fig. 1. We claim that this type of representation can be presented in mathematics classes before the use of Cartesian graphs, in order to introduce the meaning of variable and dependency between variables. Moreover, it can allow building the Cartesian plane starting from a line to

Fig. 1 An image of a DynaGraph (Goldenberg et al., 1992, p. 245)

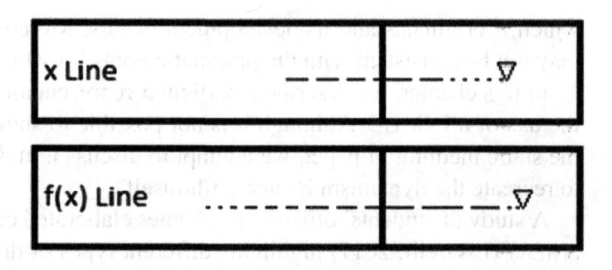

represent the real number set and following a set of steps that we explain later in this chapter.

This representation cannot be obtained without the use of a dynamic environment, such as GeoGebra, which allows objects to be moved around the screen by the dragging tool. Specifically, there are two possible movements available: direct and indirect. The *direct motion* occurs when a basic element, such as a point generated by the point tool, is dragged by acting directly on it; while the *indirect motion* occurs when a construction procedure is accomplished and the motion of the elements obtained through it can be realized only by dragging the basic points from which the construction originates.

In the case of DynaGraphs, the notion of independent variable is experienced through the possibility of freely dragging a point bounded to a line (representing the x-axis), because this motion represents the variation of the point within a specific domain. Whereas the notion of dependent variable is experienced through an indirect motion: the (direct) dragging of the independent variable along its axis causes the motion of a point, bounded to another line (representing the y-axis), which cannot be dragged directly. Indeed, the indirect motion preserves the geometrical properties defined by the construction, which in a DynaGraph consists of keeping the functional relation between the two points invariant. In other words, the use of dragging allows the user to experience functional dependency as the dependence relation between direct and indirect motions.

Moreover, movement of points experienced through the use of the dragging tool can be materialized through the trace tool, which displays the trajectory of a moving point. Although the final product of the trace tool is a static image, its use involves time, making it possible to simultaneously grasp the pointwise and the global aspects of the product of the trace tool. Concurrently, there is a twofold meaning of a sequence of positions of a moving point and of the curve consisting in the set of all such positions.

3 Pilot Study

3.1 The Design of the Dynamic Files

In this section, we provide a description of a possible development of the original idea of DynaGraph (which we call 'dynagraph' from now on), implemented within GeoGebra.

Other researchers also rearranged this representation of functions for some of their studies, for example Sinclair, Healy, and Reis Sales (2009) used Geometer's Sketchpad (Jackiw, 1991) to obtain a visualization similar to the one described above. They added a line to link together the two variables, which they called *A* and *f(A)* respectively, instead of naming the two horizontal axes.

For this study, we designed a sequence of activities aimed at making the representation of functions in the Cartesian plane rich in meaning. Starting with a sort of dynagraph, we modified its design through a sequence of activities, in order to reach the Cartesian graph. We expected that this kind of one-dimensional representation would foster the description of relative movements of the ticks and comparisons between possible walks followed by the ticks on the lines. For example, students may recognize the movements of two variables to be either in the same direction or in opposite directions, which we would identify as situated signs for monotonicity properties of functions. In addition to changes in direction, the dynagraph also provides information about the rate of motion. For example, moving the independent variable at a constant speed along the x-axis can result in constant growth, which the user actually feels while observing the dependent variable always having the same increment. Similarly, moving x at a constant speed can result in accelerated growth: the user sees $f(x)$ running off the screen. In relation to more advanced mathematical concepts, descriptions of change in speed could be read mathematically as observations on the slope of the function, which is its derivative.

As we can see in Fig. 2, the initial dynagraph used has a single horizontal line, marked with 0 and 1, and two little ticks. One of the ticks represents the independent variable and can always be dragged, while the other one represents the dependent variable and cannot be directly dragged, but moves depending on the movements of the independent tick. The two variables are represented by ticks and not by points, because a point is usually seen as a pair of coordinates, while a tick better expresses the idea of a 'value'. Moreover, the two points marked on the line determine the unit segment, highlighting that it is a real number line.

The design of our representations allows us to separate out the two variables, that is, to create a copy of the real number line in order to have one tick on each line. Therefore, what can be seen on the screen changes in this way (Fig. 3): there is a fixed horizontal line, representing the x-axis, and its copy, representing the y-axis, that can be dragged up and down maintaining the parallelism, and the alignment of the origins.

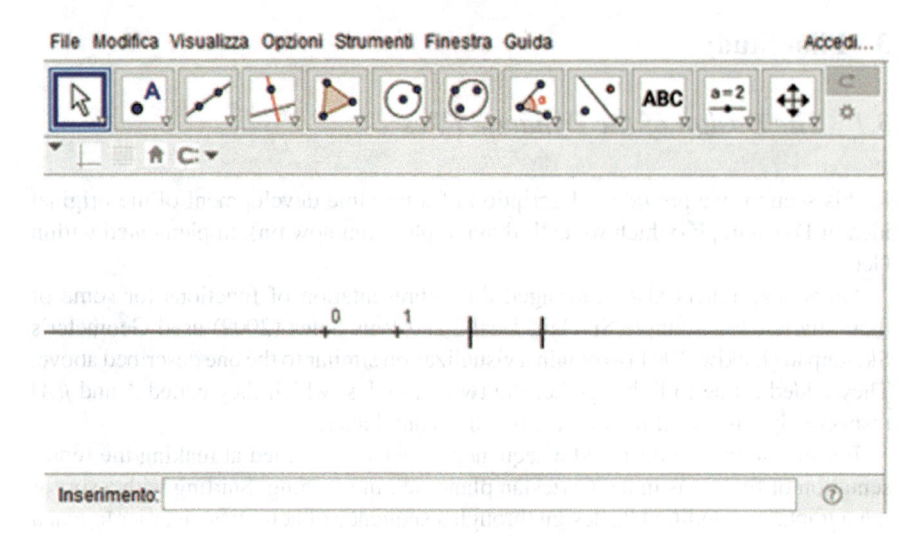

Fig. 2 Dynagraph of $f(x) = -x + 5$ with one axis

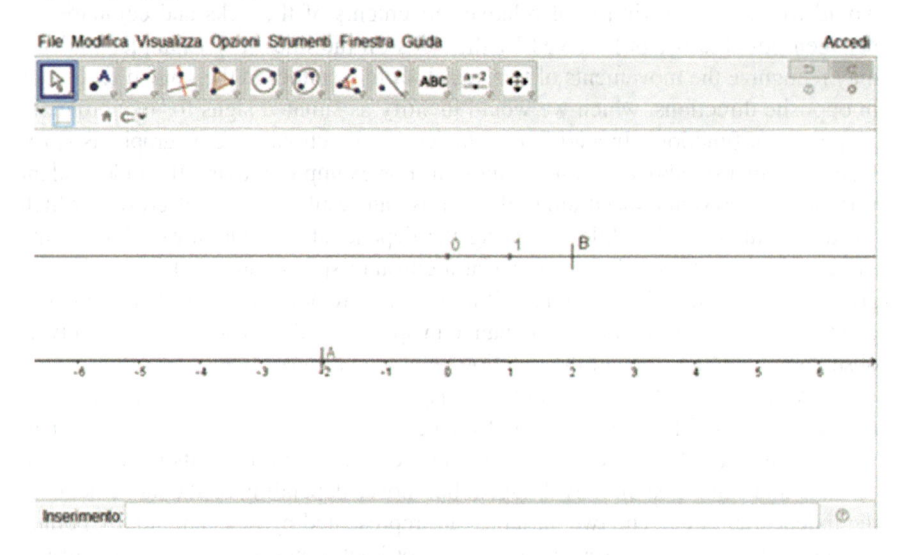

Fig. 3 Dynagraph of $f(x) = |x|$ with two axes

The user can decide to see two distinct lines on the screen or to keep them overlap. The file has been designed like this in order to address cases where it is more convenient to have two separated axes (for example to explore functions like $f(x) = \sqrt{x}$ or $g(x) = |x|$) and for other cases in which it is easier to work with just one line and both variables moving bounded to it (for example to determine x such that $f(x) = x$).

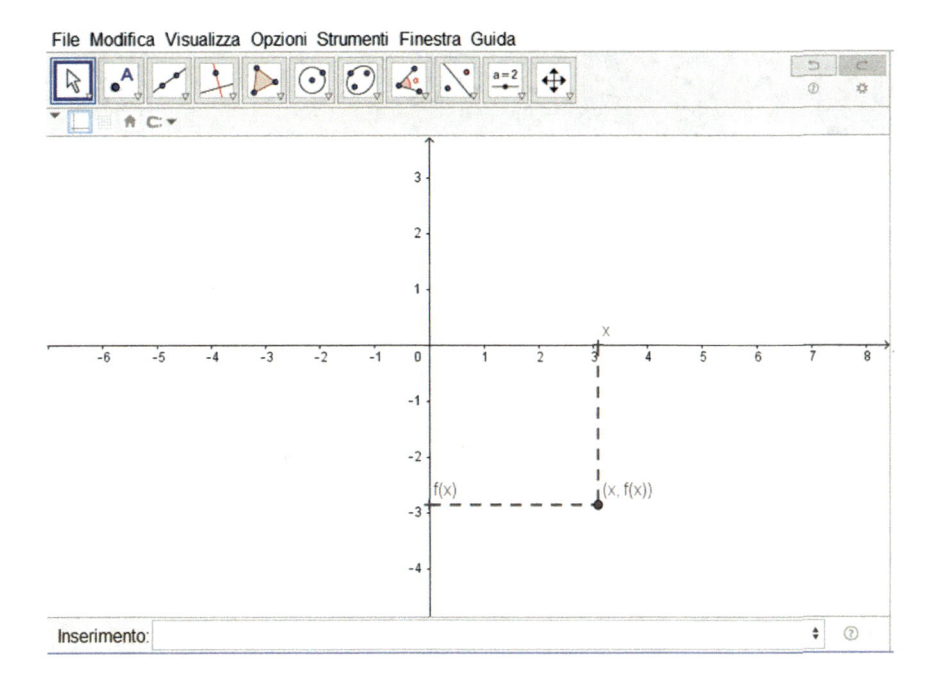

Activities Involving Dynamic Representations of Functions with … 281

Fig. 4 Dynamic graph of $f(x) = e^{x-2} - 3$ with Cartesian axes

Moreover, thanks to the dragging tool and this design of the dynamic files, there is an opportunity to rotate the y-axis: it is built by using the line tool and it is a line passing through two points that determine its unit segment. Therefore, it can be rotated by dragging these two points, joining the zeros and making it perpendicular to the x-axis, in order to obtain a representation that includes the Cartesian axes on which the two ticks can move. As described above, the tick on the x-axis can be directly dragged while the other one moves depending on it.

The following step for the construction of the Cartesian graph of a function consists of the definition of the point $(x, f(x))$, that gives the visualization in Fig. 4.

Finally, by activating the trace tool on this point and dragging the independent variable, it is possible to obtain the graph of the function in the Cartesian plane, as shown in Fig. 5.

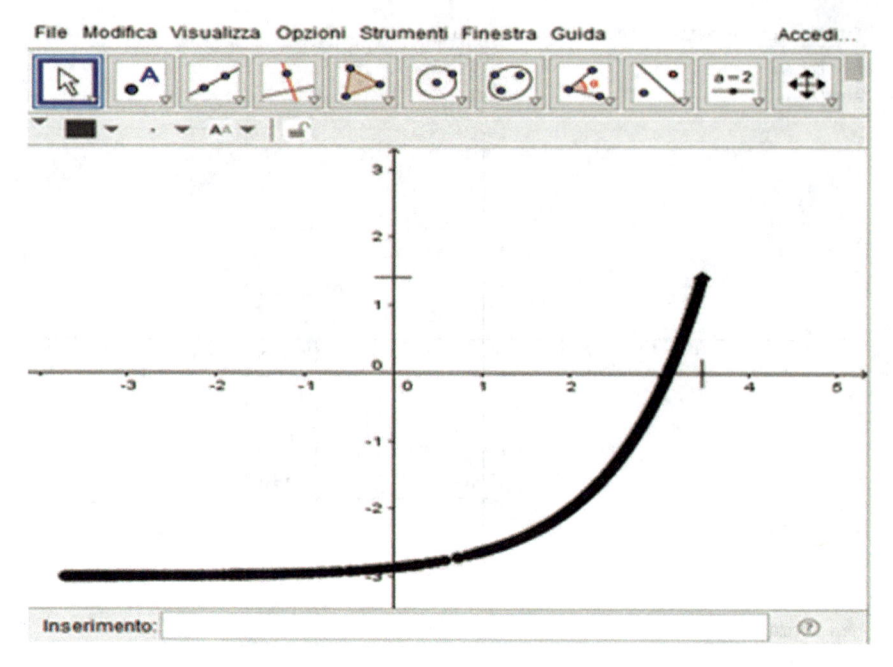

Fig. 5 Cartesian graph of $f(x) = e^{x-2} - 3$

3.2 The Experimental Context

A first experimentation was conducted in 2016 in a 10th grade of an Italian High School for Mathematics and Science, where students were introduced to the function concept through dynagraphs.

During five lessons, 27 students worked in pairs on pre-designed dynamic interactive files that they were asked to explore. The task proposed in each file was:

> Experiment with the construction, describe what you notice about possible movements. Write down your observations.

The open task is to support students' explorations, while working in pairs is to foster their speaking aloud and explaining their reasoning to each other. The students involved in the study have already worked with GeoGebra that they had used for some geometry lessons with their regular mathematics teacher.

All lessons were video-recorded through two cameras, one fixed in the back of the classroom to record the collective discussions and the other one mobile to focus on some specific processes during students' working in pairs.

The main goal of this study was to build the mathematical meaning of functional dependence as a relation between two co-varying quantities: one depending on the other one. We expected to start from the relation between the movements of the two ticks bounded to the lines.

In particular, starting from the representation of functions on one horizontal line, a sequence of activities that led to the Cartesian graph was designed, following a trajectory like the one just described. The sequence is made of several GeoGebra files with different examples of functions, including not everywhere defined functions and discontinuous functions, in order to support the production of situated signs that an expert could recognize as possible descriptions of domain, range, continuity, limit and asymptote.

The choice of students who have not met the concept of function before at school comes from the underlying assumption that this kind of dynamic representation can support a dynamic conception of functions because it draws attention to movements and variations of variables and to the relation between these variations. In addition, we wanted to study how to exploit this potential to introduce students to the idea of functional dependence. The aim of the designed activities is to promote students' verbalization to describe the moving ticks as variables that can be assigned different values and the fact that the relation between them is asymmetric: the movement (and so the value) of one tick depends on the other one, according to a specific rule that is different in each file.

In the rest of this chapter, we only discuss activities with dynagraphs with two parallel axes, investigating their semiotic and didactic potential.

3.3 Data Analysis

According to the definitions of semiotic potential and utilization schemes discussed in the first paragraph of this chapter, we analyze the semiotic potential of the representation of functions with parallel axes focusing on the embedded knowledge and the utilization schemes that students employ when exploring the dynamic files.

In the first lesson of the pilot study, after the exploration of a linear function, students are asked to explore the dynagraph of the function $f(x) = \frac{1}{x-4}$ and to write down their observations. We chose this not everywhere defined function in order to introduce them to the mathematical concept of domain, while exploring the dependence relation between the two variables. We expected that the particular behaviour of the dependent variable in a neighborhood of the vertical asymptote $x = 4$ could support the employment of a language referring to the movement and to the relation between the movements of the two ticks. Moreover, we expected that students would notice the existence of the horizontal asymptote $y = 0$ in terms of changing in speed of the tick representing the dependent variable.

After this dynagraph, students had to explore the dynagraphs of two functions: $f(x) = x^2$ and $f(x) = \begin{cases} -1, x \leq 0 \\ 1, x > 0 \end{cases}$ (without knowing the algebraic expressions). The reason why we chose these particular functions is to let students deal with cases where the range is restricted. These examples can be used to allow students to discuss the important distinction between the meaning of range and the meaning of domain, because it could result in confusing within this kind of representation of the function where the x variable always exists (it is always visible on the computer screen).

In what follows, we analyze some episodes that happened during this lesson, with the goal of recognizing instances in which the semiotic potential of dynagraphs seems to be exploited. In particular, the excerpts 1, 2, 3 and 4 contain descriptions of the same file: the dynagraph of $(x) = \frac{1}{x-4}$. Then we present the analyses of other excerpts, all of them taken from the worksheet of a student, Asia, and they refer to three different dynagraphs.

Excerpt 1

The following transcript is a dialogue between four students who are interacting with the same GeoGebra file. We chose this excerpt because during their discussion, students frequently use words that refer to movements of variables and to the relation between these movements. Moreover, as we expected from the a priori analysis, the function behavior in a neighborhood of the vertical asymptote $x = 4$ causes students' astonishment and some interesting observations.

1. Gian Oh no, it goes crazy!
2. Fra Look there, it dashes backwards
3. Gian It makes certain leaps!
4. Fra Ah, but are they three points here?
5. Dar What? Here there is back to the future!
6. Fra Eh eh, there are three points guys
7. Rob No
8. Fra Or not?
9. Gian This one doesn't move, and the meeting point is the same, it doesn't change
10. Fra No no they are two, indeed I tried to make some changes but they are equal, actually they are the same.

As we can read from the dialogue, the discontinuity of the function is something very interesting for these students, because when they drag the independent variable, they see the dependent one disappear on one side of the screen and then re-appear on the other side of the screen. They try to interpret this phenomenon by using an interpretation of continuity. It is possible that they simply do not expect to see stuttering movements in response to their continuous dragging.

Fra supposes that there could be three points (4), as if he could not accept that the same point can run off on one side and come back on the other side. However, an interesting fact is that he modifies (10) the tick representing the dependent variable (Fig. 6) in order to convince himself that the points are two and not three: while

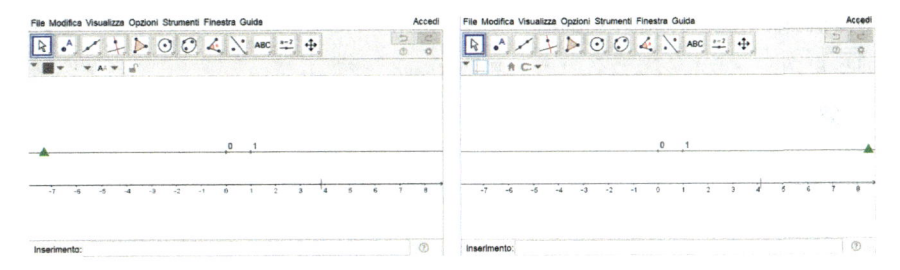

Activities Involving Dynamic Representations of Functions with … 285

Fig. 6 Fra modifies the $f(x)$-tick, which convinces him that there are two variables

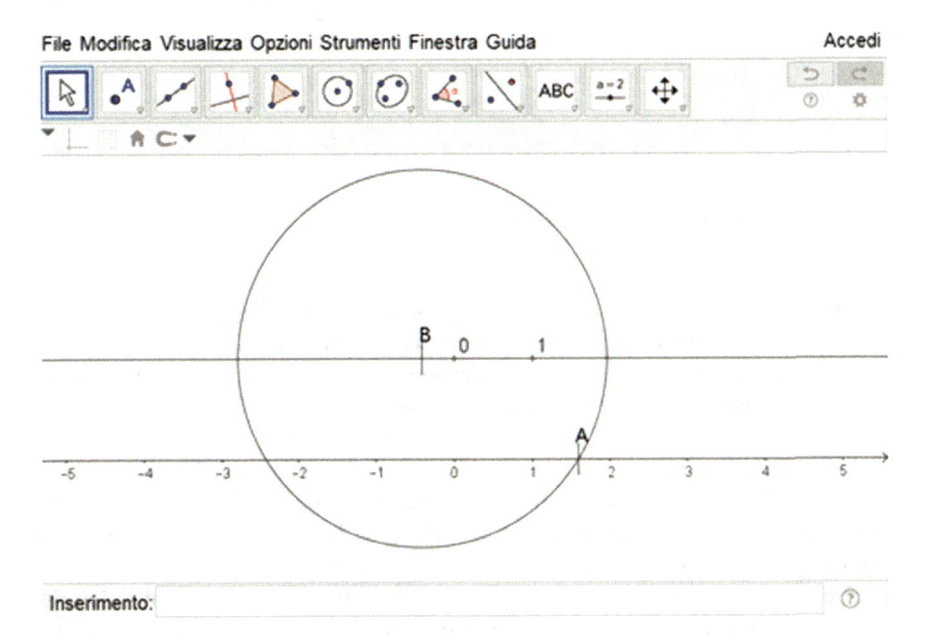

Fig. 7 Example of students' construction

dragging the independent tick in a neighborhood of 4, he always sees the same output. Therefore, feedback is directly given by the software and by useful manipulations made on the figure, which is an utilization scheme developed by this student for his purpose.

Excerpt 2

Another example of utilization scheme developed by students is the following construction (see Fig. 7): keeping the two parallel axes divided, let A and B be the independent and dependent variables respectively and build the circumference centered in B with radius AB. By dragging A along its axis, the students can observe changing in dimensions of the circumference caused by changing in length of its radius.

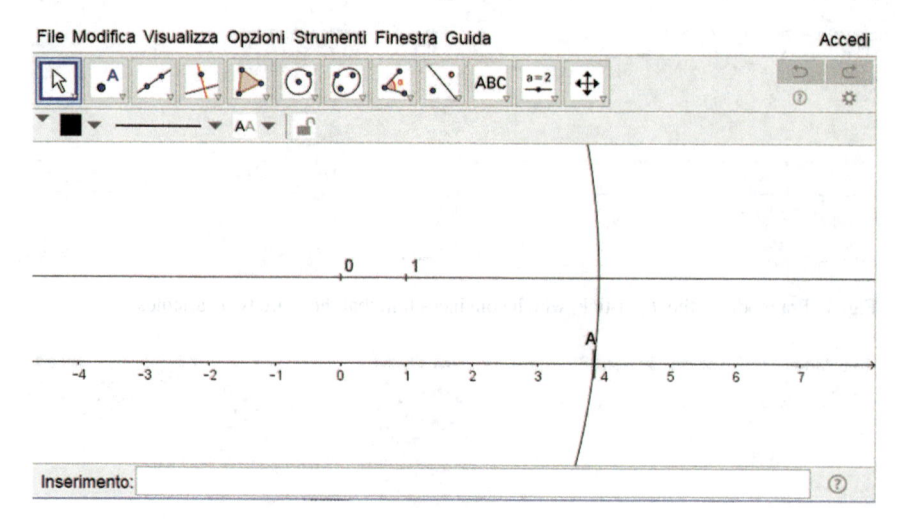

Fig. 8 The same construction for A tending to 4 from left

This manipulation of the figure, done by students, allows them to observe that the function is not defined in a certain point. Indeed, by dragging A towards A = 4, it is possible to see the radius becoming bigger and bigger (Fig. 8) until disappearing, as the circumference does, when A equals 4 and then, as A exceeds 4, the circumference changes its orientation (Fig. 9). Thus, the described construction allows the user to locate possible 'critical points' (in this case $f(x)$ is not defined for $x = 4$) by identifying the values that make the circumference disappear. This happens because the tick B does not exist when A does not belong to the domain of the function, and without B the construction of the circumference is not possible.

Probably, the change of orientation of the circumference fosters students' description of B as making some sort of circular trip, passing somehow back on the computer screen or assuming a hypothetical meeting of positive and negative infinity behind their head. We note that Goldenberg et al. (1992) and Sinclair et al. (2009) also cite very similar examples of students' narratives for functions with an asymptotic behavior like this.

For analyzing the semiotic potential of this representation, it is important to notice that the construction of the circumference can also lead to different observations. As we can read, for example, in this short excerpt taken from a student's sheet of paper:

> If I move A to the left I see B moving a little bit from the line *s*, that is perpendicular to A and B when they are in the same point. If I construct a circumference with radius AB, I see that by moving A to the right, at a certain moment the circumference has such a big radius that it becomes a line that is parallel to *s*, and so B goes to infinity.

As we can see in Fig. 10, this student adds another construction: she creates a line passing through A and B when they are aligned ("they are in the same point") on a vertical line. As discussed for the previous construction, the slope of the line changes when dragging A and the line does not exist for A = 4 since B disappears. However,

Activities Involving Dynamic Representations of Functions with ...

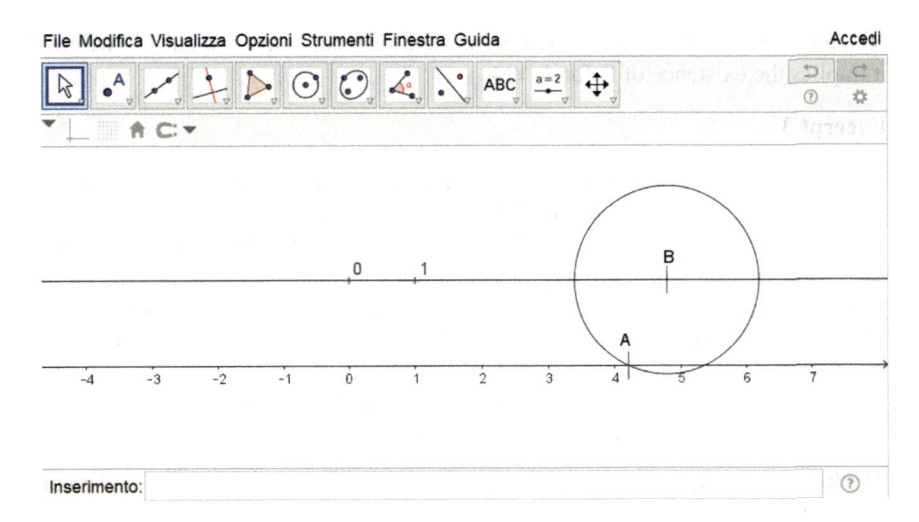

Fig. 9 The same construction for A bigger then 4

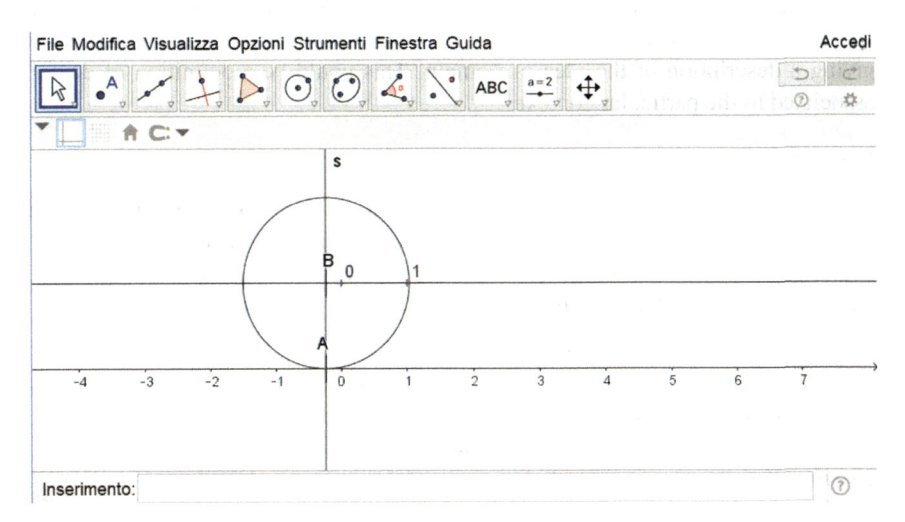

Fig. 10 Another students' construction for the same function

the student argues that the circumference becomes a line parallel to this line, as soon as A reaches the value 4 and so B, which is the center of the circumference, has to go to infinity.

From a didactical point of view, it would be interesting to investigate how she justifies the last implication ("the center of the circumference goes to infinity"), which is not a direct consequence of what she wrote previously. However, going beyond this request of clarification, it is worth to observe that in the classroom there could be students like her who describe B as going to infinity, and it is an aspect of which the teacher should be aware because this sentence is mathematically wrong if it implies the existence of B for $A = 4$.

Excerpt 3

Let us now look at a description given by a student, which we consider as a situated sign of the mathematical concept of domain of the function.

It is important to observe that the representation of the function with parallel axes requires the following interpretation of the domain: it needs to be read on the y-axis because the independent variable can always be dragged, bounded to its line. Therefore, we could say that a point on the x-axis belongs to the domain of the function if it has a corresponding output on the y-axis.

As we can read from Rob's words, there are different aspects of the semiotic potential of the representation of the function that come to light:

24 Rob After a moment, the upper point moves only in a certain range of movement of the point below

Indeed, in a very short sentence like this, he refers to time (after a moment), to space (upper, below) and to movement (a range of movement) and he succeeds in giving a description of the idea of domain. This could be a definition of domain associated to the particular context.

Excerpt 4

This dialogue between three students working on the same GeoGebra file concerns a description of the asymptotic behavior of the function for x tending to negative infinity. As in the previous excerpts, it is expressed by students without using formal mathematical terms but with many references to the specific dynamic context:

71 Fra: But do you see how it dashes away? Look!
72 Dar: Try to move a bit further backwards, look, it still moves very little.
73 Fra: It continues to move
74 Dar: Do you see? It moves a little bit
75 Fra: Yes, it is moving a little bit
76 Dar: Look, it moves here
77 Rob: Nothing is moving, where do you see that it moves?
78 Fra: It moves you're right, yes
79 Rob: No, here it does not move
80 Fra: Yes Rob it moves, look!
81 Rob: Zoom in zoom in, so we can see it. And then it makes certain leaps...
82 Fra: It leaps it leaps!
83 Rob: Look, it has leapt to one side
84 Fra: And then it stops
85 Rob: That's it, from here on it is fixed, look.

Recalling the verb (to dash away) used previously as well (see Excerpt 1), Fra underlines the unexpected acceleration of the dependent variable (71). Then the other students observe that $f(x)$ makes some leaps (81) when varying the x variable in a neighborhood of the point where the function is not defined. The semiotic potential of this dynagraph comes into play in the mathematical concept of derivative: by dragging the x-tick in a neighborhood of $x = 4$ we can observe the $f(x)$-tick leaping, which corresponds to a function having a very high slope.

Then the students discuss about the function behavior for x tending to negative infinity. Dar suggests that $f(x)$ still moves when x is dragged backwards (72), that is x tending towards negative infinity; and Fra agrees (82). However, Rob prefers to zoom in because it seems to him that the x-tick is fixed and he would convince himself of the contrary. Again, the semiotic potential of the dynagraph comes into play, supporting the mathematical concept of limit; aspects of such potential can be observed in students' words and actions. In particular, by zooming in students can observe the function behavior for smaller and smaller variations of the independent variable.

In the last sentence (85), Rob refers to x values bigger than zero and far from it, and it is possible to see it by looking at his dragging actions from the video, because he is dragging the x-tick to the right on its line.

Other Excerpts

In this section, we analyze some excerpts from Asia's worksheet. We selected them because they are very rich in reference to motion and speed. In particular, Asia describes her actions that seem to play an important role in her exploration of the dynagraph and in her description of the properties observed. Moreover, the subject of the following sentences is (almost) always a person who acts on the dynagraph and so, they give information about the utilization schemes at stake in manipulating the artifact.

Asia's description of the dynagraph with parallel axes of the function $f(x) = \frac{1}{x-4}$:

> If I drag A towards the left extremity, B moves in the opposite direction and it goes slower and slower until stopping; while if I drag A towards the right extremity, B moves in the opposite direction with an increasing speed, but if we go on the right extremity, the other point decreases its velocity. So B has the highest speed when A is in the middle of the line.

First of all, Asia speaks about the existence of two extremities of the domain, which probably are the extremities of the computer screen or they could be two spatial references that she uses to better organize her description. As Rob did in the previous excerpt, Asia sees B stopping for A tending to negative infinity, which is a way for expressing the existence of a horizontal asymptote for the function. Variables speed has a central role in this description: through it and together with some spatial references (opposite direction, in the middle of the line), Asia expresses the asymptotic behavior of the function for x tending to infinity and to 4.

From a didactical point of view, when the teacher would support a development from situated signs to mathematical signs, she should take care of the finer differences observable in students' choice of words. For example, here Asia distinguishes

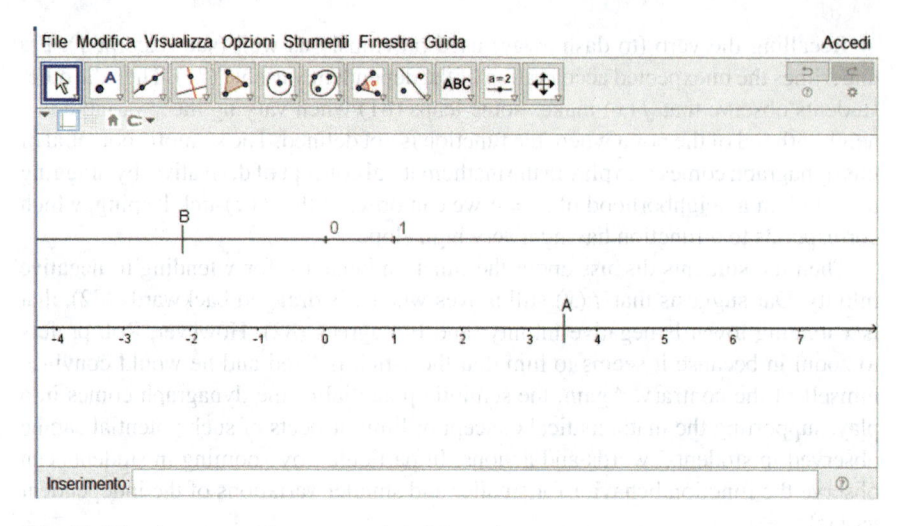

Fig. 11 What happens on the screen as Asia "drags A towards the right extremity"

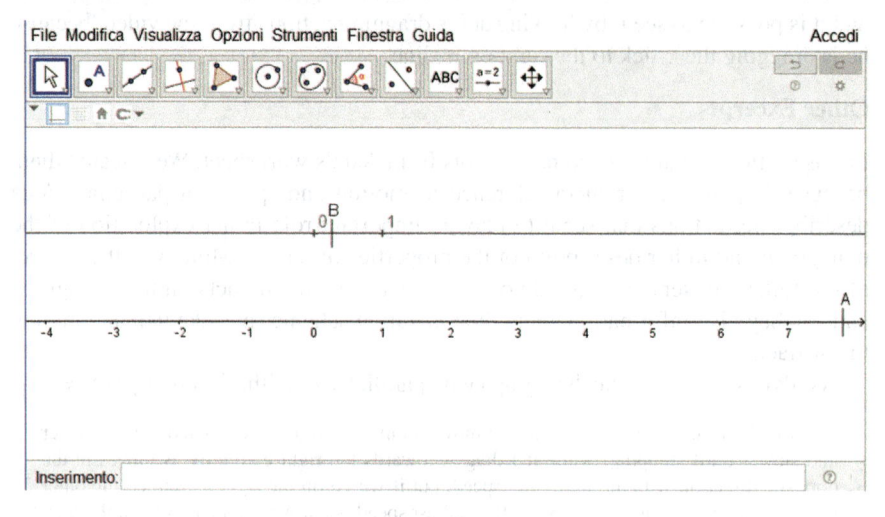

Fig. 12 What happens on the screen as Asia "goes on the right extremity"

between dragging A "towards the right extremity" (Fig. 11) and A going "on the right extremity" (Fig. 12), because she observes two different changes in the movement of B. The first expression is characterized by a sense of motion, especially conveyed through the word "towards", while the preposition "on" in the second expression evokes a more static image. Moreover, it also suggests that the right extremity of the domain is a value that the independent variable can take on.

Asia's description of the dynagraph with parallel axes of the function

$$f(x) = \begin{cases} -1, x \leq 0 \\ 1, x > 0 \end{cases} :$$

B moves when A passes below it: it is as there was a magnet between them and when they are "in the same place", even on different lines, B changes its position always within a small space.

In this short excerpt, Asia's focus seems to be on the position of A that makes B move and, again, Asia's choice of the words involves space and time ("when A passes below it"). Then, the expression "within a small space" seems to refer to the interval where A moves when "B changes its position", that is a neighborhood of zero. Therefore, it is another way to describe the speed of B, which moves very fast if A varies within a neighborhood of zero. Moreover, without explicitly referring to it in mathematical terms, Asia's description takes into account the dependence relation linking A and B and she expresses it by "when A...B...".

Asia's description of the dynagraph with parallel axes of the function $f(x) = x^2$:

When A and B are parallel, if I move A to the right, B increases its velocity to the right, but if I move A to the left, B will go to the left until it will be parallel to A, then, when A will keep going to the left, it will go to the right.

The use of the future tense suggests that Asia is not dragging A while writing, but she is probably making a prediction or remembering what she has previously seen on the screen. At a first sight, her clarification about the initial position of the two ticks ("when A and B are parallel") seems to be redundant, but it is necessary for her description that makes sense only if starting dragging A at zero. Asia's strategy (and we observed that it is commonly adopted by many students in the classroom) consists in aligning the independent tick at zero and dragging it to the right side; then, after bringing it on zero again, stopping, and dragging the same tick to the left. This strategy is different from the one we could expect that involves starting with A from a big negative number and dragging it to the right, which means following the orientation of the real numbers line or, more generally, our direction of writing and reading from the left to the right. We claim that the strategy used by students can be interpreted as a research of a point of reference, upon which to base the explorations and, more importantly, the following descriptions.

In all the three excerpts, there is evidence of the semiotic potential of the proposed representation of functions and, especially, how it allows the user to dynamically experience the dependence relation. This relation seems to be owned for example by Asia who expresses it in terms of "if I move A... B..." or "when A... B...". Moreover, it is interesting to highlight that from these brief excerpts it is possible to see the variety, the complexity and the plot of different utilization schemes. They can be exploited by researchers because after their individuation, it is possible to make a cognitive analysis of the exploration processes linked to these schemes.

4 Discussion and Conclusions

In general, studies of the interaction between humans, technology and mathematics must take into account a variety of aspects: the relation between the teacher and the technology in the mediation of mathematical knowledge, how this knowledge is influenced by constraints and actions allowed by the technological environment, and several other components that are involved. In this chapter, we presented a study of the explorations of functional dependence in a DIMLE. In particular, we have analysed aspects of the semiotic potential of the representation of functions with parallel axes, presenting some excerpts from a pilot study. The analyses of these excerpts reveal some interesting considerations consistent with the a priori analysis of the designed activities.

We noticed that students' descriptions of the different properties of functions are rich in reference to movement, time and space. This is a consequence of the choice to introduce students to functions starting from a specific dynamic one-dimensional representation, the dynagraph. Together with the dynamic environment, the focus on dynamism characterizing functional relationships between two variables is fostered by the open task that requires for exploration and description and by the possibility of dragging. Moreover, such richness in students' descriptions could also be a result of the fact that these students have no prior exposure to the concept of function (in high school), so they have not yet developed a formal mathematical vocabulary related to functions. This can be considered as an obstacle for them but, at the same time, it makes the use of such terms necessary, resulting in the production of many situated signs throughout their explorations. In particular, we have shown examples of this aspect in excerpts 3, 4 and these signs are very relevant with respect to the specific context, because the teacher can use them to mediate the movement towards mathematical signs, simply by replacing them with their corresponding mathematical terms.

From these analyses, we can infer that introducing students to functions through dynagraphs seems to promote a co-variational view of functions, seen as relations between the movements of quantities that are varying in an interval of the real number line. In the same way, some mathematical properties of functions are also conceived dynamically, for example, Rob identifies the domain of the function as a certain range of movement of the independent variable. This is consistent with the expectation that this kind of one-dimensional representation would foster the description of relative movements of the ticks and comparisons between possible walks followed by the ticks on the lines. We also noticed students' frequent use of verbs strictly related to movement and speed, which is in line with the findings of Ng (2016) about how the use of dynamic environments in particular, can support the development of co-variational reasoning.

As indicated by the analysis of Asia's discourse, we can conclude that the DIMLEs present an opportunity for students to see change directly, and to call into play all intuitions that they have developed about movement, time and speed. If we think about the Cartesian graph of a linear function, we realize that the 'behaviour' of the function is hidden to such extent that it is almost impossible to see it. In particular, it is hard for a student, especially if he is learning functions for the first time, to talk about the speed of the variables or their direction of movement in reference to the horizontal axis. However, as we saw in the analyses, these kinds of descriptions seemed to emerge spontaneously from students while working with dynagraphs in GeoGebra files.

As already highlighted by Falcade et al. (2007), the dragging tool plays the role of a potential semiotic mediator, contributing to building the meaning of asymmetric relation between the independent and dependent variables. This is realized by the possibilities of experiencing the two different kinds of motion that this tool offers: a direct motion and an indirect motion.

Finally, we highlight that the potential of this representation also emerges in students' creativity, an aspect that was studied by Sinclair et al. (2009) who found that the production of narratives was more prominent during students' explorations of dynamic representations than static ones. In our case, students' creativity was revealed in their autonomous use of the tools offered by the DIMLEs. For example, Fra changed the $f(x)$-tick visualization by modifying its dimension and its colour (Fig. 6) to verify its existence in a certain domain; Rob zoomed in to convince himself that $f(x)$ kept moving on when x tended to negative infinity. Other students built a circumference centred in B with radius AB to identify the non-definition points with a vertical asymptote. All these represent different utilization schemes developed by students to explore the dynamic representations of functions and to speak about them. After the individuation of the utilization, schemes it could be interesting to analyse the sequence of these schemes and their relationships, in order to gain insight into students' cognitive processes. For example, we have seen in excerpts 1, 2 the students trying to discover some of the features of the dynagraphs of a function having a vertical asymptote. Further research is needed in order to investigate a possible evolution of these schemes or a possible relationship with other schemes developed by students when exploring the dynagraphs of other functions having different properties.

The identification of some utilization schemes is a result of this study that has relevance at a didactical level. Indeed, the teacher could decide to promote or to avoid these kinds of students' utilization schemes, depending upon her goal. She could use the construction of the circumference centered in B with radius AB to discuss with students the difference between "B goes to infinity" and "B disappears". At the same time, if the goal of the teacher is to introduce students to the notion of domain through dynagraphs, she should consider the accuracy limits of the software that could bring students to observe or conjecture possible relations and properties even not mathematically acceptable. This fact was partially discussed in the analysis of students' definition of the circumference (Fig. 7) and the line (Fig. 10), since someone did not notice that these constructions disappeared for a certain value of the independent variable. Probably, this is due to the accuracy of the software and the velocity in using the dragging tool, which makes vanishing of the circumference or the line hard to notice. In any case, it is an example of situated sign that a teacher has to take into account when working with this kind of representation to let students talk about the concept of domain of a function.

Clearly, this study has some limitations. First of all, the limited time available did not allow us to investigate a possible evolution of the situated signs generated by students to mathematical signs. Based upon the results of this pilot study, we plan to re-design the sequence of activities with dynagraphs in order to implement them in another 10th grade Italian class, in order to add depth to these discussion points. In particular, the aim is to develop the research in the following directions. First, it would be interesting to investigate whether or not students' conceptions of the functional relationship evolve, and if so, how. This can be realized by analyzing students' frequent use of references to movement and time in their discourses, and by investigating whether the students retain or not the formal mathematical definitions after having been taught them. Another interesting issue is to explore how students deal with these dynamic terms together with the static definition of function and which terms they use to describe the static Cartesian graph of a function.

It would also be interesting to design some new interactive files by choosing functions that can support exposure of other relevant properties of functions, in order to gain a deeper insight into possible exploitation of the semiotic potential of function representation with parallel axes.

Acknowledgements GNSAGA of INdAM has partially supported this study.

References

Arzarello, F. (2000). Inside and outside: spaces, times and language in proof production. In T. Nakahara & M. Koyama (Eds.), *Proceedings of the 24th international conference psychology of mathematics education* (Vol. 1, pp. 23–38). Japan: Hiroshima University.
Bartolini Bussi, M. G., & Mariotti, M. A. (2008). Semiotic mediation in the mathematics classroom: Artifacts and signs after a Vygotskian perspective. In L. English, et al. (Eds.), *Handbook of International Research in Mathematics Education* (2nd ed., pp. 746–783). New York and London: Routledge.
Bèguin, P., & Rabardel, P. (2000). Designing for instrument-mediated activity. *Scandinavian Journal of Information Systems, 12,* 173–190.
Carlson, M., Jacobs, S., Coe, E., Larsen, S., & Hsu, E. (2002). Applying covariational reasoning while modeling dynamic events: A framework and a study. *Journal for Research in Mathematics Education, 33*(5), 352–378.
Carlson, M., & Oehrtman, M. (2005). Key aspects of knowing and learning the concept of function. *Research Sampler 9.* MAA Notes.
Confrey, J., & Smith, E. (1995). Splitting, covariation and their role in the development of exponential function. *Journal for Research in Mathematics Education, 26,* 66–86.
Dubinsky, E., & Harel, G. (1992). The nature of the process conception of function. In G. Harel & E. Dubinsky (Eds.), *The concept of function: Aspects of epistemology and pedagogy* (pp. 85–106). MAA Notes.
Falcade, R., Laborde, C., & Mariotti, M. A. (2007). Approaching functions: Cabri tools as instruments of semiotic mediation. *Educational Studies in Mathematics, 66,* 317–333.
Goldenberg, E. P., Lewis, P., & O'Keefe, J. (1992). Dynamic representation and the development of an understanding of functions. In G. Harel & E. Dubinsky (Eds.), *The concept of function: Aspects of epistemology and pedagogy* (Vol. 25, pp. 235–260). MAA Notes.
Healy, L., & Sinclair, N. (2007). If this is your mathematics, what are your stories? *International Journal of Computers for Mathematics Learning, 12*(1), 3–21.
Jackiw, N. (1991). *The geometer's Sketchpad.* Berkeley, CA: Key Curriculum Press.
Kaput, J. (1992). Technology and mathematics education. In D. Grouws (Ed.), *A Handbook of research on mathematics teaching and learning* (pp. 515–556). New York: MacMillan.

Kieran, C., & Yerushalmy, M. (2004). Research on the role of technological environments in algebra learning and teaching. In K. Stacey, H. Chick, & M. Kendal (Eds.), *The future of the teaching and learning of algebra* (pp. 99–154). Dordrecht: Kluwer Academic Publishers.

Lisarelli, G. (2017). Students' use of movement in the exploration of dynamic functions. In T. Dooley & G. Gueudet (Eds.), *Proceedings of the 10th Congress of European Research in mathematics education* (pp. 2595–2602). Dublin, Ireland: DCU Institute of Education & ERME.

Mariotti, M. A. (2002). Influence of technologies advances on students' math learning. In L. English et al. (Eds.), *Handbook of international research in mathematics education* (pp. 695–23). LEA.

Mariotti, M. A., & Bartolini Bussi, M. G. (1998). From drawing to construction: Teacher's mediation within the Cabri environment. In A. Oliver & K. Newstead (Eds.), *Proceedings of the 22nd international conference psychology of mathematics education* (Vol. 1, pp. 180–195). Stellenbosch, South Africa.

Mariotti, M. A., & Cerulli, M. (2001). Semiotic mediation for algebra teaching and learning. In M. van den Heuvel-Panhuizen (Eds.), *Proceedings of the 25th international conference on psychology of mathematics education* (Vol. 3, pp. 343–349). Utrecht, The Netherlands.

Martinovic, D., & Karadag, Z. (2012). Dynamic and Interactive Mathematics Learning Environments (DIMLE): The case of teaching the limit concept. In M. J. Alison Clark-Wilson & M. McCabe (Eds.), *Proceedings of the 10th international conference for technology in mathematics teaching* (pp. 208–213). Portsmouth, UK.

Ng, O. (2016). Comparing calculus communication across static and dynamic environments using a multimodal approach. *Digital Experiences in Mathematics Education, 2*(2), 115–141.

Rabardel, P. (1995). *Les hommes et les technologies – Approche cognitive des instruments contemporains*. Paris: A. Colin.

Sinclair, N., Healy, L., & Reis Sales, C. (2009). Time for telling stories: Narrative thinking with dynamic geometry. *ZDM—Mathematics Education, 41*, 441–452.

Tall, D. (1991). The psychology of advanced mathematical thinking. In D. Tall (Ed.), *Advanced mathematical thinking* (pp. 3–21). Dordrecht: Kluwer Academic Publishers.

Tall, D. (1996). Function and calculus. In A. J. Bishop, et al. (Eds.), *International handbook of mathematics education* (pp. 289–325). The Netherlands: Kluwer Academic Publishers.

Vinner, S., & Dreyfus, T. (1989). Images and definitions for the concept of function. *Journal for Research in Mathematics Education, 20*(4), 356–366. National Council of Teachers of Mathematics.

A Computer-Based Learning Environment About Quadratic Functions with Different Kinds of Feedback: Pilot Study and Research Design

Elena Jedtke and Gilbert Greefrath

1 Introduction

Open Educational Resources (OER) are currently of great interest in international discussions (German Education Server, 2014; OECD, 2017; UNESCO, 2017). In Germany, OER are adopting an increasingly prominent role in digital education initiatives (Federal Ministry of Education and Research, 2016; SCMECA, 2016). We shall focus on a certain class of OER known as Computer-Based Learning Environments (CBLE), and specifically the subset of CBLE based on MediaWiki software. Our research aims to expand the theoretical framework of these CBLE, with emphasis on the possible ways of integrating feedback and the effects of each approach. Feedback is believed to play a central role in CBLE (Bimba, Idris, Al-Hunaiyyan, Mahmud, & Shuib, 2017; Roth, 2015). An extensive range of studies has investigated various aspects of feedback. However, their research findings differ regarding the effectiveness of different types of feedback or the best timing—immediate or delayed feedback (e.g., Bangert-Drowns, Kulik, Kulik, & Morgan, 1991; Hattie, 2009; Hattie & Timperley, 2007; Nelson & Shunn, 2009). Some meta-analyses tend to point out elaborated feedback (EF) as an effective type with medium to high effect sizes in comparison to less complex feedback types (e.g., Van der Kleij, Feskens, & Eggen, 2015). Van der Kleij et al. (2015) added that most of the studies involved in their meta-analysis were conducted in adult education and thus there is a lack of research about primary and secondary education. In addition, "more research is needed to investigate how to provide EF effectively" (Van der Kleij et al., 2015, p. 502). These remarks in combination with the ambivalence results mentioned above

---

E. Jedtke (✉) · G. Greefrath
University of Münster, Münster, Germany
e-mail: e.jedtke@uni-muenster.de

G. Greefrath
e-mail: greefrath@uni-muenster.de

© Springer Nature Switzerland AG 2019
G. Aldon and J. Trgalová (eds.), *Technology in Mathematics Teaching*,
Mathematics Education in the Digital Era 13,
https://doi.org/10.1007/978-3-030-19741-4_13

motivate us to investigate which effects different types of feedback show, which can be included into a CBLE based on MediaWiki software. We are primarily interested in the effect of the type of feedback on the mathematical performance and self-rating capacity of students. To study this effect, we developed two equivalent versions of a CBLE on quadratic functions. Both CBLE are fully identical except for the type of integrated feedback. One version implements feedback of type knowledge-of-the-correct-response (KCR). The other version uses a form of EF that offers hints and explanations in addition to the correct response. The topic of quadratic functions was chosen as content because the multiple representations and corresponding learning obstacles associated with these functions provide an excellent opportunity to take advantage of interactive visualizations. Furthermore, research on quadratic functions within school education is currently limited. To gain insight into which of the two feedback variants (KCR or EF) allows students to achieve better performance after working with the CBLE, we plan to conduct a (quasi-) experimental study with students from the 8th and 9th grades.

This chapter presents the theoretical foundations of our research project. Afterwards, the design of the newly developed CBLE on quadratic functions is discussed. We also present the results of a qualitative preliminary study, and explain the choices of our research design plan. We begin with an overview of the international research findings on the three central pillars of our research—CBLE, feedback, and teaching and learning quadratic functions. Since the newly developed CBLE plays a key role in both the preliminary studies and the main study, we discuss the considerations underlying the design and the specific contents of the CBLE, focusing on choices regarding the type of exercise and integrated feedback, as well as other technical aspects. In preparation for the main study, several (qualitative and quantitative) preliminary studies were conducted, each with a different emphasis. The third section of this chapter presents and discusses the findings of one of these studies as an illustrative example. The objective of these qualitative pilot studies was to evaluate various specific aspects of the CBLE and improve it accordingly. Finally, we present the research design of the upcoming main study as a prospect.

2 Theoretical Background

Our research is based on three key theoretical components. Below, we give a summary of prior theoretical and empirical findings regarding CBLE, with particular attention paid to their definitions and design principles. We then present the current state of research on feedback and the relationship between feedback and performance. The third component is the didactics of quadratic functions, which is the topic that we chose for our CBLE. We explain why quadratic functions are particularly suitable for working with CBLE, and we describe the standard *Grundvorstellungen* (basic mental models) and representation-switching competencies that need to be incorporated into the CBLE. We also give a list of known examples of common learning obstacles that must be taken into consideration in the design of the CBLE and discuss possible avenues for avoiding or mitigating these obstacles.

2.1 Computer-Based Learning Environments

Searching for a formal definition of CBLE in the literature soon reveals that this term is used as a hypernym for computer- and web-based learning activities in general (Baker, D'Mello, Rodrigo, & Graesser, 2010; Balacheff & Kaput, 1996; Isaacs & Senge, 1992). Roth (2015) gives a more detailed definition of a special kind of CBLE. According to this definition, CBLE offer structured paths through a sequence of related exercises, inviting learners to work independently and autonomously (Roth, 2015, p. 8). Integrated help functions and presentations of the correct results are seen as promising ways to support independent learning processes in CBLE, alongside interactive materials such as (GeoGebra) applets, which play a key role in this context (Roth, 2015, p. 8). Wiesner and Wiesner-Steiner (2015) conducted an exploratory study to identify the most important functions of CBLE. The authors held interviews with individuals from a range of user groups including experts and students. The interviews with experts had several objectives, one of which was to develop a characterization of these central functions in technical and didactic terms. For example, on a technical level, the integration of dynamic content and the inclusion of opportunities for direct feedback received were strongly emphasized. On a didactic level, metacognitive abilities and reflective exercises were frequently cited in connection with CBLE. No prescriptive design and quality criteria for the development of OER have yet been proposed (SCMECA, 2016). A set of preliminary proposals by the SCMECA (2016) in Germany suggested that digital educational media should offer factually correct content that adheres to the curriculum but should also support competency-oriented teaching and individualized learning processes and should be designed to have the properties of multimediality, interactivity, networkability, changeability, and divisibility (SCMECA, 2016). As well as discussing the quality-related dimensions of OER, the Standing Conference also identified technical and legal aspects as potential fields of action (SCMECA, 2016). For example, hybrid and parallel applications of digital and analogue educational media were discussed. It was suggested that the latter form of media should not be fully eliminated (SCMECA, 2016). One way of applying these relatively general statements about digital educational media and OER to the specific case of CBLE based on MediaWiki software is given by a list of criteria established in 2006 by a panel of experts during the "Vienna Meeting on CBLE" and published by the Mathematik digital (2006) workgroup. This collection of criteria for evaluating the quality of a CBLE considers the dimensions of content, student orientation and student activities, user interfaces, media integration, and resources for teachers (Mathematik digital, 2006). These criteria articulate, extend, and specify the quality considerations cited above. Student-appropriate language, transparency of learning objectives, feedback, opportunities for differentiation, and suitably targeted use of media, including paper and pencil, are just some examples of the criteria included in the list. It should however be noted that none of these criteria represent fixed and prescriptive rules that must be observed when developing CBLE, and a CBLE that does not meet every criterion is not necessarily bad (Roth, 2015). This compilation of expert opinions does however offer a good starting point for CBLE development and can also be used as a basis for evaluation (Roth, 2015). CBLE based on MediaWiki software can for example be accessed on the German OER website ZUM-Wiki (https://wiki.zum.de/wiki/Hauptseite). This online resource has received acclaim for its very comprehensive and well-maintained environment (Vollrath & Roth, 2012). The learning environments proposed by the ZUM-Wiki are available under the Creative Commons CC-by-sa 3.0 licence, meaning that these CBLE can be copied and modified by any user who is logged in, provided that the original authors are cited, and any modifications are published under the same licence as the original. The ZUM-Wiki stores online copies of every version of its CBLE, which allows undesirable changes to be reverted by restoring from an older version.

2.2 Feedback

Several models and basic hypotheses have been proposed to describe the concept of feedback. For example, Boud and Molloy (2013) established a distinction between unilateral and multilateral perspectives on feedback. The unilateral perspective views feedback as a "one way transmission" (Boud & Molloy, 2013, p. 701) in which the teacher operates as the "driver of feedback" (Boud & Molloy, 2013, p. 698). By contrast, the multilateral perspective, judged by many authors to be superior to the unilateral perspective, proposes that the students themselves also play a key role in the feedback process (e.g., Nicol & Macfarlane-Dick, 2006; Sadler, 1989). For example, Sadler (1989) writes that feedback is nothing more than "dangling data" (p. 121) if we do not observe and study its effect on students. Nicol and Macfarlane-Dick (2006) also draw a connection between self-regulated working and feedback. They write that feedback can "help students take control of their own learning, i.e. become self-regulated learners" (p. 199). To support this process, they developed seven principles for good feedback practice, such as "facilitat[ing] the development of self-assessment (reflection) in learning" (p. 205). As well as research into good feedback practice, there has also been interest in the different types of feedback (e.g., Bimba et al., 2017; Fyfe, 2016; Hattie, 2009; Hattie & Timperley, 2007; Shute, 2008). The overall effect of feedback on learning processes has been demonstrated repeatedly (e.g., Azevedo & Bernard, 1995; Black & Wiliam, 1998; Hattie, 2009; Shute, 2008). At the same time, it is "generally agreed that feedback is a critical component of instruction" (Azevedo & Bernard, 1995, p. 112), especially in CBLE (Azevedo & Bernard, 1995; Fyfe, 2016). Hattie (2009) identified feedback as one of the top ten factors of academic performance in a large-scale meta-study. The strength of the effect was found to depend strongly on the type of feedback that is why it is important to examine the effectiveness of different types of feedback such as knowledge-of-the-correct-response (KCR) up to elaborated feedback (EF) (e.g., Azevedo & Bernard, 1995; Corbett & Anderson 2001; Van der Kleij et al., 2015). Whereas KCR only presents the correct response of a task without any additional information, EF provides "an explanation about why a specific response was correct or not" (Shute, 2008, p. 160) in general (Dempsey, Driscoll, & Swindell, 1993; Shute, 2008). Furthermore, Shute (2008) differentiates six types of EF, for example if the feedback is contingent on the topic or on the given response. In accordance with that, Kulhavy and Stock (1989) order EF into three key types of elaboration before: "(a) task specific, (b) instruction based and (c) extra-instructional" (p. 286). Previous studies have reported divergent findings in regard of the dimension of complexity of feedback, especially "regarding what type of feedback is most helpful and why it is helpful." (Nelson & Schunn, 2009, p. 375; cf. Mory, 2004; Shute, 2008; Van der Kleij et al., 2015). In a study published by Attali (2015), KCR produced an effect comparable to a complete absence of feedback, whereas variants that included a multiply-try function (MTF), which allows students to attempt an exercise multiple times, were found to positively influence learning (cf. Shute, 2008; Van der Kleij et al., 2015). MTFs allow students to rethink their own results and self-correct their mistakes independently. Additional hints can further support and structure these processes (Attali, 2015; Van der Kleij et al., 2015). Attali (2015) also identifies "an interesting area for future research" (p. 266), proposing that research could be conducted to measure the effect of providing additional "explanations for the correct answers" (p. 266), which means a kind of EF. Other existing meta-analysis already paid attention to the effectiveness of EF in CBLE and discovered that feedback seems to be better than no feedback in general as well as EF seems to be more effective than other types like KCR (Azevedo & Bernard, 1995; Van der Kleij et al., 2015). Thus, "ideally, feedback messages should stimulate cognitive processes and strategies so that misconceptions that jeopardize future learning attempts will not be perpetuated" (Azevedo & Bernard, 1995, p. 120). Finally, besides the feedback types described above, there are other factors that influence the effect of feedback. We do not explore them in any more detail here, since they do not align with our primary research interests and would lead us too far astray. Examples of such factors include motivation, prior knowledge and the point at which the feedback is given during the learning process (e.g. Corbett & Anderson, 2001; Hattie & Timperley, 2007; Mory, 2004; Shute, 2008). It should be noted that research on the optimal moment to provide feedback has produced inconsistent findings, although various meta-analyses have tended to favour direct feedback (Bangert-Drowns et al., 1991; Corbett & Anderson, 2001; Shute, 2008). In association with this tendency, it seems interesting that students who received feedback on demand "tended to wait until they had typed a complete solution before seeking feedback" (Corbett & Anderson, 2001).

2.3 Quadratic Functions in Mathematics Lessons

Quadratic functions play a central role in secondary-level mathematics education in Germany (SCMECA, 2003). An extensive range of literature has been published on teaching and learning quadratic functions. One especially important objective is the development of functional thinking. Quadratic functions are one of the contexts in which this type of thinking is developed at the secondary level. Functional thinking is often required when working with functions, and can be characterized by three ideas: firstly, students need to be able to apprehend and describe phenomena governed by underlying functional relationships, interpret the relationships discovered for these phenomena, and then apply these relationships to solve problems. Secondly, students need to be able to apply basic mental models (*Grundvorstellungen*) about quadratic functions as a function of context and need to develop the ability to flexibly switch between different *Grundvorstellungen*. Thirdly, the distinct representations of quadratic functions need to be understood, constructed, interpreted, converted from one to another, and applied to solve problems (Greefrath, Oldenburg, Siller, Ulm, & Weigand, 2016; Vollrath, 1989, 2014).

Parameters that change over time (such as the braking distance of a car as a function of time), physical or situational relationships (such as the kinetic energy of a body, which is a quadratic function of its speed), and inner-mathematical connections (such as the surface area of a square as a function of its side length) give examples of phenomena involving quadratic functions. These ideas can be expanded into a more precise characterization of quadratic functions in the classroom (Greefrath et al., 2016; Vollrath, 2014).

Doorman, Drijvers, Gravemeijer, Boon, and Reed (2012) identify three basic mental models (*Grundvorstellungen*) for general functions: "functions as an input-output assignment", "functions as a dynamic process of co-variation", and "functions as a mathematical object" (p. 1246). These proposed basic mental models are consistent with the work of other authors (e.g., Malle, 2000; Vollrath, 1989). The first example, the input-output model, is visualized in Fig. 1. This model adopts a local perspective on functional relationships. Each element of the domain is assigned to precisely one element of the image. Thus, it emphasizes the specific values taken by the functional relationship. Co-variation expresses the idea that one quantity changes when another quantity is varied. This perspective encompasses more than just individual values, focusing on larger regions on which a functional relationship holds. This is visualized in Fig. 2. The figure singles out an interval, in this case on the x-axis, and associates it with the corresponding interval of function values. Conceptually speaking, as the first interval runs along the x-axis, we think about how the corresponding interval of function values behaves on the y-axis.

Figures 1 and 2 thereby show explicative examples that have been chosen to visualize the models and we note that there exist different ways to visualize this. For instance, tables can also represent the input-output relation. The object perspective, on the other hand, expresses the idea that functions can be viewed as a single object that describes the entire relationship as a single entity. This is the global view of

Fig. 1 Visualization of the concept of input-output

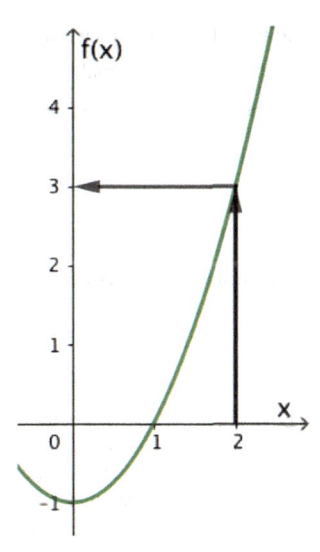

Fig. 2 Visualization of co-variation

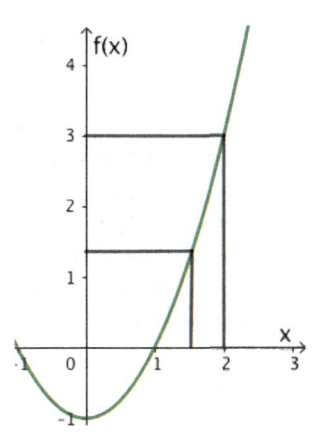

functions. This perspective can for example be used to gain intuition of how the position of a parabola can be described within a coordinate system. In Germany, this third perspective is typically only introduced from the upper secondary level onwards (Greefrath et al., 2016). Lower secondary mathematics education therefore primarily focuses on the first two of the basic mental models described above.

In terms of the distinct representations of quadratic functions (verbal, graphical, tabular, symbolic), students need to learn to apply each representation individually, but more importantly must develop the ability to flexibly switch between them. Following the approach of Swan (1982), the process of switching between representations is summarized in Table 1. The transitions between representations listed in the table provide the basis for a systematic approach to constructing appropriate exercises for students.

Table 1 Transitions between the representations of quadratic functions

From ... to ...	Verbal	Graphical	Tabular	Symbolic
Verbal	Reformulate	Sketch a graph showing the apex and any roots or other points of interest	Extract information about important points and find new values	Extract information about important points and find the equation of function
Graphical	Interpret	Translate, dilate, etc.	Read values at certain points	Read values at certain points and find the equation of the function
Tabular	Interpret the values of characteristic parameters	Draw a parabola passing through certain points	Add new rows to the table	Use suitable points to find the equation of the function
Symbolic	Interpret the meaning of parameters	Draw a parabola using the y-intercept and other parameters	Calculate pairs of values	Manipulate the equation (e.g. put in vertex form)

Nitsch (2015) reports the learning difficulties encountered by students when switching between the representations of functional relationships. One of the challenges that she identifies relates to switching from symbolic representations to graphical representations. Here, she reports that students find it difficult to understand the influence of the parameters on the graphical representations of quadratic functions. Learning resources that allow parameters to be systematically varied have been suggested as a way of supporting the understanding of this connection (Vollrath & Roth, 2012). For example, this could be implemented via sliders in dynamic geometry software (DGS).

Zaslavsky (1997) identified five "cognitive obstacles" (p. 20) or misconceptions about quadratic functions (pp. 30–33). Three of these misconceptions relate to the symbolic representation of quadratic functions, whereas the other two relate to the graphical representation. One misconception ("The relation between a quadratic function and a quadratic equation") is that the zeros are the only important features of a quadratic function, and the leading coefficient before the x^2 term can be ignored, as is the case for quadratic equations. A second misconception ("The seeming change in form of a quadratic function whose parameter is zero") also relates to the parameters of the quadratic function. If either of the other two parameters is also zero, the equation is no longer recognized as the equation of a quadratic function, or students fail to recognize that the first parameter is zero. The third misconception ("The analogy between a quadratic function and a linear function") involves incorrectly assuming that the properties of linear functions, such as interpretations of the meanings of the parameters, also apply to quadratic functions. The first misconception for the graphical representation ("The interpretation of graphical information (pictorial entailments)") is that students sometimes fail to appreciate that the graph portrays a selected extract of the function, instead assuming that the function does not exist outside of the visible region. The final cognitive obstacle ("The over-emphasis on only one coordinate of special points") is the tendency of students to assume that the x-coordinate of a single point on the graph, such as the vertex or any other significant point, is sufficient to uniquely determine the parabola.

When designing learning resources on quadratic functions, these difficulties and misconceptions can be counteracted by including exercises that target each type of representation and the transitions between them, especially graphical and symbolic representations.

3 Design of the CBLE

Since the newly developed CBLE plays a central role in our studies, this section presents various aspects of its design in more detail. We begin with a description of the concrete subject matter of the CBLE and the macro structure of this content. We also present the exercise formats used in each section of the CBLE. Finally, we discuss the types of integrated feedback, and show how both EF and KCR feedback can be implemented within the CBLE.

3.1 Structure

The CBLE consists of the ten chapters listed in Fig. 3.

Before the students begin to work with the mathematical content of the CBLE, a technical introduction on how to use the CBLE is given, and the prerequisite skills and learning objectives are transparently stated (*Welcome*). After a reminder about time management, the students can begin working with the CBLE. If a student is unsure whether he/she already possesses all prerequisite skills, an optional chapter offers the opportunity to practice these skills (*Revision*), focusing on the topics of functional thinking in general and linear functions. Alternatively, the students can choose to proceed directly to the introductory chapters *Quadratic Functions in Daily Life* and *Getting to Know Quadratic Functions*. The first of these chapters has a moti-

Discover Quadratic Functions
Welcome | Revision | Quadratic Functions in Daily Life | Getting to Know Quadratic Functions | The Parameters of the Vertex Form | Vertex Form | The Parameters of the Standard Form | Standard Form | Transform Vertex into Standard Form | Exercises

Fig. 3 Summary of the newly developed CBLE *Discover Quadratic Functions* (version: 03/05/2018)

vational emphasis, presenting a selection of everyday examples and encouraging the students to search for "parabolic curves" around them. The second introduces the simplest form of quadratic function, $f(x) = x^2$. After the students are familiar with this form, two representations of more complex quadratic functions, the vertex form and the standard form, are introduced step by step. The parameters are introduced separately (*The Parameters of the Vertex Form* and *The Parameters of the Standard Form*) and then combined together in one chapter each (*Vertex Form* and *Standard Form*). The final chapter of content (*Transform Vertex into Standard Form*) introduces the idea that the two previously encountered representations of quadratic functions are equivalent and can be converted into one another by means of simple arithmetic operations. Since the amount of time available to work on the CBLE is limited, the conversion between forms is only presented in one direction. The other direction can be introduced without the CBLE in subsequent maths lessons. The majority of students are expected to progress no further than the "*Standard Form*" chapter. However, sufficient content should be available for especially fast-working students.[1] At the end of the CBLE, there is an extra chapter with exercises on all topics (*Exercises*). The students are informed at the start that they can freely switch to this chapter at any time if they would like more practice on certain topics. These exercises can also be solved all at once at the end of the CBLE.

As a general rule, the students can decide themselves in which order they wish to complete the chapters and the exercises within each chapter. They are also free to decide how much time to spend on each exercise. Some (interactive) exercises can be completed multiple times (Fig. 4). Others have individually adjustable difficulty levels (Fig. 6). However, since the overall objective of this CBLE is to introduce new subject matter, it offers less flexibility than CBLE designed for other purposes, such as revision. In addition to working at their computer stations, the students are given an accompanying booklet. Some of the exercises explicitly ask the students to use pen and paper to solve them, providing the opportunity to develop motor skills, such as graph sketching. The booklets also summarize the formulas introduced by the CBLE and include space for planning and self-assessment activities to support the independent learning process.

3.2 Exercise Formats

The CBLE features a range of different exercise formats. Some exercises are interactive, and others are not, but other distinctions between formats can also be drawn. Some exercises allow open responses. The answers are entered into a text field in the CBLE or written down in the accompanying booklet.

Semi-open and closed responses can be implemented in a wide variety of ways within the CBLE. Semi-open responses can be presented as a text box, but also as

---

[1] In the post-test associated with the study, the students will be asked which chapters they worked on during the available time.

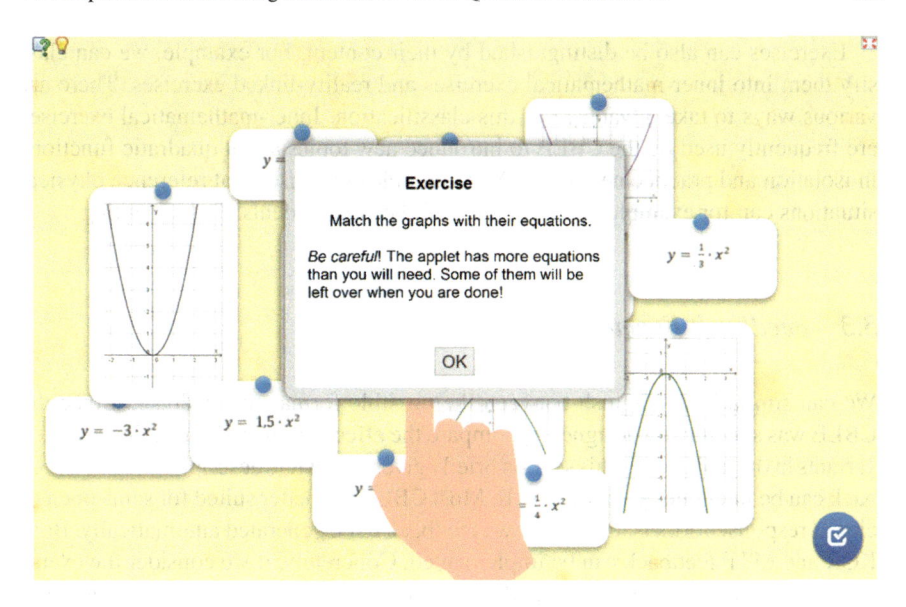

Fig. 4 Example of an inner-mathematical exercise about matching terms and graphs (translated)

crossword puzzles, fill-in-the-blank mind maps, or fill-in-the-blank sentences. Closed response formats can be implemented as multiple-choice questions or matching exercises. For example, solution elements can be matched to answer fields by interactively dragging text and image fields across the screen (Fig. 4). Multiple-choice exercises can also be grouped together as a quiz or a game (against another player or the computer) (Table 2).

Table 2 Exercise formats in the CBLE

Open exercises	Text box		
	Writing notes in the booklet		
Semi-open exercises	Text box in the CBLE		
	Crossword puzzle		
	Fill-in-the-blank mind map		
	Fill-in-the-blank sentences		
Closed exercises	Multiple-choice questions	Individual	
		Quiz	
		Game	
	Matching exercises		

Exercises can also be distinguished by their content. For example, we can classify them into inner-mathematical exercises and reality-linked exercises. There are various ways to take advantage of this classification. Inner-mathematical exercises are frequently used by the CBLE to introduce new topics about quadratic functions in isolation and practice new skills. More complex exercises that reference physical situations can for example be used to consolidate the students' understanding.

3.3 Feedback Formats

We can similarly distinguish between the possible formats of feedback. Since the CBLE was specifically designed to compare the effects of the EF and KCR feedback variants in the main study, this section briefly gives a concrete description of how feedback can be integrated into the CBLE. Most CBLE are better suited for semi-open or closed response formats, as this allows feedback to be generated automatically. Both KCR and MTF feedback can be implemented. Concretely, if we consider the example of a multiple-choice exercise, KCR formats only indicate whether the answer is correct, and then immediately proceed to the next task. MTF formats, on the other hand, allow the exercise to be repeated if the answer is incorrect—providing another round of feedback afterwards. For more complex tasks such as matching exercises, MTF formats can also display feedback before the exercise is fully solved, allowing the student to continue working on the problem with knowledge of this feedback. EF formats indicate whether the question was correctly or incorrectly solved, but also give explanations for the correct answer.

Figure 5 gives a concrete example of how feedback is presented in the CBLE. In some cases, solution feedback is directly integrated into the interactive applets and can be displayed by clicking on the blue button in the bottom-right corner (Fig. 5a). After clicking, the students are told whether their solution is correct or needs to be changed. So-called hidden solutions are another method of presenting solution feedback. This feedback can be viewed by clicking with the mouse (Fig. 5b). The authors of the CBLE can freely decide which types of feedback to integrate. Feedback can for example take the form of calculations, text, or sketches and images. To distinguish between the EF and KCR variants of our CBLE, the complexity of the hidden solutions can be varied. As well as solution feedback, hints can also be integrated into the CBLE. These hints play an important role in the EF variant of our CBLE. They can similarly be implemented by functions integrated into the interactive applets or as hidden hints. Most of the hints directly within applets are presented as text, usually as a light bulb in the top-left corner of the applet that can be clicked to expand at any time. The complexity of the hidden hints included directly in the CBLE can be freely varied in the same way as the solution feedback described above.

Fig. 5 Example of the feedback variants included in the specially developed CBLE: **a** integrated interactive applet with a control button and **b** hidden hints and solution (translated)

3.4 Examples of Exercises in the CBLE

Figure 4 shows an example of an inner-mathematical exercise in the CBLE. The exercise asks the students to match the equations of quadratic functions to the corresponding parabolas. This exercise targets the competency of switching between the symbolic and graphical representations of quadratic functions. Once the students have finished matching the elements, they can click on the button in the bottom-right to check whether their results are correct. Correct matches are indicated in green, and incorrect matches are indicated in red. No feedback is given for elements that were not matched to anything. This exercise also features a MTF.

Applied exercises require additional skills beyond the simple reproduction of acquired knowledge. Figure 6 shows an exercise that requires mathematical knowledge of parameters effects, but also encourages the students to be creative. This exercise asks students to propose an equation describing an (idealized) parabola for a ball sport of their choice. The second part of the exercise then instructs the students to switch equations with their partners and try to guess the type of ball sport chosen by the other person. This exercise targets the competency of switching between the symbolic and verbal representations of quadratic functions. In the final part of the exercise, the students compare their results and discuss how and why they arrived at their answers. This is an example of an exercise for which no feedback is provided by the CBLE.

> **a)** Think of a ball sport (or something similar) with a ball whose trajectory can be approximately modelled by a parabola. Write down a suitable equation - without telling your partner what sport you chose. You can use the GeoGebra applet below to helf you visualize this.
>
> **Show Hints**
>
> **b)** Try to guess the ball sport that your partner is trying to model. Once both think you know, compare your answers and talk about your conclusions with each other.

Fig. 6 Example of an applied exercise about quadratic functions. During the exercise, students need to work in pairs (translated)

4 Pilot Studies

This section presents one of the pilot studies in detail, including the research questions, design and a discussion of the findings. In addition, other preliminary studies are shortly demonstrated. Several qualitative pilot studies were held, as well as one quantitative pilot study. In particular, the qualitative preliminary studies pursue the objective to evaluate and enhance the designed CBLE and the quantitative pilot run shall test the materials of the main study to ensure that the latter unfolds smoothly. The upcoming paragraphs are about a first qualitative study during which students have been observed while working in a chapter of the CBLE. Afterwards they have been interviewed as well. The findings do have effects on the further design of the CBLE and on the planned procedure of the main study.

4.1 Research Questions

As part of our preparation for the main study, a qualitative preliminary study was conducted with the primary objective of evaluating and improving the CBLE on quadratic functions. The underlying research question of this preliminary study was investigated using various aspects of the CBLE that were judged to be important (e.g., Roth, 2015; Wiesner & Wiesner-Steiner, 2015). The study examined how well the students managed to work independently over certain periods of time and recorded their impressions of the support provided by the CBLE. The study also considered metacognitive aspects within the CBLE, and we recorded the students' impressions regarding the combination of digital and analogue exercise formats.

0. What did students think about working with the newly developed CBLE? Which aspects did they find helpful, and what do they think needs to be changed or expanded:

   0.1 With regard to working independently and the support provided;
   0.2 With regard to metacognitive aspects;
   0.3 With regard to the combination of digital and analogue exercises?

4.2 Research Design

In preparation for the main study, we developed a CBLE on quadratic functions as is mentioned before. The development process included multiple evaluation cycles. This section describes one of these cycles in more detail. The evaluation at stake took the form of a qualitative interview study planned and conducted in collaboration with Sur (2017).

Six ninth-graders from a high school (*Gymnasium*) in North Rhine-Westphalia participated in the study. These students had learned about quadratic functions in school lessons a few months earlier and used the CBLE as a form of revision. Since only limited time was available for the study, namely only one 45-min school period, students who already knew about the topic were deliberately hand-picked for the study. In this preliminary study, we were primarily interested to see how the students worked with each type of exercise format and how they used the hints provided, and we expected that choosing students who would not require revision would enable them to complete more tasks during the allotted time. The students worked for 45 min on the *Vertex Form* section of the CBLE.[2] Immediately after the school period ended, a structured interview was held with each student for around 15 min. Carsten Sur, who was not the students' teacher, but an unknown university student, held the interviews within the context of his master thesis (Sur, 2017). For that reason, it can be assumed that there did not appear great interviewer effects in comparison to an interview conducted by the students' teacher. The interview was organized around a series of open questions exploring the three areas identified by the research question. Figure 7 shows an excerpt of the interview transcript. Each of the interviews was transcribed and coded with MAXQDA. This coding adopted the approach of a summarizing content analysis in accordance with Mayring (2010), with emphasis placed on an inductive approach. A few deductive effects could however also be identified from pre-established anchor points.

introduction	*Getting started*: What do you think about working with the CBLE?

main part	[...]
	 5. *You have now worked on your own with the CBLE for 45 minutes.* Which parts of the CBLE supported your work?
	    a. How did you use it?
	    b. In what ways did it support your working process?

conclusion	6. Another CBLE similar to this one will be used to help teach students about quadratic functions. If you could improve anything you liked, what would it be?
	    a. Why would you make your changes?
	    b. In your opinion: why would this be better?
	 [...]

Fig. 7 Extract from the interview transcript for the qualitative preliminary study (Sur, 2017, p. XV, translated)

---

[2]See the ZUM-Wiki link: https://wiki.zum.de/wiki/Quadratische_Funktionen_erkunden/Die_Scheitelpunktform (version: 2016-11-29), available in German only.

4.3 Presentation and Discussion of the Results of the Preliminary Study

The students' answers during the structured interview were classified into multiple deductive categories according to Research Question 0. In sum, we chose to use the categories *self-reliance*, *metacognition*, and *digital and analogue*. This section gives a summary of the results. The inductively established codes were immediately assigned to one of the deductive categories to allow them to be represented by this approach. Each section begins with examples of comments by the students (translated; names have been changed). We then outline the aspects that were identified as positive, as well as the wishes and comments expressed by the students. In accordance with the purpose of the preliminary study and in order to provide an answer to the underlying research questions, each section briefly presents conclusions to improve the CBLE after reporting the research findings. It is important to note that these results simply document the individual opinions of a small group of students (n = 6) regarding one part of the CBLE. We do not claim that these opinions can be generalized. Furthermore, the students participating in the preliminary study had already studied the mathematical contents of the proposed CBLE, while during the main study students will be introduced to quadratic functions. The reasons for choosing this different test person group are shown in the paragraph before. In the future, there will be another, longer preliminary study, which will focus on the main study's target group. Nonetheless, we think the results of this first preliminary study represent a good starting point for evaluation and improvement. The results and corresponding measures for each of the three categories are summarized in Table 3.

4.4 Self-reliance

Isabell: I liked to work self-reliantly and yes it is something different from only being present in classroom and absorb things like a sponge.

Felix: This partner work. That was good; it was not working all by myself, but to have the possibility to compare how others work.

Overall, the students felt that actively engaging with the content of the CBLE was a positive experience. They also appreciated the degree of freedom afforded to them while working on certain tasks. For example, in Exercise 1 of Chapter *Vertex Form* of the CBLE,[3] students were asked to choose between multiple images. After choosing an image, they were asked to match quadratic graphs and equations to their image by varying the parameters. Similarly, Exercise 4 (Fig. 6) had an open design that allowed for some differentiation. The students also stated that it was helpful that feedback was directly available after each exercise. It should be noted that they were working with the variant of the CBLE that includes EF, as described in the

---

[3] See Footnote 2.

Table 3 Summary of results and improvements undertaken for the categories *Self-reliance*, *Metacognition*, and *Digital and analogue*

Positive aspects	Negative aspects	Measures
Self-reliance		
Actively engaging with the content	(Perceived) time pressure	Integration of time management aids
Degree of freedom for completing exercises	Desire for support from teachers	Permanent grouping into pairs
Working in pairs		Added introduction to explain technical and procedural aspects
Direct feedback on exercises		
Metacognition		
Transparent objectives	Ability to look up technical terms	Expansion of the Wiki function
Self-assessment	Some difficulties with the open format of the self-assessment exercises	More closed format for self-assessment
Suggested solutions and hints to support working independently	Risk of copying solutions without thinking about them	
Digital and analogue		
Writing things down to support the learning process	Difficulties with writing down workings in the booklet	Formulas are now provided
Written information is more easily accessible in the long term		
The DGS can be used for more than just solving each exercise		

paragraph "Feedback formats". The students also enjoyed the exercises that asked them to work with a partner rather than completely on their own (for example, see the comments made by Felix). This is consistent with the criticism expressed by the students, who noted that they would have liked more support from their teachers, especially towards the beginning of their work. One student also commented that she felt time-pressured by having to be independent for a full school period.

To address the comments raised in this category, we plan to provide students with time management resources in future trials, especially since they will be spending more than just one school period with the CBLE. These planning resources will not be integrated directly into the CBLE, but will be inserted into the booklet as a separate tab. We hope that this analogue format will allow the students to switch back and forth between the chapters of the CBLE at their leisure, and also help them to estimate the time that they require more effectively than if they were managing their time using computer-based resources in parallel to the CBLE. This way, the students can have their schedule open and available at all times and can easily modify it or add to it. This additional metacognitive activity will hopefully support the students' time management and additionally help to reduce their desire for support from their teachers, which is deliberately minimized for research-related reasons. Another idea that is currently being considered is to permanently group the students into pairs. This is motivated not only by the wishes expressed by the students, but also for the reasons listed earlier when presenting the research design. The remark that the students most strongly felt the need for support from their teachers at the start of their work with the CBLE may already have been addressed by the inclusion of a short introduction explaining what is expected from during the school periods of the intervention, as well as the roles of the CBLE and the (supporting) booklet.

4.5 Metacognition

Mia: I liked the possibility to self-control my results.
Marcus: I see slight risks because of the integrated feedback. Perhaps one looks immediately for the solution.

We view the learning objective transparency, the self-assessment prompts, and the hints and suggested solutions as the metacognitive content of the CBLE chapter considered in this preliminary study. Regarding these aspects of the CBLE, students commented that the transparency of learning objectives helped them to understand why they were being asked to work on these exercises. The students described this as a very positive experience. By contrast, their opinions on the self-assessment prompts were less uniform. The self-assessment exercise was presented as an open question, and some of the students experienced difficulties in formulating their answers, or were not sure what to write. Nevertheless, they felt that the self-assessment activities benefited their learning process and provided useful information for their teachers. Observing the students as they worked with the computer revealed that they made very little use of the hints provided. In the interviews, they commented that, while they did not require the hints themselves, they thought that they would be helpful for students who are learning about quadratic functions for the first time, instead of revising familiar content. The students' comments on the suggested solutions also varied. Most of the students described the ability to check their own results as positive. However, one student also stated that he thought it was potentially risky to include the solutions within the CBLE, as it might allow students to copy them without thinking about them. Ultimately, all of the students agreed that providing feedback made it easier to discover and correct their mistakes. An extra feature requested by the students that falls under this category is the ability to look up technical terms directly within the CBLE.

Two improvements were undertaken in response to these results. Firstly, the Wiki functionality was expanded by linking important terms to pages that briefly explain what they mean. For example, the first time that the concept of *square* is mentioned, clicking on the word opens a page that summarizes the most important properties of squares. Explanations of technical terms that are likely to be unfamiliar to a large number of students (such as the word *applet*) are also included via links. However, care was taken to limit the overall number of links to ensure that the students primarily remain focused on the mathematical content of the CBLE. Self-assessment activities are a key area of focus of our research and we view them as an important part of the content of the study. Therefore, we do not plan to remove them from the CBLE. However, their presentation was significantly revised in response to these empirical findings after performing further literature research on the subject. Specifically, the self-assessment activities were converted into a closed format. A table was added to the time management tab allowing students to check off the exercises as they complete them and rate their understanding of the exercise on a 4-point Likert-type scale. We hope that this will prove easier for students, and it will also be useful for the data analysis stage later. We fully acknowledge and agree with the point raised by Marcus about the risk of copying the suggested solutions without thinking about them. His comment precisely pinpoints one of the underlying motivations of our research questions and will be fully taken into account in our analysis of the results of the upcoming main study.

4.6 Digital and Analogue

Isabell: Writing things down again. That helps to remember them.
Fabian: I thought that it was interesting that we didn't do everything on the computer. You always have to turn on the computer to check what you've done and what you're good at. Here we have a little booklet that is easy to carry around.

Over the course of the 45 min that the students spent working with the CBLE, they used a combination of both computer work and analogue worksheets. All of the students described their experience with this combination of media as positive and beneficial. They quoted multiple aspects to support their opinions. Regarding the learning process, from the perspective of both comprehension and motivation, they felt that the written format was helpful, especially in the long run. Although they felt that working on the computer was more interesting than working with pen and paper, several students emphasized that they believed that it was positive and important to do things by hand to avoid forgetting how. The students also cited an ecological perspective, noting that DGS in the CBLE allowed them to save paper, since this software allowed them to check and visualize their attempts, as well as other similar activities. In addition to recording the comments made by the students, we also examined their workings on paper. Clear difficulties could be identified in their scripts, especially when establishing formulas and writing out justifications.

The students' comments about the combination of digital and analogue exercises were consistently positive. Still, an improvement was made for subsequent (preliminary) studies by adjusting the presentation of the formulas, which had largely been provided in an open format. This decision was motivated by several considerations. Firstly, the findings described above show that students found it difficult to establish the formulas themselves. Secondly, we feel that it is important for the students to have easy access to mathematically correct formulas. The formulas are now included in both the computer and the accompanying booklet so that the students can look them up at any time. To encourage the students to refer to the formulas, exercises were added to the CBLE that require the students to actively engage with the content by writing down suitable examples in the designated sections of their booklet.

4.7 Other Preliminary Studies

As well as the preliminary study presented in the previous paragraphs, we conducted respectively are going to conduct several other studies that we shall not discuss in detail here beyond a brief overview intended to show the overall research design of the project. The purpose of these studies is to resolve any questions that might arise before the main study and pilot the study materials.

Qualitatively, the most important results will be obtained from further interview studies. Expert interviews were conducted with the primary objective of evaluating the structure and content of the CBLE. Additional student interviews were also conducted in various settings.[4] The scope of these interviews ranges from reactions to the feedback on each type of exercise to the use of metacognitive strategies, as well as a study conducted in parallel to the quantitative pilot run asking students where they experienced difficulties and how they felt they were managing the self-guided learning process overall at several points over the course of the process. These interviews were supplemented by other elements, such as observation reports and screen captures. We hope that these qualitative preliminary studies will allow us to offer students a validated learning environment in which they can work gladly and effectively.

Quantitatively, we investigated the design of the main study by conducting a pilot study. Four classes from a high school (*Gymnasium*) in North Rhine-Westphalia participated in this pilot study (n = 119 students). The procedure was identical to that of the main study, which is described in the next paragraph "Prospect: Main study". The students were grouped into pairs, and half of the students were assigned to the control group, with the other half forming the experimental group. In each school period during the study, a university representative involved in the project was present to observe and supervise its execution. An adapted version of the CODI test developed by Nitsch (2015) with a pre-post design was used as a performance test. The composition of the test is shown in Table 4. A total of 52 unique items were used. Item groups A and B contained items on both functional thinking and linear

---

[4] We would also like to thank several students who supported our research as part of their master theses at the University of Münster.

Table 4 Distribution of test times in the pre-test and post-test. Each letter denotes a group of items. The number of items is indicated in parentheses

	Group 1	Group 2	Group 1 and 2	Group 1 and 2
Pre-test	Items A (17)	Items B (17)	Anchor items C (8)	Anchor items E (2)
Post-test	Items B (17)	Items A (17)	Anchor items D (8)	Anchor items E (2)

and quadratic functions. Item group C contained items on linear functions, and item group D contained items on quadratic functions. The items in groups C and D were used as anchor items at each given point in time. Item group E contained two items on functional thinking, which were used as anchor items across test versions and test dates.

An analysis of the quantitative and qualitative data collected during the pilot study has not yet been performed.

5 Prospect: Main Study

As a last section in terms of content a prospect of the planned main study is given. Therefore, the research questions inspired by the theoretical framework on feedback in CBLE discussed above are presented and furthermore used to give an overview of the main study's planned research design.

5.1 Research Questions

The research questions of the main study are primarily motivated by the theoretical framework established by previous research findings on the effect of specific types of feedback on student performance, which we outlined before (cf. paragraph "Feedback" in the theoretical background). More complex types of feedback, represented here by a variant of EF, are believed to have a more positive influence on mathematics performance than simpler KCR feedback (e.g., Van der Kleij et al., 2015). The findings reported by Attali (2015) even suggest that the effect of the latter type of feedback on performance is comparable to that of a complete absence of feedback. We also wish to study whether the type of feedback can also be shown to influence the students' self-rating ability, in special their ability to evaluate their own performance (e.g., Nicol & Macfarlane-Dick, 2006). These research interests can be formalized by the following two research questions, which are represented schematically in Fig. 8.

1. Does the self-rating of students (in mathematics) increase when their CBLE for the topic of quadratic functions incorporates feedback that features additional explanations and hints (EF) compared to feedback that is limited to knowledge of the correct response (KCR)?

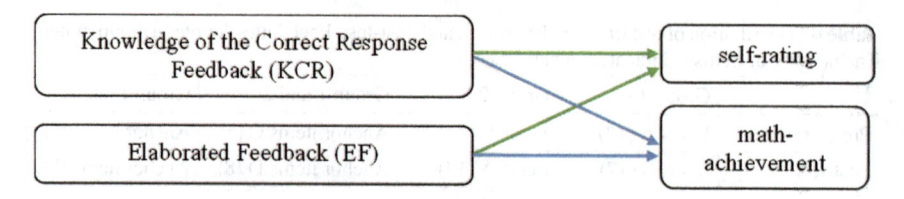

Fig. 8 Illustration of the areas targeted by the research questions

2. Does a CBLE for the topic of quadratic functions that incorporates feedback featuring additional explanations and hints (EF) have a more positive influence on the mathematical achievement of students than the same CBLE with feedback that is limited to knowledge of the correct response (KCR)?

5.2 Research Design

The research questions of the main study focus on the effect of the type of feedback integrated into the CBLE on the mathematics performance of students. The main study also investigates whether a connection between the type of feedback received and the self-rating ability of students can be demonstrated. These research interests require a quantitative study design. The main study will therefore have a pre- and post-test design with one experimental group and one control group (Fig. 9).

An adapted form of the CODI test developed by Nitsch (2015) will be used as the performance test. This test is used to diagnose learning obstacles relating to representation switching in the areas of functional thinking and linear and quadratic functions. It was modified to allow it to be applied at two distinct points in time and the content was adapted to the students' abilities, since the students participating in the study will not yet have encountered quadratic functions in the classroom, and will only be familiar with linear functions. Accordingly, the items of the pre-test prioritize linear functions and functional thinking. A few items on quadratic functions are included to check for prior knowledge. The focus of the post-test is shifted, replacing some of the items on linear functions with new items on quadratic functions. As well as mathematical items, the test includes brief questionnaire prompts, for example

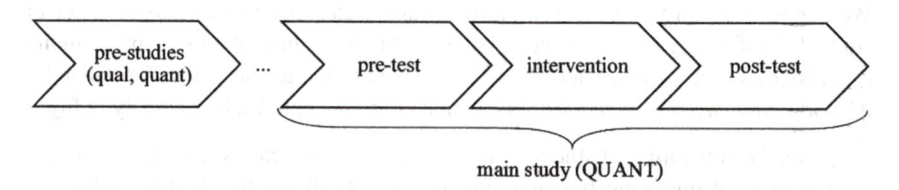

Fig. 9 Illustration of the design of the research project

asking about motivation and self-rating. The specific composition of the test will be evaluated in a quantitative preliminary study (Table 4).

The intervention between the pre-test and the post-test asks students to work independently with the newly developed CBLE on quadratic functions. In addition to the online learning environment, each student will be given a booklet to record on a four-point Likert-type scale how well they feel they have understood the content after completing each section of the CBLE. Since the experimental group and the control group receive different types of feedback, there are two versions of the CBLE. With the exception of the type of feedback, these two versions of the CBLE are completely identical. The CBLE used by the experimental group offers a form of EF. As well as feedback about whether the solution is correct, hints and explanations about the solution are given. The CBLE used by the control group features KCR-type feedback exclusively. These two feedback variants were chosen because they represent the two extreme ends of the spectrum of feedback in CBLE in terms of the expected impact on performance (cf. paragraph "Feedback"). Our objective in selecting this research design is to determine whether the feedback provided to the experimental group is indeed as effective as the theory suggests. In the section presenting the theoretical background of feedback, we noted that KCR feedback has been reported to produce effects that are comparable to a complete absence of feedback, whereas an additional combination of explanations and hints is expected to positively influence the learning process (cf. paragraph "Feedback"). However, another conceivable outcome is that the students in the experimental group only utilize the EF superficially, without thinking, simply copying the hints and solution paths provided without actively engaging with the exercise themselves. In this scenario, the control group might achieve a better learning experience, since the KCR feedback does not provide guidance on the approach and the solution, possibly encouraging the students to question their own results. To ensure that the measured effects can be attributed to the feedback provided by the CBLE, video recordings of the computer screens are taken. This allows us to monitor the extent to which the students took advantage of the feedback and at which points during the exercise they did so.

The main study will unfold over a total of eight school periods (45 min each). Performance tests will be conducted in the first and last periods. For the six periods in between, the students will work with the CBLE. They will work independently and autonomously, and we provisionally plan to group them into pairs. This is partly motivated by the layout of school infrastructure—in many schools, the computer rooms are designed for multiple students to work together at each computer station. Additionally, the CBLE includes several exercises that require a partner. In general, long-term independent learning is expected to be supported by working in pairs. During the working phase of the intervention, the teachers take on advisory role, answering the students' technical questions, but otherwise mostly holding back to allow the students to manage their own working and learning processes. To monitor teacher interventions, every question asked by a student is recorded on a form, together with the answer given by the teacher. This form can also be used to revise and modify the learning path to address frequently asked questions.

6 Summary

This chapter gives an overview of an upcoming study on the different types of feedback that can be integrated into CBLE based on MediaWiki software. The effects of two feedback types on mathematical achievement and self-rating ability will be examined by this study using a specially designed CBLE to introduce the topic of quadratic functions. After providing some context on the underlying theoretical framework, this chapter presented various design aspects and components of the CBLE on quadratic functions. In addition, findings and conclusions of a qualitative preliminary study conducted to evaluate and improve the CBLE are presented. Other qualitative and quantitative preliminary studies are currently under way to allow us to continue to improve the CBLE. We hope that these studies will yield valuable insight that can be used to adjust the research design of the upcoming main study and ensure that it unfolds smoothly. The chapter ended with a report of the prospective research questions and research designs of the main study.

References

Attali, Y. (2015). Effects of multiple-try feedback and question type during mathematics problem solving on performance in similar problems. *Computers & Education, 86*, 260–267.

Azevedo, R., & Bernard, R. (1995). A meta-analysis of the effects of feedback in computer-based instruction. *Journal of Educational Computing Research, 13*(2), 111–127.

Baker, R. S. J., D'Mello, S. K., Rodrigo, M. M. T., & Graesser, A. C. (2010). Better to be frustrated than bored: The incidence, persistence, and impact of learners' cognitive-affective states during interactions with three different computer-based learning environments. *International Journal of Human-Computer Studies, 68*(4), 223–241.

Balacheff, N., & Kaput, J. J. (1996). Computer-based learning environments in mathematics. In A. J. Bishop, K. Clements, C. Keitel, J. Kilpatrick, & C. Laborde (Eds.), *International handbook of mathematics education: Part 1* (pp. 469–501). https://doi.org/10.1007/978-94-009-1465-0_14.

Bangert-Drowns, R. L., Kulik, C.-L. C., Kulik, J. A., & Morgan, M. T. (1991). The instructional effect of feedback in test-like events. *Review of Educational Research, 61*(2), 213–238. https://doi.org/10.3102/00346543061002213.

Bimba, A. T., Idris, N., Al-Hunaiyyan, A., Mahmud, R. B., & Shuib, N. L. M. (2017). Adaptive feedback in computer-based learning environments: A review. *Adaptive Behavior, 25*(5), 217–234. https://doi.org/10.1177/1059712317727590.

Black, P., & Wiliam, D. (1998). Assessment and classroom learning. *Assessment in Education: Principles, Policy & Practice, 5*(1), 7–74. https://doi.org/10.1080/0969595980050102.

Boud, D., & Molloy, E. (2013). Rethinking models of feedback for learning: The challenge of design. *Assessment & Evaluation in Higher Education, 38*(6), 698–712.

Corbett, A. T., & Anderson, J. R. (2001). Locus of feedback control in computer-based tutoring: Impact on learning rate, achievement and attitudes. In J. Jacko, A. Sears, M. Beaudouin-Lafon, & R. Jacob (Eds.), *Proceedings of ACM CHI'2001 Conference on Human Factors in Computing Systems* (pp. 245–252). New York: ACM Press.

Dempsey, J. V., Driscoll, M. P., & Swindell, L. K. (1993). Text-based feedback. In J. V. Dempsey & G. C. Sales (Eds.), *Interactive instruction and feedback* (pp. 21–54). Englewood Cliffs, NJ: Educational Technology.

Doorman, M., Drijvers, P., Gravemeijer, K., Boon, P., & Reed, H. (2012). Tool use and the development of the function concept: From repeated calculations to functional thinking. *International Journal of Science and Mathematics Education, 10*(6), 1243–1267.

Federal Ministry of Education and Research. (2016). *Bildungsoffensive für die digitale Wissensgesellschaft. Strategie des Bundesministeriums für Bildung und Forschung*. Retrieved from https://www.bmbf.de/files/Bildungsoffensive_fuer_die_digitale_Wissensgesellschaft.pdf.

Fyfe, E. R. (2016). Providing feedback on computer-based algebra homework in middle-school classrooms. *Computers in Human Behavior, 63*, 568–574. https://doi.org/10.1016/j.chb.2016.05.082.

German Education Server. (2014). *Open Educational Resources (OER). An Overview of Initiatives Worldwide*. Retrieved from http://www.bildungsserver.de/Open-Educational-Resources-OER-an-Overview-of-Initiatives-Worldwide-6998_eng.html.

Greefrath, G., Oldenburg, R., Siller, H.-S., Ulm, V., & Weigand, H.-G. (2016). *Didaktik der Analysis. Aspekte und Grundvorstellungen zentraler Begriffe*. Berlin and Heidelberg: Springer Spektrum.

Hattie, J. (2009). *Visible learning: A synthesis of 800+ meta-analyses on achievement*. Abington: Routledge.

Hattie, J., & Timperley, H. (2007). The power of feedback. *Review of Educational Research, 77*(1), 81–112. https://doi.org/10.3102/003465430298487.

Isaacs, W., & Senge, P. (1992). Overcoming limits to learning in computer-based learning environments. *European Journal of Operational Research, 59*(1), 183–196. https://doi.org/10.1016/0377-2217(92)90014-Z.

Kulhavy, R. W., & Stock, W. A. (1989). Feedback in written instruction: The place of response certitude. *Educational Psychology Review, 1*(4), 279–308.

Malle, G. (2000). Zwei Aspekte von Funktionen: Zuordnung und Kovariation. *mathematik lehren, 118*, 57–62.

Mathematik-digital. (2006). *Was ist ein guter Lernpfad? - Qualitätskriterien. Hrsg. vom Arbeitskreis Mathematik digital*. Retrieved from http://wiki.zum.de/Mathematik-digital/Kriterienkatalog, version of 18.11.2017.

Mayring, P. (2010). *Qualitative Inhaltsanalyse. Grundlagen und Techniken* (11 revised ed.). Weinheim and Basel: Beltz.

Mory, E. H. (2004). Feedback research revisited. In D. H. Jonassen (Eds.), *Handbook of research on educational communications and technology* (2nd ed., pp. 745–783). Taylor & Francis.

Nelson, M. M., & Schunn, C. D. (2009). The nature of feedback: How different types of peer feedback affect writing performance. *Instructional Science, 37*(4), 375–401. https://doi.org/10.1007/s11251-008-9053-x.

Nicol, D. J., & Macfarlane-Dick, D. (2006). Formative assessment and self-regulated learning: A model and seven principles of good feedback practice. *Studies in Higher Education, 31*(2), 199–218.

Nitsch, R. (2015). *Diagnose von Lernschwierigkeiten im Bereich funktionaler Zusammenhänge. Eine Studie zu typischen Fehlermustern bei Darstellungswechseln*. Wiesbaden: Springer Spektrum.

OECD. (2017). *Open Educational Resources (OER)*. Retrieved from http://www.oecd.org/edu/ceri/open-educational-resources-oer.htm.

Roth, J. (2015). Lernpfade: Definition, Gestaltungskriterien und Unterrichtseinsatz. In J. Roth, E. Süss-Stepancik, & H. Wiesner (Eds.), *Medienvielfalt im Mathematikunterricht. Lernpfade als Weg zum Ziel* (pp. 3–25). Wiesbaden: Springer Spektrum.

Sadler, D. R. (1989). Formative assessment and the design of instructional systems. *Instructional Science, 18*, 119–144.

SCMECA—Standing Conference of the Ministers of Education and Cultural Affairs. (2003). *Bildungsstandards im Fach Mathematik für den Mittleren Schulabschluss*. Wolters Kluver.

SCMECA—Standing Conference of the Ministers of Education and Cultural Affairs. (2016). *Bildung in der digitalen Welt. Strategie der Kultusministerkonferenz*. Retrieved from https://www.kmk.org/fileadmin/Dateien/pdf/PresseUndAktuelles/2016/Bildung_digitale_Welt_Webversion.pdf.

Shute, V. J. (2008). Focus on formative feedback. *Review of Educational Research, 78*(1), 153–189. https://doi.org/10.3102/0034654307313795.

Sur, C. (2017). *Selbstgesteuertes Lernen mit Lernpfaden im Mathematikunterricht aus Sicht der Lernenden – Eine qualitative Untersuchung auf Basis von Interviews*. Unpublished master thesis, University of Münster.

Swan, M. (1982). The teaching of functions and graphs. In *Proceedings of the Conference on Functions* (pp. 151–165). Enschede, The Netherlands: National Institute for Curriculum Development.

UNESCO. (2017). *Open Educational Resources*. Retrieved from http://www.unesco.org/new/en/communication-and-information/access-to-knowledge/open-educational-resources/.

Van der Kleij, F. M., Feskens, R. C. W., & Eggen, T. J. H. M. (2015). Effects of feedback in a computer-based learning environment on student's learning outcomes: A meta-analysis. *Review of Educational Research, 85*(4), 475–511. https://doi.org/10.3102/0034654314564881.

Vollrath, H.-J. (1989). Funktionales Denken. *Journal für Mathematikdidaktik, 10*(1), 3–37.

Vollrath, H.-J. (2014). Funktionale Zusammenhänge. In H. Linneweber-Lammerskitten (Ed.), *Fachdidaktik Mathematik*. Friedrich: Seelze.

Vollrath, H.-J., & Roth, J. (2012). *Grundlagen des Mathematikunterrichts in der Sekundarstufe* (2nd ed.). Heidelberg: Springer Spektrum.

Wiesner, H., & Wiesner-Steiner, A. (2015). Einschätzungen zu Lernpfaden – Eine empirische Exploration. In J. Roth, E. Süss-Stepancik, & H. Wiesner (Eds.), *Medienvielfalt im Mathematikunterricht. Lernpfade als Weg zum Ziel* (pp. 27–45). Wiesbaden: Springer Spektrum.

Zaslavsky, O. (1997). Conceptual obstacles in the learning of quadratic functions. *Focus on Learning Problems in Mathematics Winter Edition, 19*(1), 20–44.

Concluding Remarks

One of the aims of the biyearly International Conference on Technology in Mathematics Teaching (ICTMT) is, to provide a state of the art regarding research in the field of mathematics education with technology. This 13th edition of the conference does not break the rule.

Contributions of the 13th ICTMT

This 13th edition of ICTMT focused specifically on the issue of assessment with technology, which is mirrored in the first four chapters of the book (Part 1). These contributions and discussions during the conference provide evidence that technology is an extraordinary tool for teachers to implement formative assessment strategies in mathematics classrooms. Therefore, technology for mathematics teaching needs to be considered broadly, beyond specific pieces of mathematical software, such as dynamic geometry, spreadsheet or CAS, that tend to be over-emphasized in the field. Indeed, as the four chapters in Part 1 of this book show, technology that facilitates communication, like polls (Cusi et al.), platforms embedding analytical tools (Olsher) or self-assessment applications (Ruchniewicz & Barzel; Barzel et al.), in mathematics classrooms has also great potential for improving mathematics teaching and learning by allowing to deploy innovative pedagogical methods.

An important aim of the ICTMT conference is to allow a direct dialogue between teachers, researchers, software developers, technologists, engineers, whose synergy is required for the design of cutting-edge and didactically sound technological tools and innovative technology-based instructional approaches. The 13th edition of ICTMT witnessed a few examples reported in the chapters in Part 2 of the book: new ways of representing the world through virtual reality (Dimmel & Bock), technology as a vector of creativity (El Demerdash et al.), or an innovative pedagogical use of existing technology—WIMS (Kobylanski).

© Springer Nature Switzerland AG 2019
G. Aldon and J. Trgalová (eds.), *Technology in Mathematics Teaching*,
Mathematics Education in the Digital Era 13,
https://doi.org/10.1007/978-3-030-19741-4

The use of (old and new) technology in the service of mathematics education supposes an appropriate teacher education and professional development (TPD). The chapters in Part 3 of the book show that technology has a lot to offer to this field as well. Technology makes it possible to reach hundreds, or even thousands teachers and engage them in new forms of professional development, based on collaboration and peer-learning, while benefiting from specific interactions with teacher trainers (Aldon et al.; van den Bogaart et al.). MOOCs and other online TPD programmes aiming at the development of professional knowledge and skills teachers need to use efficiently technology in their classrooms (Tabach & Trgalová) are wide spreading all over the world, opening new avenues for research.

Finally, ICTMT is also a place of sharing experiences with teaching, learning or designing digital technologies. Outstanding examples are reported in Part 4 of the book: a thorough analysis of student's mathematical conceptualisations while working with 3D modelling software (Uygan & Turgut), an in-depth exploration of the potential of a particular representation of functions enabled by a dynamic environment to learning of a difficult concept of co-variation (Lisarelli) and a design of particular feedback in a computer-based learning environment and an analysis of its effects on students' learning (Jedtke & Greefrath).

Perspectives for Future Research on Technology in Mathematics Education

The chapters present in this book, as well as the rich discussions that occurred during ICTMT13, outline possible orientations for future research on technology in mathematics education.

Coming back to the acronym ICT, standing for Information and Communication Technology, highlights three keywords: *information*, *communication* and *technology*. Considering *information*, nowadays trends are towards new teachers', but also students', systems of resources (see for example a recent international conference on this issue, Res(s)ources 2018, https://resources-2018.sciencesconf.org/). Profound evolutions of these systems imply modifications of interactions between students, teacher and knowledge in a technological era. This raises the need for specific research methodologies making possible monitoring these evolutions.

Regarding *communication*, technology enables new ways of teaching, learning and training, engaging students and trainees in collaborative work, project-based pedagogy and self-assessment. Peer-learning and new forms of tutorial intervention are just two among many other possible topical research issues, as highlighted by Aldon et al.:

> Our analysis shows that a real involvement of trainees in collaborative work needs to be triggered and supported by suitable tools added to the platform. The availability in the platform of tools consonant with the social networks used in everyday life increases the triggering of what Manlove et al. (2006) call co-regulated learning, in the sense that the trainees themselves regulate their tasks and collaboration (Aldon et al. this book)

The advances of *technology* open ways toward:

- new approaches to and new representations of mathematical concepts. For example, augmented and virtual reality is a step between real world and mathematical abstraction: "we are attempting to ensure that research-based ideas about the nature of productive mathematical activity are represented in this next generation of virtual learning environments" (Dimmel & Bock);
- developing creativity and giving mathematics education a new dimension: *"Innovative technology has been designed allowing for producing resources offering to students a rich exploratory environment with carefully devised scaffolding supporting students' learning mathematics as well as their creative approach to problems at hand"* (El Demerdash et al.)

The interrelatedness of technological, mathematical, didactic and pedagogical aspects in research on technology in mathematics education raises a necessity of networking frameworks from these fields: ergonomy and didactics, semiotic and content knowledge, etc.

The research agenda is still rich, promising an exciting 14th edition of ICTMT in July 2019 at the University of Duisburg- Essen!